Energy and Power Systems

Energy and Power Systems

Edited by **Helena Walker**

Callisto Reference
New York

Published by Callisto Reference,
106 Park Avenue, Suite 200,
New York, NY 10016, USA
www.callistoreference.com

Energy and Power Systems
Edited by Helena Walker

International Standard Book Number: 978-1-63239-685-3 (Hardback)

Printed in China.

Contents

Preface

In my initial years as a student, I used to run to the library at every possible instance to grab a book and learn something new. Books were my primary source of knowledge and I would not have come such a long way without all that I learnt from them. Thus, when I was approached to edit this book; I became understandably nostalgic. It was an absolute honor to be considered worthy of guiding the current generation as well as those to come. I put all my knowledge and hard work into making this book most beneficial for its readers.

Energy systems can consist of: electric, bioenergetic and thermodynamic systems whereas power systems are used for the transmission of energy. Power and energy are overlapping fields as they both focus on meeting the current energy demand of the world. The various studies that are constantly contributing towards advancing technologies and evolution of these fields are examined in detail. While understanding the long-term perspectives of the topics, the book makes an effort in highlighting their impact as a modern tool for the growth of the discipline. Those in search of information to further their knowledge in this area will be greatly assisted by this book.

I wish to thank my publisher for supporting me at every step. I would also like to thank all the authors who have contributed their researches in this book. I hope this book will be a valuable contribution to the progress of the field.

Editor

1

Performance Enhancement of the Proton Exchange Membrane Fuel Cell Using Pin Type Flow Channel with Porous Inserts

Vaibhav Pal[1], P. Karthikeyan[2*], R. Anand[1]

[1]Department of Mechanical Engineering, National Institute of Technology, Tiruchirappalli, India
[2]Department of Automobile Engineering, PSG College of Technology, Coimbatore, India
Email: vp5830@gmail.com, *apkarthipsg@gmail.com, anandachu@nitt.edu

Abstract

The design of the flow field is highly responsible for the performance of the Proton Exchange Membrane Fuel Cell (PEMFC). In this study, pin type flow channel is numerically analyzed by arranging carbon made porous material in uniform and zigzag manner on the rib surface of the flow field. The study focuses on enhancing the performance of PEMFC by reducing liquid flooding in the interface between the rib and Gas Diffusion Layer (GDL). A single PEMFC having an active area of 25 cm^2, with three flow channel designs (conventional serpentine, pin type flow channel with 2 mm cubical porous inserts in zigzag and uniform pattern) are modeled for the numerical analysis. The effect of porosity of the carbon inserts on the cell performance is studied by varying its value from 0.6 to 0.9. The results show that the performance of the flow channel with zigzag and uniformly positioned porous inserts is more than the conventional serpentine flow channel by 20.36% and 16.87% respectively. The reason for this increase is the removal of the accumulated water from the rib surface due to the capillary action of the porous carbon inserts. This helps in eliminating the stagnant water regions under the rib and thereby helps in reducing liquid flooding.

Keywords

Flooding, Uniform and Zigzag Pin Type, Porosity, Porous Carbon Inserts

1. Introduction

With the increasing problem of environmental degradation due to the harmful emissions from the existing fuels

*Corresponding author.

and the shortage of the non-renewable fossil fuels, alternative fuels and non-polluting sources of energy are the need of the hour. Fuel cells are a viable option for the future as they are portable and have almost no harmful emissions [1]-[4]. Out of the various available fuel cells, Proton Exchange Membrane Fuel Cell (PEMFC) is capable of operating at a temperature close to the atmospheric temperature and has a rapid start up time, thus making it a potential candidate for automobile application. But there are various obstacles for the commercialization of the PEMFC such as high cost of its materials, short cycle life, and lack of hydrogen infrastructure for its generation, and storage.

In a typical PEMFC, hydrogen gas is used as a fuel and is passed through the anode region, whereas an oxidant either oxygen or air is passed through the cathode region. Water is obtained as a by-product at the cathode end due to the electrochemical reaction between hydrogen and oxygen. The water obtained as a by-product has to be precisely balanced for the better performance of PEMFC. At the cathode side, oxygen reduction reaction (ORR) occurs in the catalyst-membrane interface and due to this water is generated. For the hydration of the membrane, inlet fuel and oxidant are supplied in a fully humidified state. Due to this, the gases in the fuel cell may get oversaturated with the water vapour and get condensed to form liquid water, thereby the amount of water content in the cathode region increases. Apart from this, the cathode region also gets water through electro-osmotic drag, wherein the positively charged protons tend to attract and drag the water molecules along with them, when migrating from anode to cathode flow channels. The water transport also takes place due to pressure drop as well as concentration difference. Hence, managing the produced water towards the cathode end becomes important. The water content towards the cathode end must be judiciously maintained in order to achieve high efficiency. At present, the biggest problem researchers are facing is the water management at the cathode side. If the water removal rate is greater than the generation rate, the membrane gets dehydrated and therefore results in the poor performance due to excessive ohmic losses [5] [6]. But if there is excess water, it can retard the transport of reactants to the active reaction site by blocking the pores in the catalyst layer as well as in the GDL. This problem is called flooding and it reduces the PEMFC's performance to a huge extent and makes it unreliable and unpredictable [7] [8].

Different methods have been proposed for managing the water in an effective manner and the most productive method found is to effectively designing the flow channel [1] [9] [10]. Flow patterns with less pressure drop (straight parallel, Z type parallel, cascade, and pin type) allow preferential flow path and there is uneven distribution of gases in the membrane electrode assembly (MEA), which decreases the performance of the PEMFC [11] [12]. Serpentine and interdigitated flow patterns distribute the reactants uniformly over the catalytic layer and thus increases its efficiency [13] [14]. On the other side, higher pressure fall in the flow pattern increases the energy required for the gas flow and this reduces the cell's efficiency. But the higher pressure fall in these two patterns also promote the under rib convection which helps in the removal of water and thus increases the efficiency of the PEMFC [15] [16].

For solving this problem of cathode flooding, different researchers have suggested different techniques. Nguyen [17] initially studied the interdigitated flow field pattern and found that the dead end profile of the flow channel forced the gases to go through the GDL and gas pushed out the water droplets entrapped in the layers of electrode. This also increased the amount of reactants reaching the catalyst layer from the channels leading to increase in performance and the problem of flooding was also reduced. Wang *et al.* [18] investigated the baffle blocked interdigitated flow field and concluded that the gas convection got increased by the use of baffle blocks which further enhanced the reactant transport to the electrode and the water removal. Qin *et al.* [19] proposed the use of hydrophilic plate flow field which increased the water removal rate. In this study, a novel flow field design towards the cathode side of the PEMFC is used for efficient water removal. The effect of three flow patterns (conventional serpentine, pin type flow channel with adoption of porous carbon insertsin uniform and zigzag pattern) have been studied numerically and their performances have been compared. Also, the porosity of carbon inserts have been varied and then its influence on the PEMFC's performance has been analysed.

2. Model Development

A three-dimensional model with three flow field patterns (conventional serpentine, pin type flow channel with the adoption of porous carbon inserts in uniform and zigzag pattern) has been evaluated on an active area of 25 cm^2. As the problem of water flooding occurs towards the cathode side of the PEMFC, a modification is made in the cathode flow channel only. The modification is done by inserting porous carbon inserts in the uniform and zigzag pattern on the rib surface of the pin type flow channel. By placing these porous carbon inserts in a uniform and zigzag manner on the rib surface, the pin type flow channel becomes similar to a serpentine flow channel (dimension of porous inserts 2 mm × 2 mm × 2 mm). Therefore, it gains the merits of the serpentine flow channel. **Table 1** shows the geometric dimensions used in the conventional serpentine flow channel. The geometric dimensions of the other two modified patterns remain the same, except for the change in the flow field pattern at the cathode side as shown in **Figure 1(a)** and **Figure 1(b)**. The configuration of the landing to channel (L: C) 2 mm × 2 mm is used throughout the three patterns.

2.1. Numerical Model Development

The commercially available Fluid Dynamics software FLUENT 14.5 which is based on a control volume approach was used to solve the various governing equations. 3-dimensional, steady state, laminar, double precision, serial processing modes were used for this simulation. A special add-on module called "Battery and Fuel Cell" was used for this study. In order to get accurate results, a grid independency test was done and finally the results were found grid independent at about 650,000 elements. In order to simplify the analysis, the gases used were considered to be ideal. The flow was taken as steady, laminar, incompressible, and the system was considered as isothermal at 325 K. The various thermo physical properties were considered as constant and the effect of gravity was neglected. Also, the GDL, the catalyst layer, carbon inserts and the membrane were considered to be isotropic.

Table 1. Dimensions of the fuel cell.

Part	Length (mm)	Width (mm)	Height (mm)
GDL	50	50	0.3
Catalyst Layer	50	50	0.06
Membrane	50	50	0.15
Serpentine/Pin Type Flow Channel	50	50	2
Current Collector	80	80	10

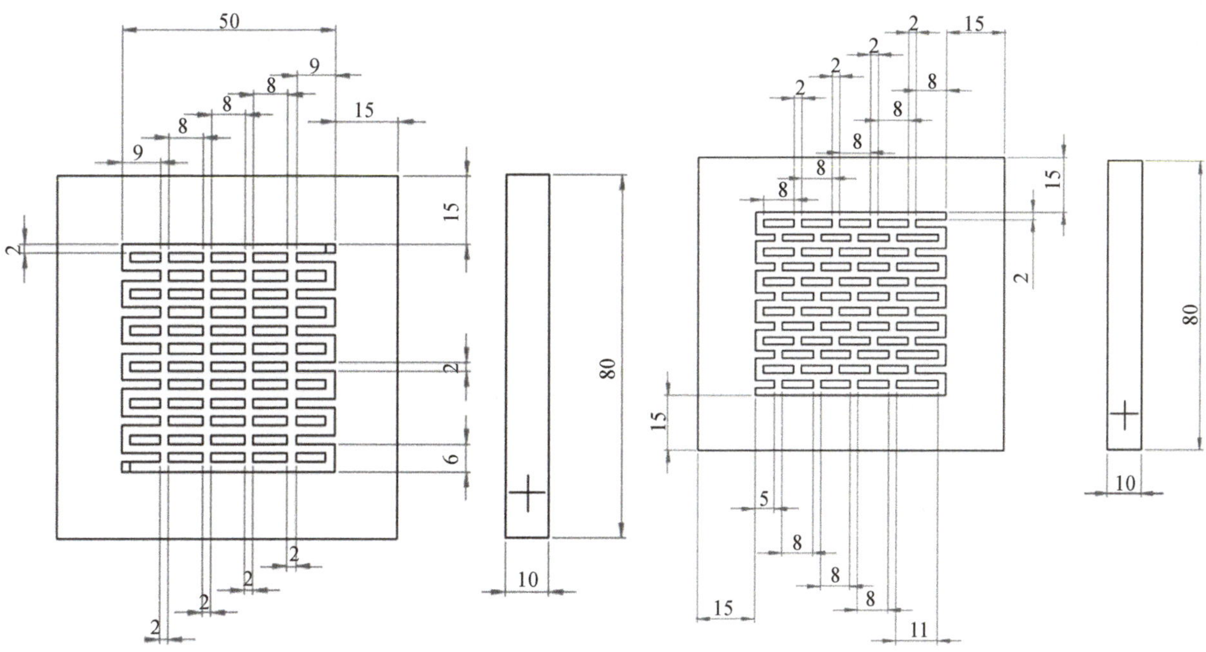

(a) (b)

Figure 1. Dimensions of the modified flow channels (a) Uniform pin type; (b) Zigzag pin type (All dimensions are in mm).

2.2. Boundary Conditions

In this model, the anode and cathode channel inlet zone were set as mass flow inlet respectively, the anode and cathode channel outlet zone were set as pressure outlet, exhaust-fan, outlet-vent whereas the other left surfaces were set as wall. The operating conditions for all the three patterns were set at a pressure of 1 bar and temperature of 325 K. Water produced was assumed to be fully liquid (100% humidified) and the mass fractions towards the anode side were set at 0.8, 0.2, 0 for H_2, H_2O, O_2 respectively [20]; for the cathode side mass fractions were set at 0, 0.1, 0.2 for H_2, H_2O, O_2 respectively [20]. The summary of various reaction parameters and material properties used in this work is shown in **Table 2**. For the uniform and zigzag pattern of porous carbon inserts the porosity value was varied from 0.6 to 0.9.

The various equations which are solved in this study (discussed in sections 2.3 to 2.5) are taken from the Falcao *et al.* [21] and are as follows:

2.3. Electrochemistry Equations

In this study Fuel Cell and Electrolysis Model is used to solve the two potential equations and they are as follows [21]:

$$\nabla \cdot (\sigma_{sol} \cdot \nabla E_{sol}) + R_{sol} = 0 \quad (1)$$

$$\nabla \cdot (\sigma_{mem} \cdot \nabla E_{mem}) + R_{mem} = 0 \quad (2)$$

Table 2. Reaction parameters and material properties.

Reaction Parameters		Value	Units	Reference
Open circuit voltage		1.1	V	[20]
Reference anode concentration		1	$k \cdot mol \cdot m^{-3}$	[20]
Anode charge transfer coefficient		2		[20]
Anode exchange current density		10,000	$A \cdot m^{-2}$	[20]
Reference cathode concentration		1	$k \cdot mol \cdot m^{-3}$	[20]
Cathode charge transfer coefficient		2		[20]
Cathode exchange current density		20	$A \cdot m^{-2}$	[20]
Physical Parameters		**Value**	**Units**	**Reference**
GDL	Porosity	0.5		[20]
	Thermal conductivity	10	$W \cdot m^{-1} \cdot K^{-1}$	[20]
	Density	2719	$kg \cdot m^{-3}$	[20]
	Electrical conductivity	5000	$ohm^{-1} \cdot m^{-1}$	[20]
Catalyst Layer	Porosity	0.5		[20]
	Surface/volume ratio	200,000	m^{-1}	[20]
	Thermal conductivity	10	$W \cdot m^{-1} \cdot K^{-1}$	[20]
	Electrical conductivity	5000	$ohm^{-1} \cdot m^{-1}$	[20]
Membrane	Thermal conductivity	2	$W \cdot m^{-1} \cdot K^{-1}$	[20]
	Dry membrane density	1980	$kg \cdot m^{-3}$	[20]
	Equivalent weight	1100	$kg \cdot K^{-1} \cdot mol^{-1}$	[20]
	Electrical conductivity	1×10^{-16}	$ohm^{-1} \cdot m^{-1}$	[20]

where σ, E and R represent electrical conductivity, electric potential and volumetric transfer current respectively. The indices sol, mem, ano and catho represent solid phase, membrane phase, anode region and cathode region respectively. The transfer currents or the source terms in the Equation (1) and Equation (2) are not zero at the catalyst layers only and are calculated as [21]:

if phase is solid, $R_{\text{sol}} = -R_{\text{ano}} \ (<0)$ at the anode region and $R_{\text{sol}} = +R_{\text{catho}} \ (>0)$ at the cathode region.

if phase is membrane, $R_{\text{mem}} = +R_{\text{ano}} \ (>0)$ at the anode region and $R_{\text{mem}} = -R_{\text{catho}} \ (<0)$ on the cathode region.

Source terms for Equations (1) and (2) are also termed as exchange current density and have the generic Tafel formulation [21]:

$$R_{\text{ano}} = j_a^{\text{ref}} \left(\frac{C_{\text{H}_2}^{\text{ACL}}}{C_{\text{H}_2}^{\text{ref}}} \right)^{\gamma_a} \left(e^{\alpha_a F \frac{\eta_{\text{ano}}}{GT}} \right) \quad (3)$$

$$R_{\text{catho}} = j_c^{\text{ref}} \left(\frac{C_{\text{O}_2}^{\text{CCL}}}{C_{\text{O}_2}^{\text{ref}}} \right)^{\gamma_c} \left(e^{-\alpha_c F \frac{\eta_{\text{catho}}}{GT}} \right) \quad (4)$$

where G, T, and F are the gas constant, the temperature of the fuel cell and the Faraday's constant respectively. $C_{\text{H}_2}^{\text{ACL}}$ is the hydrogen concentration in the catalytic layer, $C_{\text{H}_2}^{\text{ref}}$ is the reference hydrogen concentration, $C_{\text{O}_2}^{\text{CCL}}$ is the oxygen concentration in the catalytic layer, $C_{\text{O}_2}^{\text{ref}}$ is the reference oxygen concentration in the catalytic layer, j_a^{ref} is the anode volumetric reference exchange current density, j_c^{ref} is the cathode volumetric reference exchange current density; α_a is the coefficient of anode transfer, α_c is the coefficient of cathode transfer, γ_a is the anode concentration dependence, γ_c is the cathode concentration dependence, η_{ano} is the anode over potential and η_{catho} is the cathode over potential. The anode over potential is the difference between the solid potential (E_{sol}) and membrane potential (E_{mem}) whereas the cathode over potential is the difference between the solid potential (E_{sol}), membrane potential (E_{mem}), and the open circuit voltage (E_{O}) [21]:

$$\eta_{\text{ano}} = E_{\text{sol}} - E_{\text{mem}} \quad (5)$$

$$\eta_{\text{catho}} = E_{\text{sol}} - E_{\text{mem}} - E_{\text{O}} \quad (6)$$

2.4. Current and Mass Conservation

The source terms (S) for the equation of species and energy are denoted in Equations (7)-(9) [21]:

$$S_{\text{H}_2} = -\frac{M_{\text{H}_2}}{2F} R_{\text{ano}} \quad (7)$$

$$S_{\text{O}_2} = -\frac{M_{\text{O}_2}}{2F} R_{\text{catho}} \quad (8)$$

$$S_{\text{H}_2\text{O}} = \frac{M_{\text{H}_2\text{O}}}{2F} R_{\text{catho}} \quad (9)$$

where M_{H_2} is the molecular weight of hydrogen, M_{O_2} is the molecular weight of oxygen, $M_{\text{H}_2\text{O}}$ is the molecular weight of water, S_{H_2} is the volumetric source term for H_2, S_{O_2} is the colume tric source term for O_2, and $S_{\text{H}_2\text{O}}$ is the volumetric source term for H_2O. So, in this study the following equation has been used for current conservation [21]:

$$\int_{\text{anode}} R_{\text{ano}} \mathrm{d}V = \int_{\text{cathode}} R_{\text{catho}} \mathrm{d}V. \quad (10)$$

2.5. Liquid Water Formation and Transport

A saturation model is used in FLUENT for modeling the formation and transport of the liquid water. The following conservation equation governs the formation of the water (in liquid form) and its transport [21]:

$$\frac{\partial(\varepsilon\rho_l s)}{\partial t}+\nabla\cdot\left(\rho_l\vec{V}_l s\right)=r_w \tag{11}$$

where s is the volume fraction of liquid water or the water saturation, the subscript l stands for liquid water, ε is the porosity, ρ is the density and r_w is the rate of following conservation equation which is defined as [21]:

$$r_w=c_r\max\left\{\left[(1-s)\frac{P_{wv}}{P_{\text{sat}}}M_{\text{H}_2\text{O}}\right],[-s\rho_l]\right\} \tag{12}$$

where c_r is the condensation rate constant and is defined as $c_r=100\ \text{s}^{-1}$, P_{wv} is the water vapour pressure and P_{sat} is the saturation pressure in terms of atm (standard atmosphere). The liquid velocity (V_l) is also assumed to be equal to the velocity of gas in the flow channel. The convective term in Equation (11) isreplaced by the capillary diffusion termto form the following conservation equation (used for high resistant porous zones) [21]:

$$\frac{\partial(\varepsilon\rho_l s)}{\partial t}+\nabla\cdot\left(\rho_l\frac{Ks^3}{\mu_l}\frac{\text{d}p_c}{\text{d}s}\nabla s\right)=r_w \tag{13}$$

The capillary pressure (p_c) is calculatedas following depending upon the wetting phase [21]:

$$\text{for } \theta_c<90^\circ,\quad p_c=\frac{\sigma\text{Cos}\theta_c}{\left(\frac{K}{\varepsilon}\right)^{0.5}}\left(1.417(1-s)-2.12(1-s)^2+1.263(1-s)^3\right) \tag{14}$$

$$\text{for } \theta_c>90^\circ,\quad p_c=\frac{\sigma\text{Cos}\theta_c}{\left(\frac{K}{\varepsilon}\right)^{0.5}}\left(1.417s-2.12s^2+1.263^3\right) \tag{15}$$

where σ is the surface tension, K is the absolute permeability and θ_c is the angle of contact. Equation (11) depictsthe different physical phenomenon like surface tension, capillary diffusion, condensation, and vaporization.

3. Results and Discussion

3.1. Effect of Porosity Variation on Cell Performance

On comparing the peak power density for pin type flow channel with uniformly positioned porous carbon inserts with different porosities, it was found that there was not much effect on the cell performance with the porosity variation (peak power density for 0.9 porosity was 0.512 W/cm^2 whereas for 0.6 porosity peak power density obtained was 0.511 W/cm^2). Similarly, on comparing the peak power density for pin type flow channel with zigzag positioned porous carbon inserts with different porosities, it was found that there was not much effect on the cell performance with the porosity variation (peak power density for 0.9 porosity was 0.527 W/cm^2 whereas for 0.6 porosity peak power density obtained was 0.526 W/cm^2).

3.2. Comparison between the Performance of Uniform and Zigzag Pattern with Porous Carbon Inserts

The performance of the uniform and zigzag pattern with porous carbon inserts is analyzed by varying the porosity from 0.6 to 0.9. **Figures 2(a)-(d)** show the P-I (Power Density-Current Density) and the V-I (Voltage-Current Density) curves for both the patterns used, at porosity equals to 0.6, 0.7, 0.8 and 0.9 respectively. The above mentioned figures show that the performance of zigzag pin type channel with porous inserts is better than the uniform pin type channel with porous carbon inserts since the plot for zigzag pattern is higher than the uniform pattern for all the different porosity used. Also, **Table 3** depicts the peak values of power density and corresponding current densities for the uniform and zigzag pattern with porous carbon inserts and there seems to be an

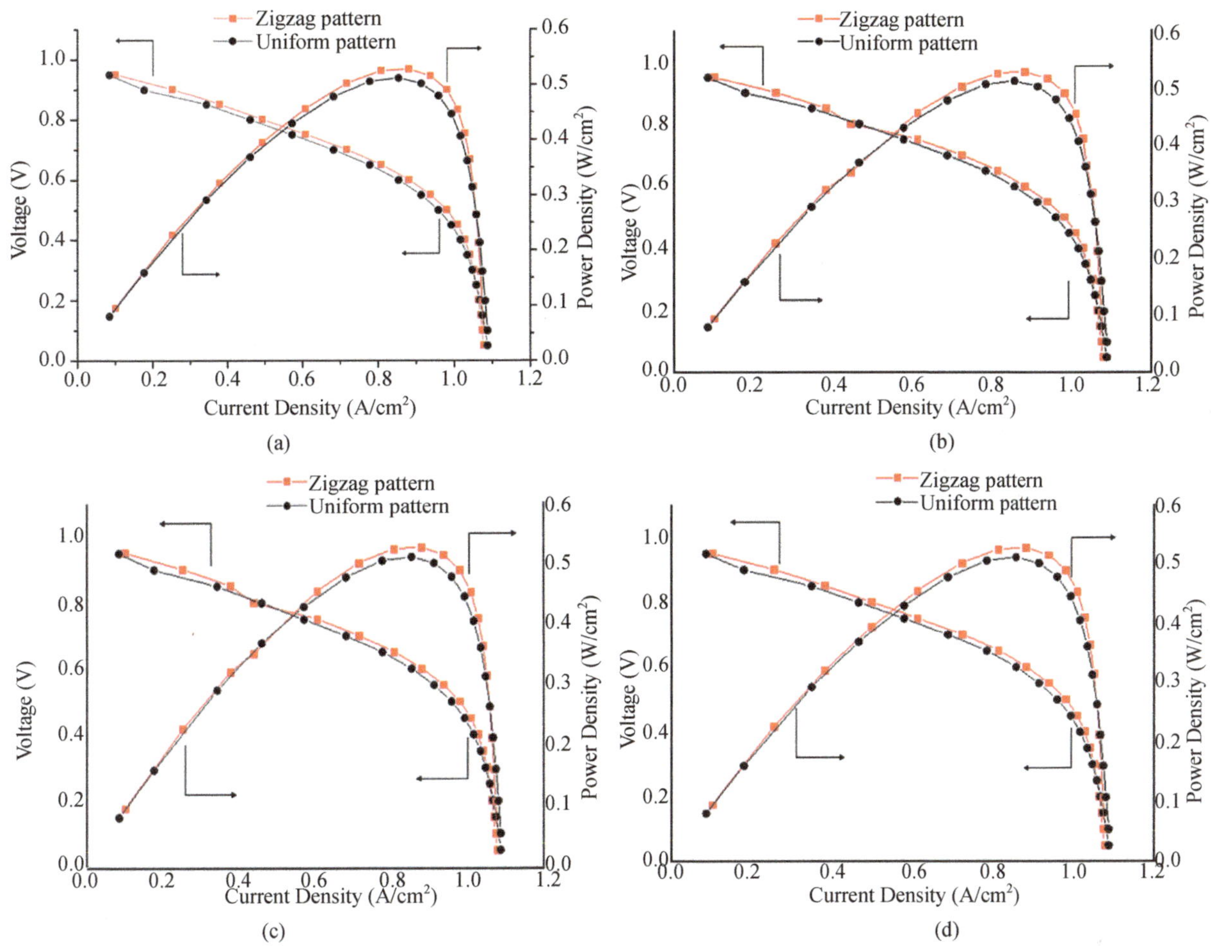

Figure 2. Performance comparison of modified patterns at different porosities (a) 0.6; (b) 0.7; (c) 0.8; (d) 0.9.

Table 3. Comparison of peak power density and current density for all the three flow patterns analyzed.

Flow channel Type/pattern	Peak Power Density (W/cm^2)	Current Density (A/cm^2)	Voltage(V)
Conventional serpentine	0.438336	0.73056	0.6
Uniform pin type flow channel with porous carbon inserts (porosity 0.9)	0.512296	0.853824	0.6
Zigzag pin type flow channel with porous carbon inserts (porosity 0.9)	0.527584	0.879306	0.6

increase in the values of power density for zigzag pattern over the uniform pattern by 2.98%. In the case of uniform pattern, the arrangement of porous inserts becomes localized, whereas for the zigzag pattern the arrangement is widespread. As a result, the water absorbed by the zigzag pattern in the interfacial region between the gas diffusion layer and the membrane becomes more globalized and hence eliminates the stagnant water more effectively. This effective water elimination results in the higher performance of the zigzag pattern.

3.3. Comparison of Cell Performance for Conventional Serpentine Pattern with the Uniform and Zigzag Pattern of Porous Carbon Inserts in Serpentine Flow Field

Numerical studies have been done on the three flow patterns (conventional serpentine, pin type flow channel with the adoption of porous carbon inserts in zigzag and uniform pattern) in order to analyze the performance of PEMFC. **Figure 3(a)** shows the performance comparison between the conventional serpentine and pin type flow channel in uniform pattern with porous carbon inserts (at porosity 0.9 as it shows maximum performance among the different porosities) using the P-I and the V-I curves. **Figure 3(b)** shows the P-I and the V-I curves for performance comparison between the conventional serpentine and pin type flow channel in zigzag pattern with porous carbon inserts (at porosity 0.9 as it shows maximum performance among the different porosities). For conventional serpentine flow channel, the maximum current density and power density are obtained as 0.731 A/cm^2 and 0.438 W/cm^2 respectively. For uniform and zigzag pattern with porous inserts the peak power density and corresponding current density are 0.512 W/cm^2, 0.854 A/cm^2 and 0.527 W/cm^2, 0.879 A/cm^2 respectively. The peak values of power and corresponding current densities for the three patterns used are mentioned in **Table 3**. It is evident that the uniform and zigzag pattern with porous carbon inserts show higher peak power density than the conventional serpentine channel pattern by 16.87% and 20.36%. The reason for the increase in performance is due to the elimination of stagnant region under the rib surface. Due to the capillary action of the porous inserts water is absorbed from the interfacial region between the rib and GDL, thereby reducing the obstacles for the incoming reactant gases. **Figure 4** compares the water generation rate for both the modified pin type flow patterns (porosity 0.9) with the conventional serpentine channel pattern. This comparison shows that in case of

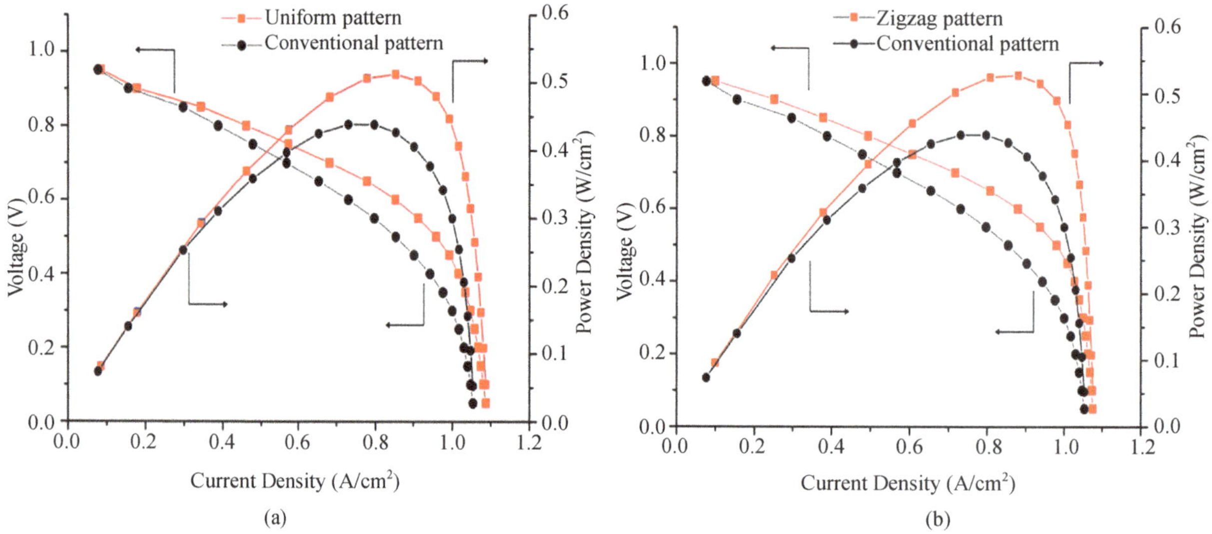

Figure 3. Performance comparison of conventional pattern with (a) Uniform pattern; (b) Zigzag pattern at 0.9 porosity.

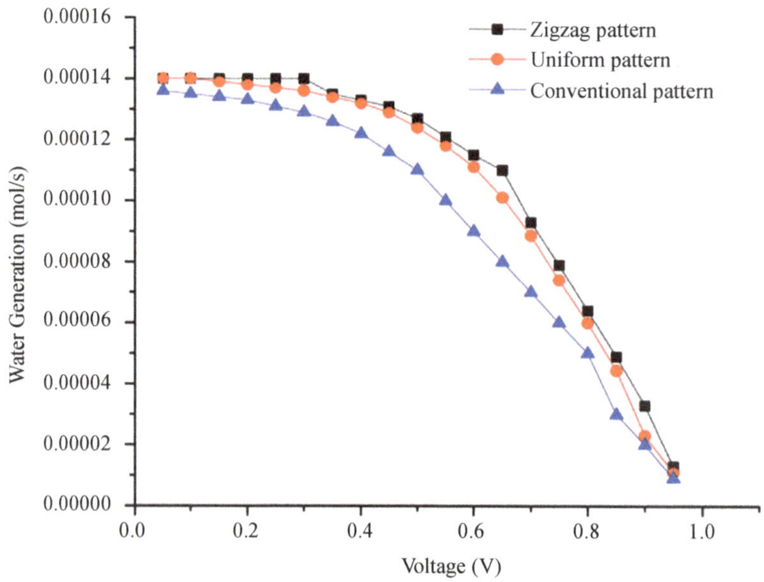

Figure 4. Comparison of water generation rate for conventional pattern with (a) Uniform pattern; (b) Zigzag pattern (both at 0.9 porosity).

both the modified pin type flow patterns there is more water generation. The reason for this higher water generation rate is the effective distribution of the reactants throughout the cathode channel (as there is no water logging in both the modified flow pin type flow channel). This validates that the use of both the modified pin type flow patterns enhances the performance by reducing the water accumulation in the flow channel.

4. Conclusion

Analysis was done on 25 cm^2 PEMFC with three flow channel configurations (conventional serpentine, pin type flow channel with the adoption of porous carbon inserts in zigzag and uniform pattern on the rib area) in order to study the cell performance. From the analysis, it was found that the uniform and zigzag pattern with porous inserts show 16.87% and 20.36% increase in performance when compared to the conventional serpentine flow channel. While comparing pin type channels (uniform and zigzag) with porous carbon inserts, zigzag pattern pin type flow channel showed 2.98% more performance than the uniform. It is also found that there is not much effect on the performance of the two new adopted designs with the variation in porosity level of the porous inserts from 0.6 to 0.9. The analysis also shows that the water absorption in case of the zigzag pattern flow channel is more than the uniform porous carbon insert channel pattern.

Acknowledgements

The work done for this paper was due to the support of the Computer Support Group Department, National Institute of Technology, Tiruchirappalli and Department of Automobile Engineering, PSG College of Technology, Coimbatore.

References

[1] Li, X.G. (2006) Principles of Fuel Cells. Taylor & Francis, New York.

[2] Li, X.G. (2005) Diversification and Localization of Energy Systems for Sustainable Development and Energy Security. *Energy Policy*, **33**, 2237-2243. http://dx.doi.org/10.1016/j.enpol.2004.05.002

[3] Zamel, N. and Li, X.G. (2006) Life Cycle Analysis of Vehicles Powered by a Fuel Cell and by Internal Combustion Engine for Canada. *Journal of Power Sources*, **155**, 297-310. http://dx.doi.org/10.1016/j.jpowsour.2005.04.024

[4] Granovskii, M., Dincer, I. and Rosen, M.A. (2006) Life Cycle Assessment of Hydrogen Fuel Cell and Gasoline Vehicles. *International Journal of Hydrogen Energy*, **31**, 337-352. http://dx.doi.org/10.1016/j.ijhydene.2005.10.004

[5] Zawodzinski Jr., T.A., Derouin, C., Radzinski, S., *et al.* (1993) Water Uptake by and Transport through Nafion^R 117 Membranes. *Journal of Electrochemical Society*, **140**, 1041-1047. http://dx.doi.org/10.1149/1.2056194

[6] Li, H., Tang, Y., Wang, Z., Shi, Z., *et al.* (2008) A Review of Water Flooding Issues in the Proton Exchange Membrane Fuel Cell. *Journal of Power Sources*, **178**, 103-117. http://dx.doi.org/10.1016/j.jpowsour.2007.12.068

[7] Yang, X.G., Burke, N., Wang, C.Y., Tajiri, K. and Shinohara, K. (2005) Simultaneous Measurements of Species and Current Distributions in a PEFC under Low-Humidity Operation. *Journal of Electrochemical Society*, **152**, 759-766. http://dx.doi.org/10.1149/1.1864492

[8] Pierre, J.S., Wilkinson, D.P., Knights, S. and Bos, M. (2000) Relationships between Water Management, Contamination and Lifetime Degradation in PEFC. *Journal of New Materials for Electrochemical Systems*, **3**, 99-106.

[9] Cavalca, C., Homeyer, S.T. and Walsworth, E. (1997) Flow Field Plate for Use in a Proton Exchange Membrane Fuel Cell. US Patent No. 5686199.

[10] Li, X.G. and Sabir, I. (2004) Review of Bipolar Plates in PEM Fuel Cells: Flow-Field Designs. *International Journal of Hydrogen Energy*, **30**, 359-371. http://dx.doi.org/10.1016/j.ijhydene.2004.09.019

[11] Manso, A.P., Marzo, F.F., Barranco, J., Garikano, X. and Mujika, M.G. (2012) Influence of Geometric Parameters of the Flow Fields on the Performance of a PEM Fuel Cell—A Review. *International Journal of Hydrogen Energy*, **37**, 15256-15287. http://dx.doi.org/10.1016/j.ijhydene.2012.07.076

[12] Ferng, Y.M. and Su, A. (2007) A Three-Dimensional Full-Cell CFD Model Used to Investigate the Effects of Different Flow Channel Designs on PEMFC Performance. *International Journal of Hydrogen Energy*, **32**, 4466-4476. http://dx.doi.org/10.1016/j.ijhydene.2007.05.012

[13] Liu, H., Li, P., Robles, D.J., Wang, K. and Hernandez-Guerrero, A. (2014) Experimental Study and Comparison of Various Designs of Gas Flow Fields to PEM Fuel Cells and Cell Stack Performance. *Frontiers in Energy Research*, **2**, 2. http://dx.doi.org/10.3389/fenrg.2014.00002

[14] Ji, M. and Wei, Z. (2009) A Review of Water Management in Polymer Electrolyte Membrane Fuel Cells. *Energies*, **2**,

1057-1106. http://dx.doi.org/10.3390/en20401057

[15] Lee, B., Park, K. and Kim, H.M. (2013) Numerical Optimization of Flow Field Pattern by Mass Transfer and Electrochemical Reaction Characteristics in Proton Exchange Membrane Fuel Cells. *International Journal of Electrochemical Science*, **8**, 219-234.

[16] Nama, J.H., Lee, K.-J., Sohnb, S. and Kimb, C.-J. (2009) Multi-Pass Serpentine Flow-Fields to Enhance Under-Rib Convection in Polymer Electrolyte Membrane Fuel Cells: Design and Geometrical Characterization. *Journal of Power Sources*, **188**, 14-23. http://dx.doi.org/10.1016/j.jpowsour.2008.11.093

[17] Nguyen, T.V. (1996) A Gas Distributor Design for Proton-Exchange-Membrane Fuel Cells. *Journal of Electrochemical Society*, **143**, 103-105. http://dx.doi.org/10.1149/1.1836666

[18] Wang, X.-D., Duan, Y.-Y. and Yan, W.-M. (2007) Novel Serpentine-Baffle Flow Field Design for Proton Exchange Membrane Fuel Cells. *Journal of Power Sources*, **173**, 210-221. http://dx.doi.org/10.1016/j.jpowsour.2007.08.037

[19] Qin, Y.Z., Li, X.G., Jiao, K., Du, Q. and Yin, Y. (2014) Effective Removal and Transport of Water in a PEM Fuel Cell Flow Channel Having a Hydrophilic Plate. *Applied Energy*, **113**, 116-126. http://dx.doi.org/10.1016/j.apenergy.2013.06.053

[20] ANSYS, Inc. ANSYS FLUENT (2007) Modeling a Single-Channel, Counter-Flow Polymer Electrolyte Membrane Fuel Cell.

[21] Falco, D.S., Gomes, P.J., Oliveria, V.B., Pinho, C. and Pinto, A.M.F.R. (2011) 1D and 3D Numerical Simulations in PEM Fuel Cells. *International Journal of Hydrogen Energy*, **36**, 12486-12498. http://dx.doi.org/10.1016/j.ijhydene.2011.06.133

2

Power System Harmonics Study for Unbalanced Microgrid System with PV Sources and Nonlinear Loads

Zhanhe Liu[1], Xufeng Xu[1], Hany A. Abdelsalam[1,2], Elham Makram[1]

[1]Electrical and Computer Engineering Department, Clemson University, Clemson, USA
[2]Electrical Engineering Department, Kafrelshikh University, Kafrelshikh, Egypt
Email: zhanhel@clemson.edu, xufengx@clemson.edu, hanyahmed2008@yahoo.com, makram@clemson.edu

Abstract

Harmonics distortion is a crucial problem in microgrid. Harmonic sources can be categorized as two main factors: renewable energy integration and nonlinear loads. Both factors are investigated in this paper. For renewbale energy, photovoltaic (PV) power is one of the most effective solutions for energy crisis and it is showing great potential for serving customers in microgrid. A three-phase PV source model is establised and integrated at different locations in order to observe the impact of harmonics on a microgrid and power quality (PQ). A composite load is modeled using Crossed Frequency Admittance Matrix theory. A practicdal microgrid loacated at GA, USA is used as a study system. The microgrid, PV model and nonlinear load model are simulated in MATLAB/ Simulink environment. The results show the impact of installing PV sources at two types of locations considering linear and composite nonlinear loads. In addition, three PQ indices are discussed to show the numerical impacts with various perspectives.

Keywords

Crossed Frequency Admittance Matrix, Nonlinear Load, Harmonics Distortion, PV, Power Quality Indices

1. Introduction

Apart from conventional sources of power generation, photovoltaic (PV) is a promising zero-pollution renewable energy resource. PV transforms solar energy into electrical power. Investments are more willing to focus on PV energy especially with the fast developments in PV cell technology. As expected, inverters and intermittent characteristic of PV bring negative impacts on micrigrid [1]-[5]. Harmonics distortion is one of the main reasons which leads to these negative impacts. Previous research on PV integration provides a view on PV impacts during integration. In order to consider the intermittent characteristic of solar generation plant, it is necessary to look into weather condition variation during a certain time period [6]. Likewise related simulations have been developed in [7] to see total harmonics distortion (THD) at point of common coupling (PCC). In microgrid, while it is operating in grid connected mode, voltage profile is less distorted and relatively stable within a small range of fluctuation compared to currents. Hence most of negative impacts of PV injection suffer from current distortion. Simulations in [8] also verified that compared to voltages, current distortion was the main problem for PV source. Similar type of approach has been likewise developed in [9], which mostly focuses on voltage profiles and PV output using FFT analysis.

In addition to the PV sources of microgrid, house appliances or small scale of business buildings are the majority of customers, which are considered as nonlinear loads. Therefore, nonlinear loads is another source of harmonics that should be investigated with PV injection. In [10], personal computer and compact fluorescent light (CFL) have been included while modeling the residential system. In [11], a frequency-domain harmonic model has been introduced only for compact fluorescent lamps (CFL). However, a composite model, which consists of multiple types of nonlinear load, can present a more accurate performance on simulating harmonics distortion. In [12], a mathematical approach has been introduced through crossed frequency admittance matrix. This method is able to represent interactions between voltages and currents of different harmonic orders. Two simplified load models (diode bridge rectifier load (DBR) and phase angle AC voltage controller (PAVC)) have been studied in [13].

Power quality (PQ) indices are used to quantify the quality of the system and comparing the negative impacts of different disturbances/modeling on power networks [14] [15]. PQ indices provide different approaches to compare and distinguish power disturbances, which have great significance for microgrid monitoring and operation. Therefore, PQ indices can work as references, especially for microgrid with renewable energy.

This paper studies the impact of PV sources on a microgrid system considering composite linear/nonlinear loads. The crossed frequency admittance matrix method is used for nonlinear load modeling. In addition to the study of harmonic distortion, three PQ indices are applied and compared in order to give a clear view about the power distortion, waveform distortion and system unbalance. The paper is organized as follows: Section 2 describes the PV source and its control scheme, Section 3 describes the composite nonlinear load model based on cross-frequency admittance matrix method, Section 4 introduces the harmonic distortion method and PQ indices, Section 5 presents a description for the microgird study system and the corresponding data, Section 6 presents the simulation results, and Section 7 concludes the study and future direction.

2. PV Source

PV source consists of PV panel (a set of PV cells), DC-DC boost converter with MPPT, coupling capacitor, three phase inverter and its control and filter. In this paper, KC200GT is chosen as a model of PV cell [16]. The KC200GT data is determined by its own manufacture process. The data can be obtained from the manufacturer data sheet. There are two main methods for inverter control with PV integration: one is the active power control, and the other is the voltage frequency control. Since the harmonic current influences on grid-connected mode are being studied, active power control is the better option for lowering the reactive power output of PV sources. **Figure 1** shows the configuration of three phase active power inverter control. The dq transformation block converts the three components (i_a, i_b, i_c) of AC currents into two components (i_d, i_q). Active and reactive power signals (P_{ref}, Q_{ref}) are used to obtain the reference signal (i_{dref}, i_{qref}) using the matrix solver in (1) [17].

$$L\frac{\mathrm{d}}{\mathrm{d}t}\begin{bmatrix} i_d(t) \\ i_q(t) \end{bmatrix} = \begin{bmatrix} -R & \omega L \\ -\omega L & -R \end{bmatrix}\begin{bmatrix} i_d(t) \\ i_q(t) \end{bmatrix} - \begin{bmatrix} u_d(t) \\ u_q(t) \end{bmatrix} + \begin{bmatrix} e_d(t) \\ e_q(t) \end{bmatrix} \tag{1}$$

where L and R represent inductance and resistance of the impedance Z_a, Z_b and Z_c between three phase inverter and voltage feedback in **Figure 1**. Since the purpose of this control scheme is to force the reactive power output close to zero, the reference on q-axis set to zero. The u_d and u_q are the output control signals of the current control block.

3. Nonlinear Load Modeling

The crossed frequency admittance matrix model [12] is based on the relationship between voltages and currents as

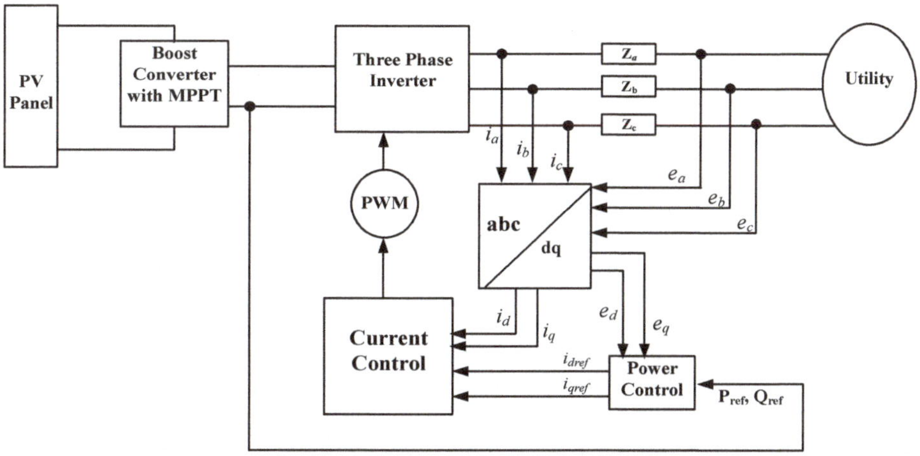

Figure 1. PV inverter control scheme.

$$\begin{bmatrix} \dot{I}_1 \\ \dot{I}_2 \\ \dot{I}_3 \\ \vdots \\ \dot{I}_m \end{bmatrix} = \begin{bmatrix} \dot{Y}_{11} & \dot{Y}_{12} & \dot{Y}_{13} & \cdots & \dot{Y}_{1m} \\ \dot{Y}_{21} & \dot{Y}_{22} & \dot{Y}_{23} & \cdots & \dot{Y}_{2m} \\ \vdots & \vdots & \ddots & & \vdots \\ \vdots & \vdots & & \ddots & \vdots \\ \dot{Y}_{m1} & \cdots & \cdots & \cdots & \dot{Y}_{mm} \end{bmatrix} \begin{bmatrix} \dot{V}_1 \\ \dot{V}_2 \\ \dot{V}_3 \\ \vdots \\ \dot{V}_m \end{bmatrix} \tag{2}$$

where voltages and currents are represented as complex numbers. The subscript m stands for the harmonic order. For pure fundamental frequency voltage sources, voltage vector elements are zeros except V_1. But currents flowing into loads still have harmonic distortion due to the fact that Y_{21}, Y_{31} and Y_{m1} are not all zeros for nonlinear load. On contrary, for linear load, the crossed frequency admittance matrix becomes a diagonal matrix and the off diagonal elements are zeros. The crossed frequency matrix models the load as a harmonic currents source. In this paper, a composite load consists of a linear load and three types of nonlinear loads. Linear load is very clearly built in software platform. Three different crossed frequency admittance matrixes are built for the three nonlinear load models, which are DBR, CFL and PAVC loads. DBR, CFL and PAVC cover major appliances for both residential and commercial building in microgrid system. Circuit diagrams used in this research are given in **Figure 2**.

For the purpose of composite load simulation, with given voltages, output currents can be monitored. When monitoring output currents, time domain current waveform is analyzed by the Fast Fourier Transformation (FFT) so that it can be expressed as a complex number matrix. Each row of this matrix stands for a current on a distinct harmonic order. With only fundamental voltage, 60 Hz, first column elements in crossed frequency admittance matrix can be calculated. Then by superimposing 3rd, 5th, 7th, 9th, 11th and 13th order harmonic voltage sources, each one at a time, the other columns can be calculated likewise. Therefore, three 7×7 admittance matrixes are built for the three types of nonlinear loads. **Figure 3** shows the block diagram of the composite load model used in the Simulink simulation.

4. THD and PQ Indices

THD is the measure of the harmonic distortion at each bus in the system. In order to look at the harmonic distortion in the whole system, whole system harmonics distortion level (WSHDL) is defined as:

$$WSHDL = \sqrt{\frac{\sum_{i=1}^{n} THD_i}{n}} \tag{3}$$

$$THD = \sqrt{\sum_{h=1}^{m} \left(\frac{I_h}{I_1}\right)^2} \tag{4}$$

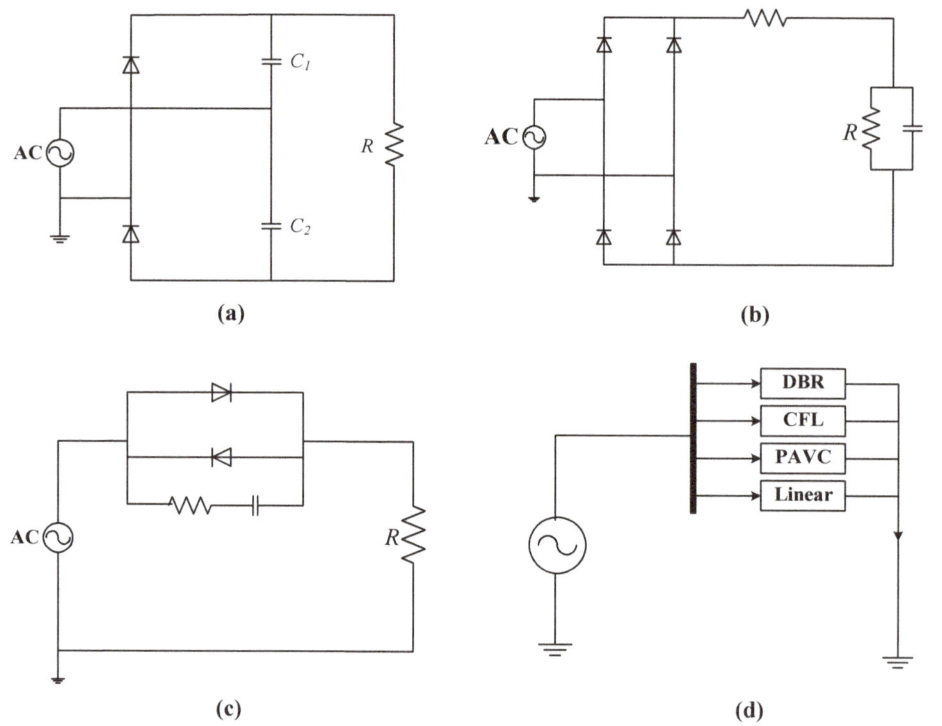

Figure 2. Circuit diagrams of nonlinear loads: (a) CFL; (b) DBR; (c) PAVC; (d) Composite load.

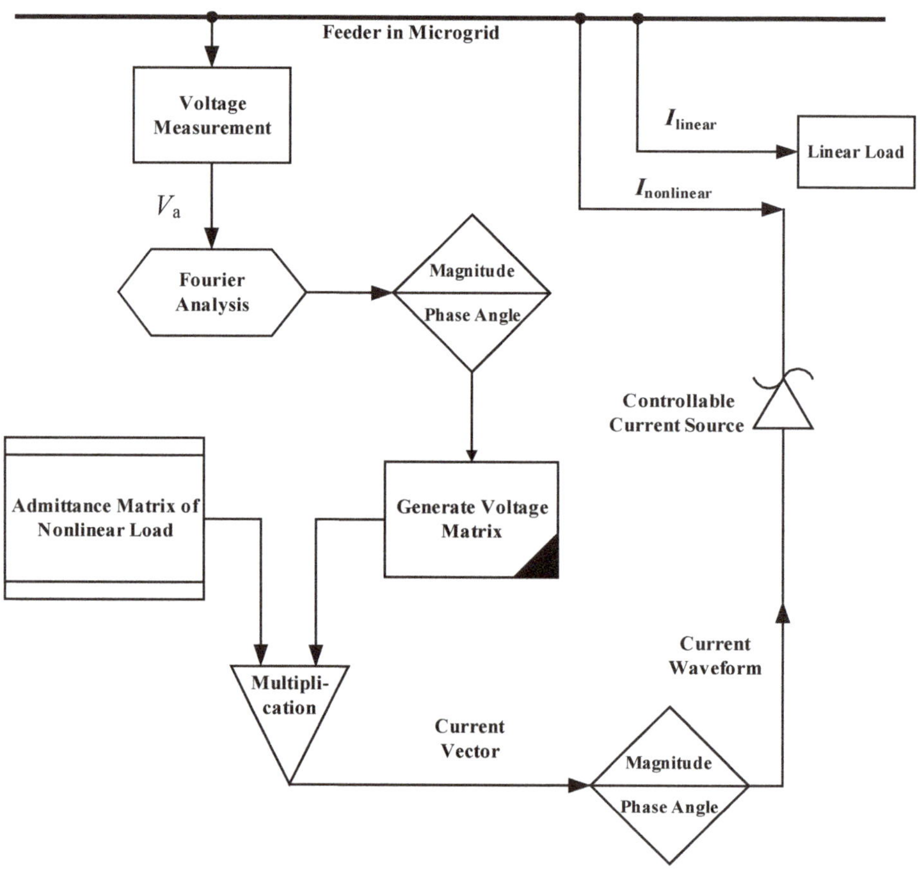

Figure 3. Composite load diagram.

where n is the number of nodes; h represents the harmonic order; m is the considered maximum harmonic order. I_1 is the absolute value of fundamental (60 Hz) current.

In addition to the *WSHDL*, there are three types of PQ indices are introduced and discussed along with simulation results. Distortion power *DP* index is defined in (5) [15]. The total apparent power, fundamental active power and fundamental reactive power are defined in (6)-(8).

$$DP = \sqrt{S^2 - P_1^2 - Q_1^2} \approx V_1 \cdot I_1 \cdot \sqrt{THD_V^2 + THD_I^2 + \left(THD_V \cdot THD_I\right)^2} \tag{5}$$

$$\text{where the total apparent power } S = \sqrt{\sum_{n=0}^{m-1} V(n)^2} \times \sqrt{\sum_{n=0}^{m-1} I(n)^2}, \tag{6}$$

$$\text{fundamental active power } P_1 = V_1 \cdot I_1 \cdot \cos(\theta), \tag{7}$$

$$\text{fundamental reactive power } Q_1 = V_1 \cdot I_1 \cdot \sin(\theta) \tag{8}$$

By normalizing *DP* to unity, normalized DP^{norm} index can be obtained. DP index has the ability to show contributions of distortion power from individual customers to PCC. Waveform distortion WD index is defined as [14]:

$$WD = \left(\sqrt{\left|I_{m1} - I_1\right|^2 + \sum_{i=2}^{M} I_{integ-h,i}^2 + \sum_{j=1}^{N} I_{inter-h,j}^2} \right) \Big/ I_1 \tag{9}$$

where I_1 is the rated current magnitude and I_{m1} is the measured fundamental current magnitude. $I_{integ-h,i}$ is the i_{th} integer harmonic component and $I_{inter-h,j}$ is the j_{th} inter-harmonic component. WD index expresses how much a component, AC current, is distorted or deviated from ideal sinusoidal waveform. WD index includes inter-harmonic components, which can take a large part of harmonics when different types of inverters involved. Also inter-harmonics cannot be presented by THD which only include integer order of harmonics. Instead of an average value, WD index gives an instantaneous distortion ratio and it can be depicted along with time axis to be monitored. Symmetrical components deviation SCD index is defined as [15]:

$$SCD = \left(\sqrt{\left|I_{mp} - I_1\right|^2 + I_{mn}^2 + I_{mz}^2} \right) \Big/ I_1 \tag{10}$$

where I_{mp}, I_{mn}, I_{mz} are the measured currents at positive, negative and zero sequences. SCD index has a significant meaning for microgrid because it can give the level of unbalance on currents. SCD index can help staff who work in substations to recognize the unbalance in each node and its impact on the whole system.

5. Study System

Study system shown in **Figure 4** is a 14-bus 4.16 kV microgrid (located in GA, USA). Bus 10 is the slack bus and the balance point. The slack bus voltage is set to 1.05 pu to keep all node voltages at least 0.95 pu. In **Figure 4**, three lines represent three phase feeder. Two lines represent two phase feeder and single line means one phase feeder. Load information is given in **Table 1**. The PV sources are integrated to the system as shown in **Figure 4**. The inverter control inputs e_a, e_b, e_c, i_a, i_b, i_c are obtained from monitoring the integration point in microgrid. If these values are away from ideal balanced condition and changing along with time, then PV sources will provide more harmonics than running under ideal balanced condition. **Figure 5** shows the solar irradiance and temperature during 24 hours (average value of typical day of July within 10 years). In this work, the simulation only runs during the sunniest period (7 am to 6 pm).

6. Simulation Results

The system in Section 5 is simulated in MATLAB/Simulink and analyzed considering the PV model in Section 2 and the composite load model in Section 3 to study the harmonic distortion and PQ indices described in Section 4. The PV data are set as in **Table 2**. After running the power flow in the basic microgrid (without PV integration and without nonlinear loads), voltages on each node are in the acceptable fluctuation range (0.95, 1.05 pu).

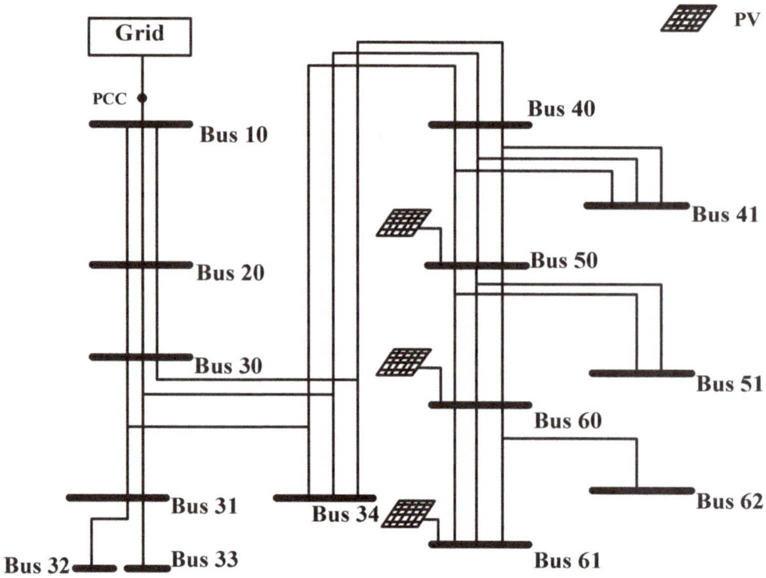

Figure 4. Microgrid study system.

Table 1. Load data of the study system.

Bus	Load Data for the System		
	Note	Active Power (kW)	Reactive Power (kVar)
10	—	—	—
20	Three phase	630	212
30	Three phase	412	112
31	—	—	—
32	One phase	37	12
33	One phase	24	8
34	Three phase	343	122
40	Three phase	175	100
41	Three phase	133	51
50	Three phase	298	151
51	Two phase	68	12
60	Three phase	200	70
61	Three phase	74	28
62	One Phase	32	11

6.1. Impact of PV Sources with Linear Loads

Two cases are studied based on PV locations. In case I, PV locations shown in **Figure 4** are considered. Applying Equation (2) gives the harmonics distortion data for each bus from 7:00 am to 6:00 pm. WSHDL values demonstrates the harmonics distortion in the microgrid system as shown in **Figure 6**. In **Figure 6**, the harmonic distortion is significant at noon time. In other words, as the power supplied by PV is going higher, the WSHDL is going higher. THD provides analysis in each bus as shown in **Figure 7**. According to the standards in [18], when voltage level is under 69 kV and I_{sc}/I_L is smaller than 20, THD of current is regulated under 5% by standards.

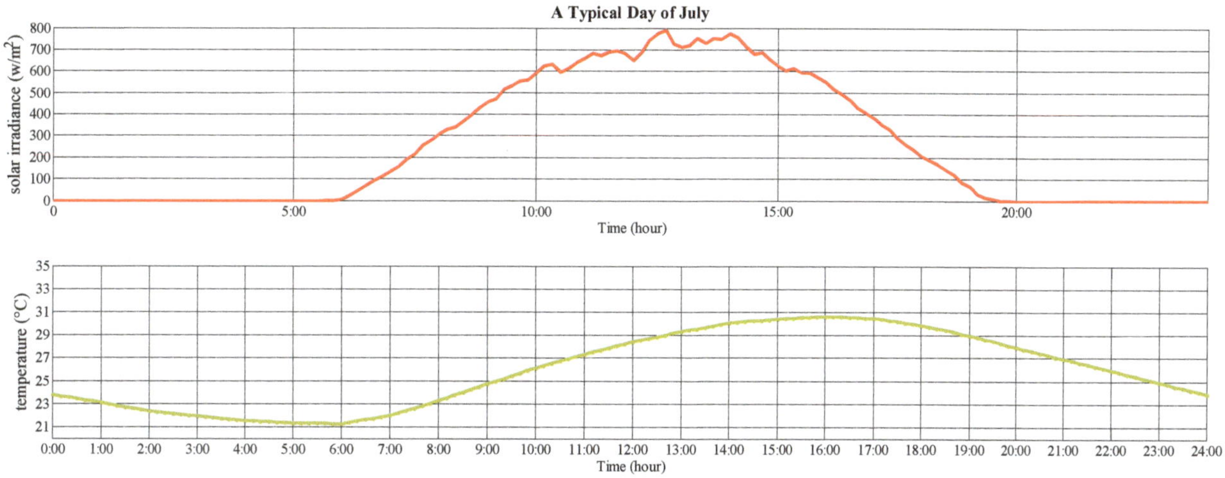

Figure 5. Meteorological data on the typical day of july.

Table 2. PV system data.

Element	Data
PV array	600 kW maximum power at 1000 W/m^2 irradiance
DC-DC boost converter	1 kHz, 300 V boost converter with MPPT controller.
DC filter	0.001 F, 100 μH for output of the PV array. 0.003 F capacitor for inverter input.
3-PHASE INVERTER system	1 MW, 720 V DC/300V AC with active power control
AC filter	600 μH inductor
Coupling transformer	1 MVA, 60 Hz, 300 V/4.16kV.

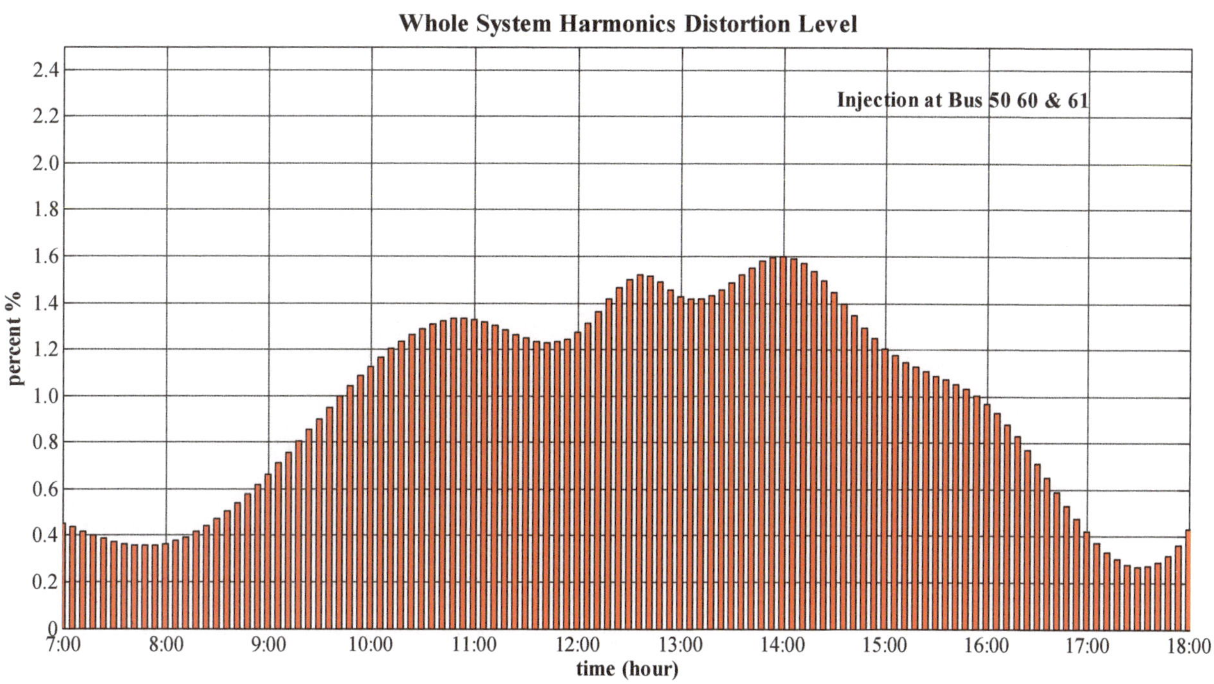

Figure 6. WSHDL from 7:00 AM - 6:00 PM (Case I).

So the THD values above 5% represent nodes that negatively influenced by PV integration. In case II, the PV sources are integrated at buses 20, 40 and 60. **Figure 8** shows the WSHDL of case II. In **Figure 9**, the red line (phase A of bus 50) is the most affected. However, THD value on all these nodes are under 5%, which is acceptable for microgrid system. Compared to case I, the integration locations affect the THD. In case I WSHDL is large when the penetration level is high; however in case II WSHDL is not affected much by the penetration level change. This results give a clear demonstration about how the integrations of PV affect the microgrid distortion. However, when THD is over 5% the power quality is required to be improved. Methods of decreasing current harmonics are being studied by scholars, for example innovative filter and compensation techniques (which are out of focus of this work).

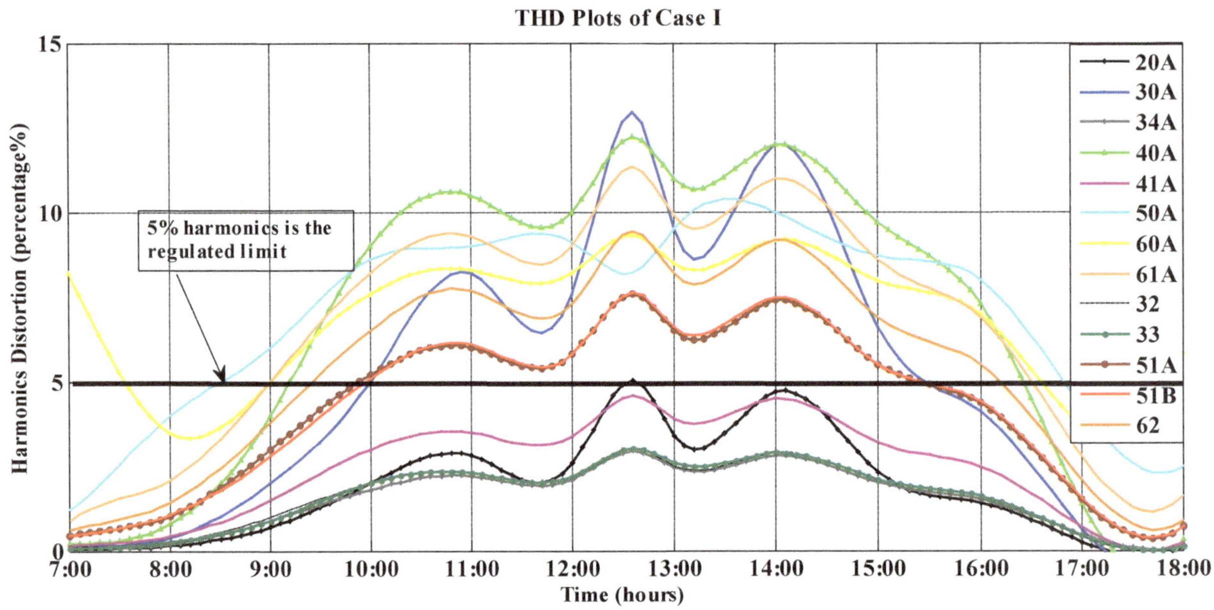

Figure 7. THD (Case I).

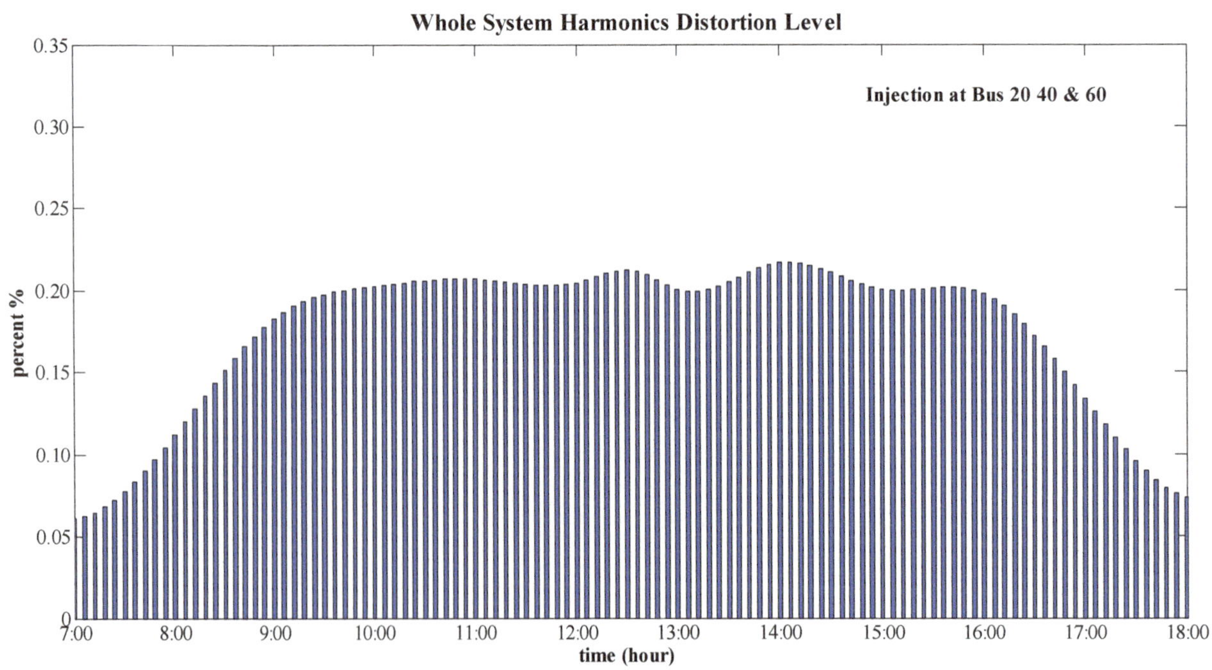

Figure 8. WSHDL from 7:00 AM to 6:00 PM (Case II).

6.2. Considering PV with Composite Load

This section investigates the effect of nonlinear loads. According to experiences from Grainger Industrial Supply [19], a normal business building can withstand up to 15% of nonlinear load without apparent negative influences. But when it comes larger than 15%, necessary devices should be installed to improve power quality. Nonlinear loads are added at different percentage w.r.to the linear loads. Blue solid line in **Figure 10** shows the current waveform with 10% of nonlinear load at Bus 61, phase A. At this time, the THD is 4.0% as sown in **Figure 11**. Red solid line in **Figure 10** represents the 15% of nonlinear load at the same node and the corresponding THD is 5.6%. **Figure 11** shows how the current harmonic distortion increase with increasing the percentage of nonlinear load.

6.3. Impact of PV and Composite Load on the PQ Indices

For the purpose of PQ indices investigation, the microgrid system in **Figure 4** is simplified as shown in **Figure 12**. An equivalent PV source is considered at the PCC point. The simplified system consists of four main loads. The composite load information is shown in **Table 3**. Bus 51 (considered as a small scale factory and because of

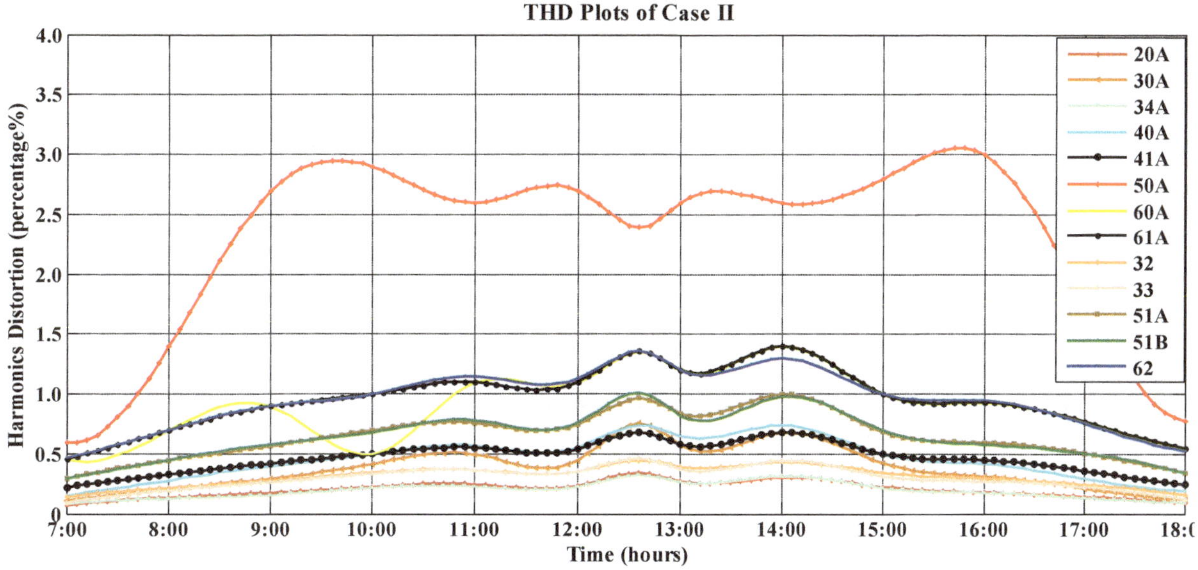

Figure 9. THD (Case II).

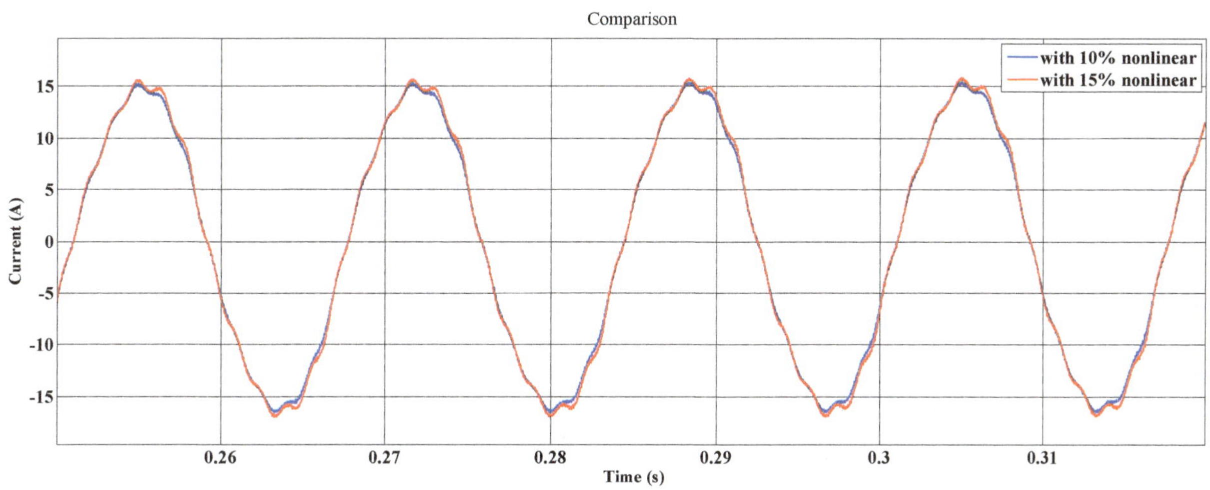

Figure 10. Load current of Bus 16A with nonlinear load.

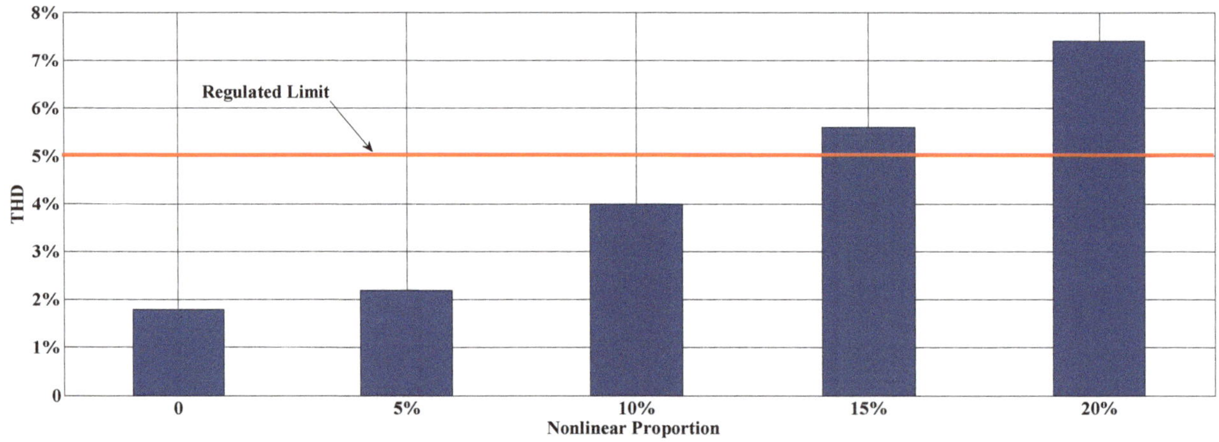

Figure 11. THD comparison of different loads with PV injection.

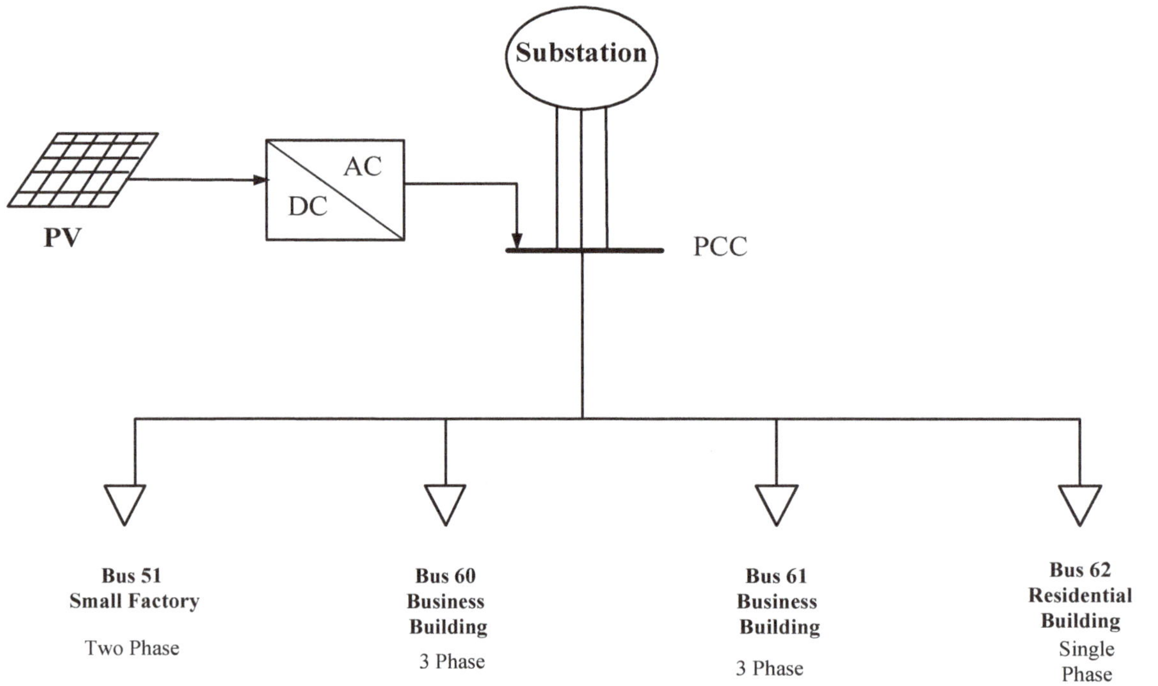

Figure 12. Simplified system configuration.

Table 3. Composite load information.

Bus Number	Phase	DBR	CFL	PAVC	Linear Load
51	A&B	15%	15%	30%	40%
60	ABC	20%	20%	10%	50%
61	ABC	25%	25%	10%	40%
62	C	15%	10%	15%	60%

large amount of motors and heating pumps) is a PAVC type of nonlinear loads. Buses 60 and 61 (two business buildings which contain many devices like computers) are classified as DBR. Load of bus 62 is considered residential. PQ indices section 4 are applied on the simplified system. Harmonics distortion ranking (HDR) can be found through descending sorting of DP and WD indices. The simulation results are shown in **Table 4** and **Ta-**

ble 5. All these results are based on currents measurement because voltage distortion is not dominant compared with currents. **Table 4** lists the DP index which implies that the large amount of load contributing more to DP and the ranking gives the same information. However in **Table 5**, the WD index reflects that without considering power but the waveform of currents. Clearly, due to large use of motors, heating pumps and other heavy duty nonlinear loads, even load on Bus 51 does not consume the most power but it still has the worst impact on current waveform. In **Table 5**, the HDR is different with that in **Table 4**; this can give substation engineers multiple views of power quality. The SCD index can indicate how much unbalance the system have from data obtained at PCC. But before applying the SCD equation, one assumption has to be made that in this case study mutual coupling between lines can be neglected due to the short length of each feeder. According to simulation results, the magnitude of current sequences running through PCC is [25.87 12.64 10.23] Amps ([positive negative zero]). By applying Equation (10), the SCD index is 0.54. SCD index could weight from 0 to 1 and larger number shows more unbalanced. The contribution of harmonics from PV inverter, different types of nonlinear loads on each phase and system configuration have determined that this microgrid is very unbalanced. Engineers who work in substations can also use SCD as power quality index for regulating the system.

7. Conclusion

This paper discussed the impact of two harmonic sources in a microgrid. First, the PV source integration is studied using linear loads. Then PV sources are combined with nonlinear loads. While considering only linear loads, this study provides a comparison between two types of PV integrating locations. Then the whole system

Table 4. Distortion power index.

Bus Number	Phase	DP^{norm}	HDR
51	A	0.0791	7
51	B	0.0787	8
60	A	0.1417	3
60	B	0.1798	1
60	C	0.1645	2
61	A	0.0793	6
61	B	0.1072	4
61	C	0.0977	5
62	C	0.0720	9

Table 5. Waveform distortion index.

Bus Number	Phase	WD	HDR
51	A	0.482	2
51	B	0.511	1
60	A	0.301	9
60	B	0.336	7
60	C	0.324	8
61	A	0.382	6
61	B	0.453	3
61	C	0.419	5
62	C	0.437	4

harmonics distortion level as well as the THD are considered as the main measures to evaluate the PQ and impact of PV integrations. Instead of single PV source, this study investigates mutual influences with multiple PV sources. A composite load model helps to analyze the harmonic distortion compared to linear loads. An actual microgrid with historical-based weather conditions is used as a study system. Simulation results can indicate a discernible look at these negative impacts on the microgrid PQ. In order to make numerical comparison and ranking PQ indices are used to estimate the current distortion from several different perspectives. Three PQ indices are applied and compared to investigate the power distortion, waveform distortion and system unbalance. Further research may develop an algorithm to find the optimized integration plan for unbalanced microgrid system, with multiple PV sources and nonlinear loads.

References

[1] Katiraei, F. and Aguero, J.R. (2011) Solar PV Integration Challenges. *IEEE Power and Energy Magazine*, **9**, 62-71. http://dx.doi.org/10.1109/MPE.2011.940579

[2] Papaioannou, I.T., Alexiadis, M.C., Demoulias, C.S., Labridis, D.P. and Dokopoulos, P.S. (2009) Modeling and Measurement of Small Photovoltaic Systems and Penetration Scenarios. *PowerTech of the IEEE Bucharest*, Bucharest, 28 June-2 July 2009, 1-7. http://dx.doi.org/10.1109/PTC.2009.5281836

[3] Yazdani, A., Di Fazio, A.R., Ghoddami, H., Russo, M., Kazerani, M., Jatskevich, J., *et al.* (2011) Modeling Guidelines and a Benchmark for Power System Simulation Studies of Three-Phase Single-Stage Photovoltaic Systems. *IEEE Transactions on Power Delivery*, **26**, 1247-1264. http://dx.doi.org/10.1109/TPWRD.2010.2084599

[4] Kulkarni, A. and John, V. (2013) Mitigation of Lower Order Harmonics in a Grid-Connected Single-Phase PV Inverter. *IEEE Transactions on Power Electronics*, **28**, 5024-5037. http://dx.doi.org/10.1109/TPEL.2013.2238557

[5] Blaabjerg, F., Teodorescu, R., Liserre, M. and Timbus, A.V. (2006) Overview of Control and Grid Synchronization for Distributed Power Generation Systems. *IEEE Transactions on Industrial Electronics*, **53**, 1398-1409. http://dx.doi.org/10.1109/TIE.2006.881997

[6] Shirek, G.J. and Lassiter, B.A. (2012) Solar Plant Modeling Impacts on Distribution Systems PV Case Study. *Rural Electric Power Conference of the IEEE*, Milwaukee, 15-17 April 2012, B5-1-B5-10. http://dx.doi.org/10.1109/REPCon.2012.6194569

[7] Kalbat, A. (2013) PSCAD Simulation of Grid-Tied Photovoltaic Systems and Total Harmonic Distortion Analysis. *Electric Power and Energy Conversion Systems of* 3rd *International Conference on of the IEEE*, Istanbul, 2-4 October 2013, 1-6. http://dx.doi.org/10.1109/EPECS.2013.6713002

[8] Schlabbach, J. and Kammer, L. (2006) Prediction of Harmonic Currents of PV-Inverters Using Measured Solar Radiation Data. *Electrotechnical Conference of the IEEE Mediterranean*, May 2006.

[9] Gallo, D., Langella, R., Testa, A., Hernandez, J.C., Papic, I., Blazic, B. and Meyer, J. (2013) Case Studies on Large PV Plants: Harmonic Distortion, Unbalance and Their Effects. *Power and Energy Society General Meeting of the IEEE*, Vancouver, 21-25 July 2013, 1-5. http://dx.doi.org/10.1109/PESMG.2013.6672271

[10] Munir, S. and Li, Y.W. (2013) Residential Distribution System Harmonic Compensation Using PV Interfacing Inverter. *IEEE Transactions on Smart Grid*, **4**, 816-827. http://dx.doi.org/10.1109/TSG.2013.2238262

[11] Yong, J., Chen, L., Nassif, A.B. and Xu, W. (2010) A Frequency-Domain Harmonic Model for Compact Fluorescent Lamps. *IEEE Transactions on Power Delivery*, **25**, 1182-1189. http://dx.doi.org/10.1109/TPWRD.2009.2032915

[12] Fauri, M. (1997) Harmonic Modelling of Non-Linear Load by Means of Crossed Frequency Admittance Matrix. *IEEE Transactions on Power Systems*, **12**, 1632-1638. http://dx.doi.org/10.1109/59.627869

[13] El-Saadany. E.F., Salama, M.M.A. and Chikhani, A.Y. (2000) Passive Filter Design for Harmonic Reactive Power Compensation in Single-Phase Circuits Supplying Nonlinear Loads. *IEE Proceedings-Generation on Transmission and Distribution*, **147**, 373-380. http://dx.doi.org/10.1049/ip-gtd:20000654

[14] Lin, T. and Domijan, A. (2005) On Power Quality Indices and Real Time Measurement. *IEEE Transactions on Power Delivery*, **20**, 2552-2562. http://dx.doi.org/10.1109/TPWRD.2005.852333

[15] Jo, S.H., Son, S. and Park, J.W. (2013) On Improving Distortion Power Quality Index in Distributed Power Grids. *IEEE Transactions on Smart Grid*, **4**, 586-595. http://dx.doi.org/10.1109/TSG.2012.2222943

[16] Villalva, M.G. and Gazoli, J.R. (2009) Modeling and Circuit-Based Simulation of Photovoltaic Arrays. *Power Electronics Conference of COBEP'*09 *Brazilian of the IEEE*, 27 September-1 October 2009, 1244-1254. http://dx.doi.org/10.1109/COBEP.2009.5347680

[17] Cao, H., Zhang, H., Jiang, W. and Wei, S. (2012) Research on PQ Control Strategy for PV Inverter in the Unbalanced Grid. *Power and Energy Engineering Conference of the Asia-Pacific IEEE*, Shanghai, 27-29 March 2012, 1-3.

http://dx.doi.org/10.1109/APPEEC.2012.6307159

[18] Halpin, S.M. (2005) Comparison of IEEE and IEC Harmonic Standards. *Power Engineering Society General Meeting of the IEEE*, 12-16 June 2005. http://dx.doi.org/10.1109/PES.2005.1489688

[19] http://www.grainger.com/content/safety-electrical-power-system-harmonics

3

Voltage Stabilization of Hybrid Micro-Grid Using Super Capacitors

Jamshed Khan[1], **Usman Nasir**[2]

[1]Goldwind Science and Technology Co. Ltd., Beijing, China
[2]North China Electric Power University, Beijing, China
Email: 2480516495@qq.com, 09beeunasir@seecs.edu.pk

Abstract

Scarcity of fossil fuel resources has motivated the researchers to develop renewable energy based power projects. Instead of using a single or independent renewable energy source, it is preferable to use the combination of such energy sources in a distributed way to compensate the power fluctuations of the system and this leads to the concept of hybrid micro-grid energy. Voltage stability is an important parameter for the secure operation of the hybrid-micro grid, and IEEE 1547 Standard defines the limit of the voltage for the successful operation of the micro-grid. Although Vanadium Redox Batteries (VRBs) can help the system to stabilize the voltage when voltage sag occurs when a heavy load is suddenly connected to the system, this stabilization process takes some time. This paper discusses the application of super capacitors to the hybrid micro-grid system, as a higher energy density element, to help the system quickly recover its transient voltage.

Keywords

Super Capacitors, Micro-Grid, Voltage Sags, Voltage Stabilization, Hybrid Micro-Grid

1. Introduction

The centralized power generation system has been used for decades and usually the system utilizes fossil fuel as a resource, but unfortunately, the use of fossil fuel in the power generation poses a threat to human life and more specifically to those people who are situated near the power generation plants. Therefore, due to environmental issues *i.e.* emission of CO_2 and other hazardous gases, the centralized power generation plants are located remotely [1]. In addition to the mentioned problem, the centralized system also faces the problems related to system stability and high transmission line losses because the system is usually located in the remote area [2].

Nowadays, the demand of electricity is also increasing and to overcome the shortage of electricity, development of new centralized systems is not a wise decision. Therefore, instead of increasing the quantity of energy there is a need to improve the quality of energy by employing renewable energy sources *i.e.* wind and solar energy. Since renewable energy technologies unlike fossil fuel based power generation systems are "eco-friendly", locating these "eco-friendly" power generation systems in remote areas is not required. It not only helps in the reduction of the transmission line losses, but also decreases the risk of large blackouts. Hence, this idea leads to the concept of Distributed Generation (DG) using micro-grids because the large scale applications of the DG are possible by employing micro-grid. Micro-grid is developed by grouping a cluster of loads and parallel DG units in local distribution network.

It must be noted that the micro-grid is connected to the grid via Point of Common Coupling (PCC), which, in case of a fault, will disconnect from the grid, and then the micro-grid will operate in the islanded mode. So, micro-grid must be able to operate in grid-connected as well as in islanded mode [3].

In order to keep the micro-grid in the grid-connected mode, certain constraints must be satisfied such as maximum voltage sag and swell, and the limits for maximum and minimum voltages are defined in ANSI C84.1-2006 [4] and referred in IEEE 1547 Standard for Interconnecting Distributed Resources with Electric Power System [5]. For the micro-grid to operate in grid-connected mode, it must maintain a particular level of the voltage for its successful operation because according to the mentioned standards, if the frequency and the amplitude of the voltage exceed the defined value, then PCC breaker will open instantly [6]. Therefore, the transient stability of the voltage must be considered as an important parameter for the stable operation of micro-grid. The removal of a heavy load to the system will cause voltage swells, and similarly, voltage sags will occur when a heavy load is connected [7]. Therefore, it is necessary to have the voltage as sine wave and also it must be among certain defined limits.

This paper presents the practical study of a micro-grid model under the influence of load change, and this is due to the fact that the transient voltage of the system is largely affected by the increase of load. The voltage sag due to the increase of load will increase the amplitude of current, which consequently can cause damage to the power electronic converter. This paper proposes to employ super capacitors in a DG hybrid micro-grid system. Therefore, the inclusion of super capacitors to the micro-grid system becomes necessary.

2. A Hybrid Distributed Generation Micro-Grid System

Today, independent PV based renewable energy system is being employed but due to changing intensity of sun, the power of the PV system is not constant.

Therefore, in order to avoid fluctuations, instead of using an independent PV system, a hybrid system, integrating wind energy system (WES) and energy storage (ESS) elements with the PV system, is usually used. Similarly, the power of an independent WES will adversely affect the power system. Hence, a hybrid system formed due to the integration of ESS *i.e.* battery energy storage system (BESS) and flywheel energy storage system (FESS) with renewable energy technologies *i.e.* PV and wind energy is best possible solution to avoid power fluctuation problems in a power system. A hybrid micro-grid system which is shown in **Figure 1** is developed at Goldwind Science & Technology Co., Ltd. 2010.

The micro-grid hybrid system shown in Fig. 1 includes three PV Energy Systems (PVES) with two of them adding about 400 kW to the system *i.e.* PV1 and PV2 while another PV panel system *i.e.* PV3 contributing 100 kW to the system. The mentioned system also includes a wind energy system, with a capacity of 2.5 MW, connected to the 400 V AC channel via step-down transformer.

As discussed before, in order to secure the system from power surges from WECS and power fluctuations from PV system, two different technologies of EES, each with a capacity of 800 kWh, have also been integrated via bi-directional power electronic converters. One of the EES being used in the system employs Vanadium Redox Battery Energy Storage System (VRBESS). VRB's are usually used in the system to due to its capability of providing high energy density, long cycle life of 10,000 cycles (which is 3 times more than that of lithium ion battery), increased operating temperature range and a long time scale of minutes to hours [8]. The complete modeling of the VRB's has been presented in [9] [10]. The capacity of the Iron Phosphate based Lithium Ion Battery Energy Storage System (LIBESS) was same as of VRB's. It must also be noted that the studied micro-grid operates in grid-connected mode. **Table 1** shows the capability and the specifications of the mentioned experimental system.

The mentioned hybrid micro-grid system also includes two units of 65kW Natural Gas Micro Turbine. C65

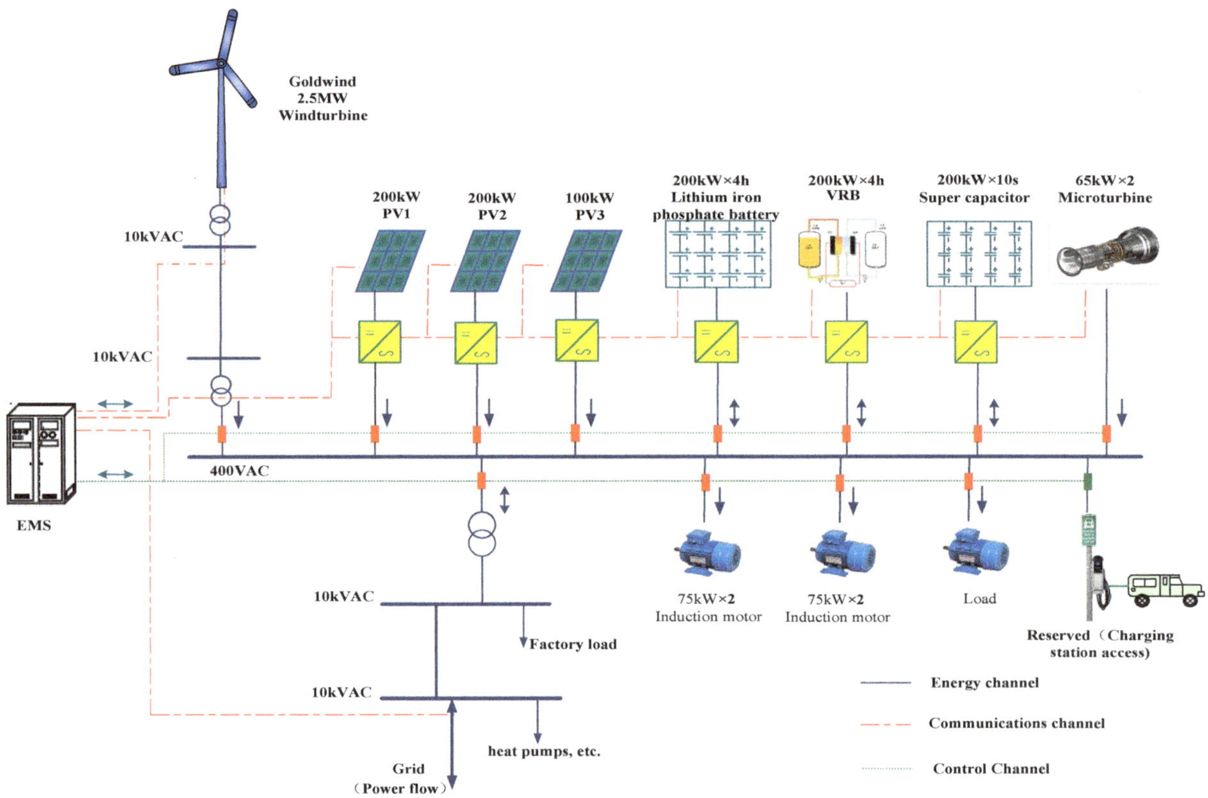

Figure 1. A hybrid micro-grid system topology at Gold wind Co., Ltd.

Table 1. Energy capacity and load specifications for the experimental hybrid micro-grid system at Goldwind Co. Ltd.

Energy Sources/Energy Storage Elements/Load	Number of Units	Specifications Per Unit		Total Power
	3	PV1 = 200 kW		
PV Energy System (PVES)		PV2 = 200 kW	200 kW × 2 +100 kW	500 kW
		PV3 = 100 kW		
		WT1 = 1.5 kW		
Wind Energy System (WES)	2			2.5 MW
		WT2 = 1.5 kW		
		MT1 = 65 kW		
Micro Turbine Energy System (MTES)	2		65 kW × 2	130 kW
		MT2 = 65 kW		
Vanadium Redox Battery Energy Storage System (VRBESS)	1	200 kW for a duration of 4 hours		800 kWh
Lithium Ion Battery Energy Storage System (LIBESS)	1	200 kW for a duration of 4 hours		800 kWh
Super Capacitor Energy Storage System (SCESS)	1	200 kW for a duration of 10 seconds		0.55 kWh
		IM1 = 75 kW		
		IM2 = 75 kW		
Three Phase Induction Motor (TPIM)	4		75 kW × 4	300 kW
		IM3 = 75 kW		
		IM4 = 75 kW		

ICHP combined heat and power (CHP) natural gas based micro turbine has a capability of providing 100A in grid-connected as well as in islanded mode operation of micro-grid whereas, operational voltage range is between 400VAC and 480VAC. One of the advantages is that C65 ICHP has ultra low NOx emissions. **Figure 2** shows the two units of the mentioned micro turbine.

The purpose of this research is to experimentally check the effect of super capacitor units on micro-grid under different load conditions. However, there are different control schemes that have been developed for grid-connected mode and islanded mode of the micro-grid system which has been presented in [11] while [12] discusses the modeling of the PVES and EES.

3. Modeling of the Super Capacitor

The Super Capacitor or Ultra-Capacitor has very higher value of energy density in comparison to the normal capacitors. Unlike a conventional capacitor, the super capacitor doesn't use the dielectric material and it uses "plates".

The equivalent model of the super capacitor is shown in **Figure 3**. The given model has two resistance values and one inductance value. R_S represents the series resistance which is responsible for the internal heating losses in the super capacitor whereas, the parallel resistant *i.e.* R_P represents the self discharge effect and the long term performance is dependent on R_P. The actual behavior of the super capacitor is non-linear and its capacitance value can be considered as a constant value if the significant variations are not present in the parameters such as temperature, voltage and current.

The super capacitor is connected to the grid via bi-directional power electronic converter that can be controlled using a decoupled d-q axis control algorithm [13] in order to compensate the power surges or fluctuations from PVES and WES or load variations. The topology of the super capacitor used is shown in **Figure 4**.

AC filter improves the quality of the current and decreases THD value and then power is fed to the step-up transformer that step ups the voltage from 150VAC to the required grid voltage of 400VAC. The parameters for experiment, transformer and SCESS's converter are given in **Table 2**.

Figure 2. Natural gas based micro turbine units (65 kW × 2) at Goldwind Co. Ltd.

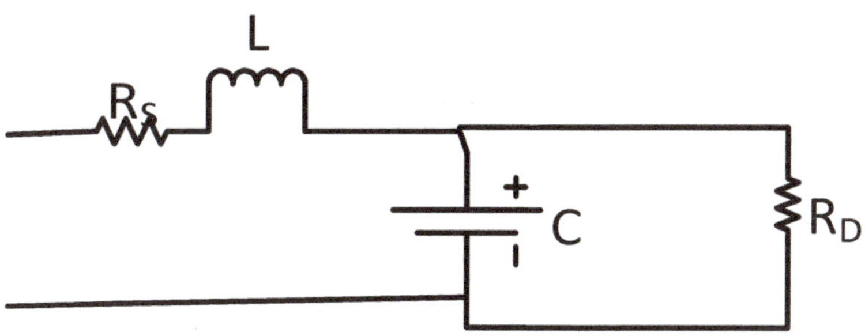

Figure 3. Equivalent model of a super capacitor.

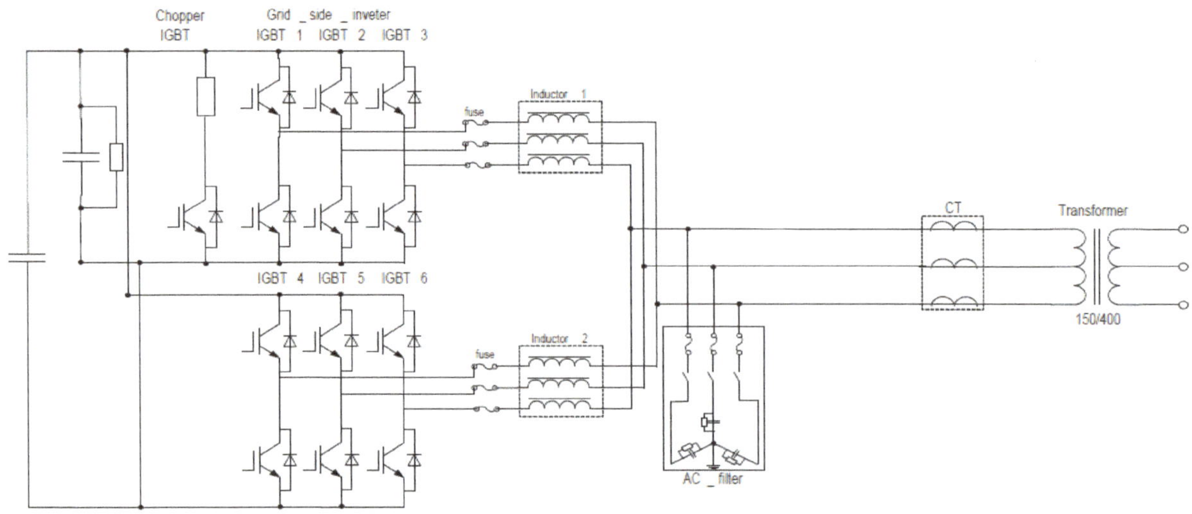

Figure 4. Bi-directional converter topology for SCESS.

Table 2. Parameters.

Parameters for Experiment	Parameters for Converter System	Parameters for Grid Side Transformer	Parameters for Super Capacitor
Altitude (<2000 m)	Rated Voltage (150 V ± 10% Fluctuation)	Type of Transformer (Y-connection)	Rated Capacitance (Approx. 30F)
Protection Type (IP22)	Active Power Capacity (200 kW)	Capacity in Apparent Power (500 kVA)	Operating Temperature Range (−40˚C - +60˚C)
Humidity (10% - 100%)	Rated Frequency (50 Hz ± 5%)		System Working Time (10 s)
Pressure (77,500 Pa - 106,500 Pa)	Switching Frequency (Approx. 2.7 kHz - 3.0 kHz)		
Storage Temperature Range (−40˚C - +60˚C)	Approx. Power Factor (0.9)		
Operating Temperature Range (−30˚C - +50˚C)			

4. Experimental Results

4.1. Transient Response of Micro-Grid under Load Increase of 12 KVA

The transient response of the grid is experimentally studied with and without the super capacitor. In order to check the effect of super capacitor on the grid, an inductive load of about 12 KVA was added to the initial resistive load at t = 80 ms. **Figure 5** shows one phase of the transient voltage response. The response in "red" is the transient voltage with super capacitor whereas the other one represents that of without super capacitor.

It can be noted from **Figure 5** that when the load changes at t = 80 ms approx., the system without super capacitor will take some time *i.e.* approximately 50 milliseconds, to recover the voltage sag while the system with super capacitor has quickly recovered the voltage transient response. The system without super capacitor will take many cycles to restore the voltage to its required level.

4.2. Transient Response of Micro-Grid under Load Increase of 60 KVA

In order to experimentally check the response of the system under the influence of super capacitor, the load is increased to about 5 times the previous value *i.e.* 60 KVA and the transient response of the voltages are shown in **Figure 6**.

Figure 6 shows the three phases of the voltage with phase A in (a), phase B in (b) and phase C in (c). The load change is applied at t = 40 ms approx. and the response show that the system with super capacitor doesn't

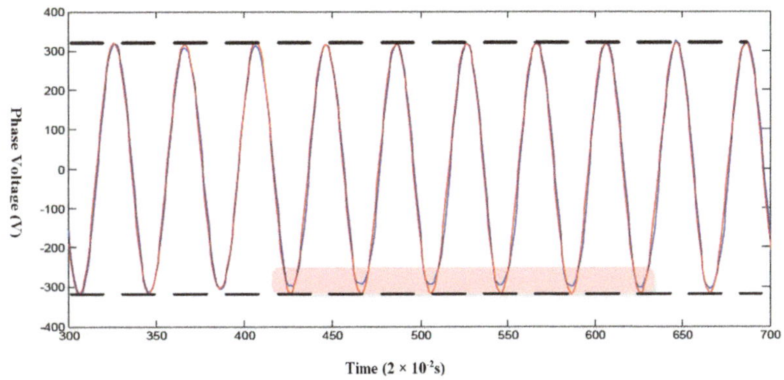

Figure 5. Phase Voltage of the system when load 12 kVA load is added at t = 80 ms ("Blue" without SCESS and "Red" with SCESS).

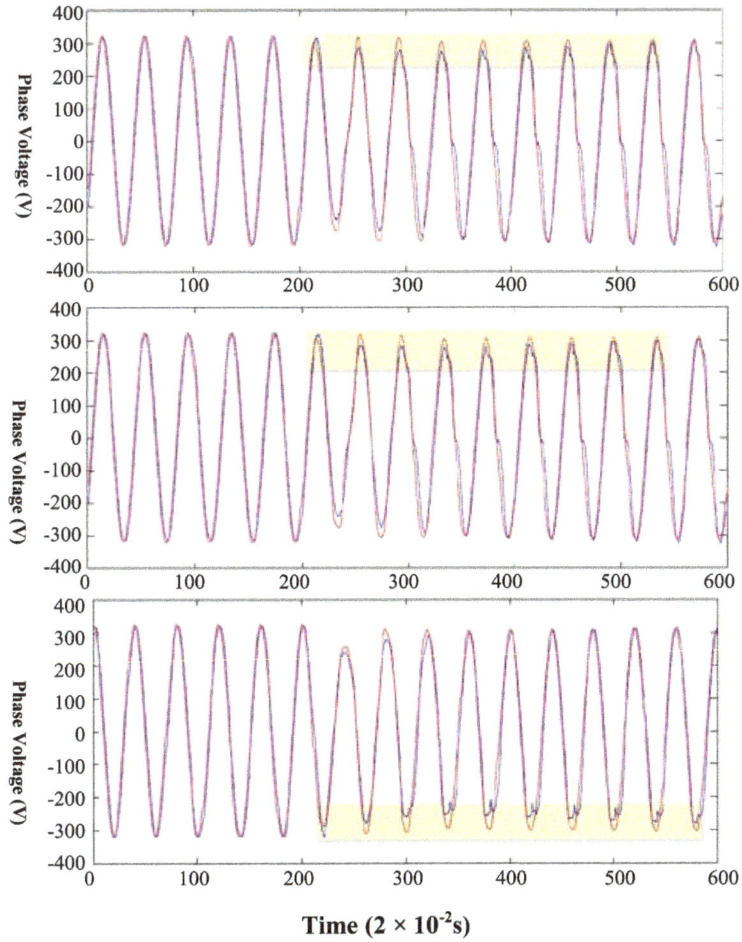

Figure 6. Phase Voltages of the system when load 60 kVA load is added at t = 40 ms ("Blue" without SCESS and "Red" with SCESS) (a) Phase voltage 1; (b) Phase voltage 2; (c) Phase voltage 3.

need 90 milliseconds to recover the voltage and it also shows that it has successfully recovered the voltage sag within a short time of one to two cycles and super capacitor has helped the system to quickly stabilize the response.

4.3. Transient Response of Micro-Grid Due to Sudden Start-Up of an Induction Motor

The starting current for an induction motor has usually a very high value and therefore, the effect of a sudden start of an induction motor on system with and without super capacitor is studied.

Figure 7 shows the voltage and current response comparison for the system with and without the inclusion of super capacitors in the micro-grid. **Figure 7(a)** and **Figure 7(b)** show the three phase transient voltages and currents of the system without super capacitor. For the system without super capacitor, it can be seen that when an induction motor suddenly starts-up at t = 0.27 s approx., voltage sags till t = 0.45 s and then it gets stabilized

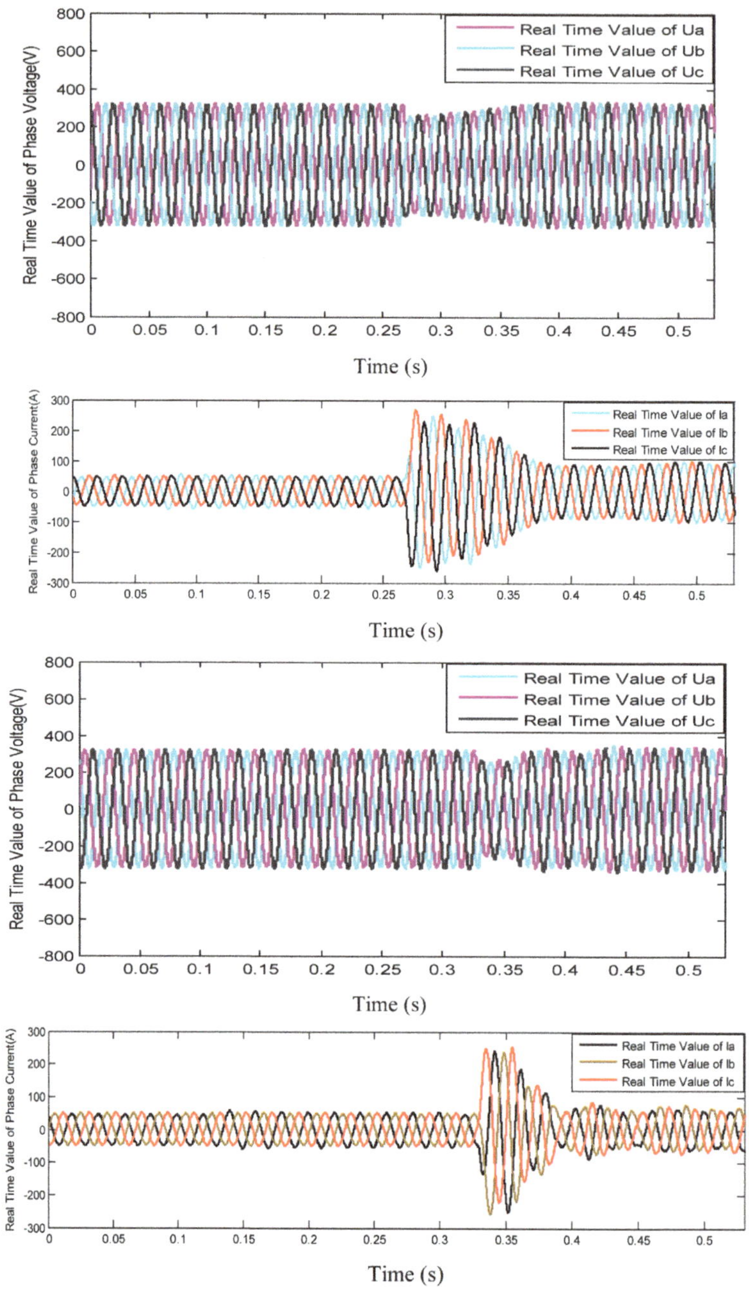

Figure 7. Phase voltages of the system when an induction motor start-ups (a) Three Phase Voltages (without SCESS); (b) Three Phase Currents (without SCESS); (c) Three Phase Voltage (with SCESS); (d) Three Phase Currents (with SCESS).

in about 180 ms but for the system with super capacitor, the induction motor starts-up at t = 0.33 s approx., the voltage sets stabilized in less than 70ms. Similarly, the amplitude of three phase currents increase when the induction motor is connected and current takes less time to stabilized for the system with super capacitor as shown in **Figure 7(d)**.

5. Conclusion

The concept of hybrid micro-grid technology is emerging because of its stable and eco-friendly behavior. Transient analysis of the system is very important in this regard for its successful and stable operation. The parameters such as upper and lower limits of the system's voltages must be taken into consideration, and a careful observation is required to check the transient response of the system under different loads because the hybrid micro-grid must develop a robust system that can work in grid-connected as well as islanded mode. The transient response of the voltage is disturbed for some cycles when a particular load change occurs in the micro-grid system. Although vanadium redox batteries (VRBESS) have higher energy density than lithium-ion batteries (LIBESS), they are not fast enough to stabilize the response. Therefore, in addition to VRBESS and LIBESS, an introduction of a super capacitor (SCESS), as very high energy-density energy storage equipment, will help the system to recover and stabilize three phase voltages rapidly. Hence, using a super capacitor in a hybrid micro-grid system is a requirement of today's micro-grid technology.

Acknowledgements

We would like to thank Goldwind Science & Technology Co. Ltd. Beijing, China for providing us an opportunity to enhance our research potential by co-operating with us and helping us to use their equipment for experimental verification of stabilizing voltage using super capacitors. Mr. Jamshed Khan is associated with Goldwind Science & Technology Co. Ltd. Beijing, China.

References

[1] Hassanzahraee, M. and Bakhshai, A. (2011) Transient Droop Control Strategy for Parallel Operation of Voltage Source Converters in an Islanded Mode Microgrid. *IEEE 33rd International Telecommunications Energy Conference* (*INTELEC*), Amsterdam, 9-13 October 2011, 1-9.

[2] Liang, Y.W., Hu, Z.J. and Chen, Y.P. (2003) A Survey of Distributed Generation and its Application in Power System. *Power System Technology*, **27**, 71-75.

[3] Han, M.X., Su, X.L., Lin, S.B. and Zhao, Z.K. (2013) Transient Analysis and Control for Microgrid Stability Controller. 2013 *IEEE PowerTech* (*POWERTECH*), Grenoble, 16-20 June 2013, 1-6.

[4] ANSI C84.1-2006 Electric Power Systems and Equipment—Voltage Ratings (60 Hertz). http://wenku.baidu.com/view/1554580d6c85ec3a87c2c563.html

[5] IEEE Standard for Interconnecting Distributed Resources with Electric Power Systems, IEEE Standard 1547, 2003.

[6] Xu, Z., Zhang, P. and Wang, Q. (2012) The Key Technology Research on Transient Stability Control of Micro-Grid. *China International Conference on Electricity Distribution* (*CICED*), Shanghai, 10-14 September 2012, 1-5. http://dx.doi.org/10.1109/CICED.2012.6508621

[7] Deng, W., Pei, W. and Qi, Z.Q. (2008) Impact and Improvement of Distributed Generation on Voltage Quality in Micro-Grid. *3rd International Conference on Electric Utility Deregulation and Restructuring and Power Technologies*, Nanjing, 6-9 April 2008, 1737-1741. http://dx.doi.org/10.1109/DRPT.2008.4523687

[8] Li, W. and Joós, G. (2008) A Power Electronic Interface for a Battery Supercapacitor Hybrid Energy Storage System for Wind Applications. *IEEE Power Electronics Specialists Conference* (*PESC*), Rhodes, 15-19 June 2008, 1762-1768. http://dx.doi.org/10.1109/pesc.2008.4592198

[9] Qiu, X., Nguyen, T.A., Crow, M.L. and Elmore, A.C. (2014) Modeling of Vanadium Redox Battery by Field Analysis and Neural Network Approach. *Power and Energy Conference at Illinois* (*PECI*), Champaign, 28 February-1 March 2014, 1-7.

[10] Chahwan, J., Abbey, C., Chamberland, M. and Joos, G. (2007) Battery Storage System Modelling, Design and Operation for Wind Energy Integration in Power Systems. *CIGRE Canada Conference on Power Systems*, Calgary, 26-28 August 2007, 1-8.

[11] Peas Lopes, J.A., Moreira, C.L. and Madureira, A.G. (2005) Defining Control Strategies for Analysing Microgrids Islanded Operation. *IEEE Russia Power Tech*, St. Petersburg, 27-30 June 2005, 1-7.

http://dx.doi.org/10.1109/ptc.2005.4524548

[12] Li, M.H., Funaki, T. and Hikihara, T. (2007) A Study of Output Terminal Voltage Modeling for Redox Flow Battery Based on Charge and Discharge Experiments. *4th Power Conversion Conference*, Nagoya, 2-5 April 2007, 221-225. http://dx.doi.org/10.1109/pccon.2007.372971

[13] Nasir, U., Iqbal, Z., Minxiao, H. and Rasheed, M.T. (2015) Active and Reactive Power Control of a Variable Speed Pumped Storage System using DFIM's Stator. The Paper Is Accepted and Will Be Presented at Rome, Italy, 10-13 June 2015.

4

Dual Layered Architecture for Multi Agent Based Islanding and Load Management for Microgrids

Al Kulasekara[1], Ktmu Hemapala[2], Rarc Gopura[1]

[1]Department of Mechanical Engineering, University of Moratuwa, Katubedda, Sri Lanka
[2]Department of Electrical Engineering, University of Moratuwa, Katubedda, Sri Lanka
Email: udayanga@elect.mrt.ac.lk

Abstract

This paper proposes a novel dual layered multi agent system (MAS) based control system for the use in microgrid operations. In developing a smarter grid capable of withstanding disturbances and/or outages and providing quality service to the consumers, reliable microgrid control architecture is vital. The innovative microgrid control system proposed, makes the microgrid capable of isolating the local grid from effects of any upstream disturbances in the main utility grid by operating disconnected from the main utility via islanding, and it allows the most critical local loads to be supplied by any, available, local power source during such islanded operation. The proposed MAS control architecture is developed using the JADE platform and it is used to control a test network simulated in MATLAB. The results of these simulations show the capability of developing MAS based reliable control mechanism for islanding and load management of microgrids based on the proposed concept.

Keywords

Autonomous Agents, Load Management, Microgrids, Multi-Agent Systems

1. Introduction

The modern world moves forward with an ever increasing demand for electricity. This growth in demand for electrical power is becoming a major challenge for electrical supply authorities worldwide with the present energy crises. Dwindling oil reserves, consequent rise in electricity costs and the mounting need to address concerns regarding climate change due to the emission of greenhouse gases (GHG) are forcing utilities to look at new concepts in electricity generation, transmission, distribution and utilization [1].

The present grid systems which have mostly been built over the last century are rapidly aging. Legacy power distribution systems are facing a tough challenge in meeting the needs of a rapidly evolving electricity market. Time is ripe for a smarter grid, capable of meeting the present and future power delivery requirements [1]-[3]. The need to reduce GHG emissions have ushered in the need for greener electricity generation and incorporating renewable energy sources into the energy mix. These complications demand a paradigm shift from large scale fossil fuel based sources towards small scale lower emission systems. The incorporation of smaller distributed generation (DG) units with maximum renewable energy penetration becomes vital in meeting these demands.

Centralized control systems are more powerful in controlling available resources in traditional power systems. However the current power network, with more renewable power sources with intermittent nature, dynamic behavior has increased considerably. In this context literature has recognized multi agent based systems gives more flexibility to the power network [4].

Grids with such DG systems require the ability to handle active generation and loads, which existing legacy system designs are unable to address [2]. Smaller "micro" grids comprising local generation is a concept put forward to cater this requirement [2]. These microgrids can be operated either in island mode, where the local loads are fully supplied by the local generation, or in grid connected mode, where the microgrid is either exporting or importing power from the main grid. With the local generation and loads coming into the picture, local distributed control is essential for the grid stability.

Therefore, these power systems have to become smarter with local distributed control capabilities and more reliable and robust in taking on renewable energy sources without losing stability and efficiency. Smarter microgrids are an answer to this situation. As an innovative method in providing this flexible distributed control requirement, Multi Agent Systems (MAS) are stepping forward [3]. MAS based applications can be found in several areas in the power system, such as electricity market applications [5] [6], smart grid related areas [7] [8] and power control applications [9]. Under power system control, load restoration [10] and load shedding [11] and are the key areas for microgird applications.

Agents in the MAS can use artificial intelligent (AI) techniques to attain the proper control of the microgrid to keep the system voltage and the frequency within the acceptable level [12] [13]. MAS architecture play a vital role in these applications [14] [15]. This paper proposes a novel MAS architecture which provides scalable and robust distributed control architecture for microgrid applications.

The MAS model is designed using the JADE [16] [17] platform and is implemented on a simulated test-bed in MATLAB/SIMULINK.

Rest of the paper is organized as follows. A brief introduction to microgrids and MAS is given in Section 2. A novel dual layered MAS architecture is proposed in Section 3. Simulation results of a conceptual microgrid operation based on the proposed architecture is discussed in Section 4. The conclusions are presented in Section 5.

2. Microgrids and Multi Agent Systems

Centralized, bulk generation facilities are giving way to more distributed and small scale generation systems. Distributed generation (DG) introduces the capability of embedding a wide range of low emission and potentially lower cost generation options. Such options can include internal combustion (IC) engines, micro-turbines and non-conventional renewable energy sources such as solar photo voltaic, wind, biomass, mini/micro-hydro and even waste-to-energy systems. These smaller generation technologies allow the sources to be placed optimally with respect to the loads, thus reducing emissions and transmission losses, while also allowing for local control instead of centralized dispatching [2].

Such DG systems provide the potential to take a sub-system approach in working with microgrids. During a disturbance in the main utility grid, local generation and their loads can be separated from the utility isolating the local loads from the disturbance, while maintaining the integrity of the main utility grid. This isolation is carried out at a single point of connection to the main utility known as point of common coupling. This capability, known as islanding, requires high reliability and flexibility from the microgrid. These requirements can be met by having peer-to-peer control and plug-and-play capabilities in each entity within the microgrid.

A conceptual microgrid model is shown in **Figure 1**. This can be either a small scale distribution level network or an electrical system of a building. Certain loads, which are critical, have local power sources. Non-critical

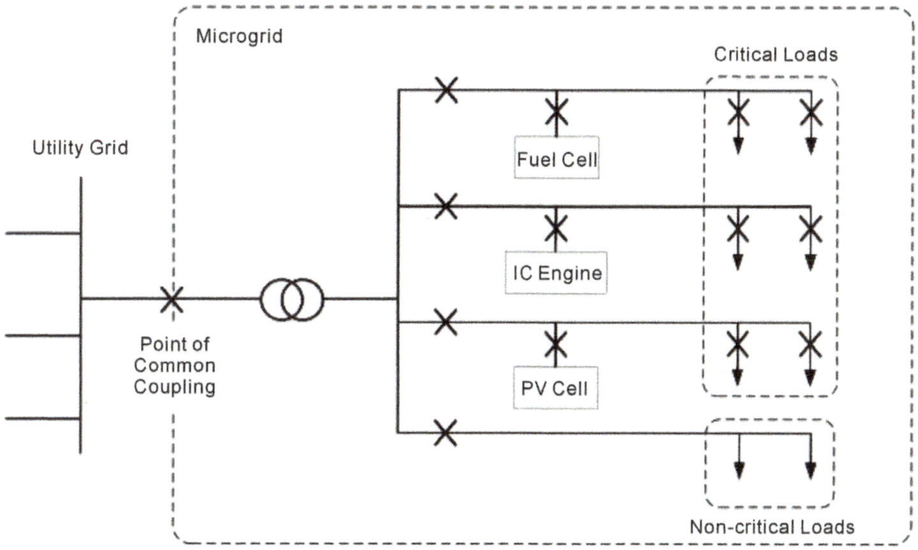

Figure 1. Conceptual microgrid model.

loads have no local supply.

Once an upstream disturbance takes place on the main grid, the microgrid can be separated from the main grid at the point of common coupling and isolate the local loads from any outage, providing a higher level of service, with minimum effect to the stability and the integrity of the main grid. This capability of intentional islanding delivers a higher reliability than legacy power systems [2]. Thereafter, the microgrid has to manage the local loads to make sure that the most critical loads are supplied by the available limited local capacity. This requires priority based shedding of non-critical loads.

2.1. Multi Agent Systems (MAS)

With the need of peer-to-peer control, plug-and-play capabilities required in successfully implementing micro-grids; more robust distributed control architectures are being required. Multi Agent Systems (MAS) have been successfully used to fulfill this requirement [18]-[21]. MAS are a collection of intelligent agents with limited local knowledge, but with enhanced communication capabilities, that can cooperate to achieve a global objective [22].

The notion of agent or agency had been developed for use mainly in computer science applications. A comprehensive comparison of the numerous definitions alongside with their benefits and drawbacks has been presented in [23]. The authors prefer the definition proposed by Wooldridge [22], as found in many of the recent literature. In [22], an agent is defined as a software entity inside an environment able to autonomously react to any changes. Here the notion of an agent, an environment and the ability of autonomy is defined. The environment suggests everything external to the agent. In order for an agent to behave in an environment it should, at least in part, be observable or alterable by the agent. In the microgrid paradigm, the environment and the agent are situated in the power system, which is observable through power-electronic sensors, and alterable through power-electronic devices *i.e.* closing or opening a breaker to reconfigure the power flow.

Definition in [22] differentiates agents from existing systems and objects in programming with the concept of flexible autonomy. Characteristics of agents are: Reactivity—ability to react to sensible changes in the environment; Pro-activity—make intuitive decisions and display goal-oriented behavior; and Social-ability—be able to communicate with other agents using a common language and initiate conversations.

Some existing control systems and software objects exhibit reactivity, but an agent with flexible autonomy differs from them by displaying a degree of intuition, alongside social ability [22]. Integration of two or more such intelligent agents coordinating to achieve a global objective by trying to fulfil local goals is a MAS.

Readers interested in a detailed guide to the concepts, methodologies, technical problems and the applicability of MAS in power systems applications are referred to [18]. An in depth discussion on the standards, tools, ontologies, design requirements and specifications can be found in [19].

2.2. MAS and Distributed Control of Microgrids

Agent systems are inherently distributable owing to their separation from their environment. An agent can be copied to a different location without losing reasoning capabilities or the goals. This allows for an agent to react in different ways to two different environments, given both environments support the intended tasks.

A basic issue for distributed generation is the technical difficulties related to control of a significant number of micro-sources. Research into implementing MAS for microgrid applications has been rising in the recent years and has mostly recently centered on distributed control, electricity trading, and power system restoration [20].

Intelligent agents can be assigned to each entity in a microgrid; *i.e.* circuit breaker, electricity generation unit, loads, users etc. This assignment will determine the MAS architecture. Several different architectures for micro-grid applications have been proposed up to now *i.e.* [23]-[25]. Single layered architectures are the most pre-dominant type in the literature. For example, single layered architectures have been proposed for microgrid applications such as resolving real and reactive power mismatch [25] and load restoration [26]. In these single layered agent topologies, agents are tasked with multiple objectives. For instance, main controlling agent in these architectures is commonly tasked with islanding; load shedding, load management and process control simultaneously. This becomes a shortcoming in improving the scalability and in building robust distributed control models. Therefore, the appropriate delegation of objectives to secondary level agents improves the robustness of the model. Such secondary agents can be tasked separately with islanding control and load shedding, while overseen by a primary agent on the primary layer.

Several multiple level control architectures have been presented in the literature [27]-[29]. However, in these multi layered topologies, the layers address different strata of the distribution grid and not of the control architecture.

3. Proposed Dual-Layered Mas

Figure 2 shows the conceptual design of the proposed dual layered control architecture for microgrids [30]. The

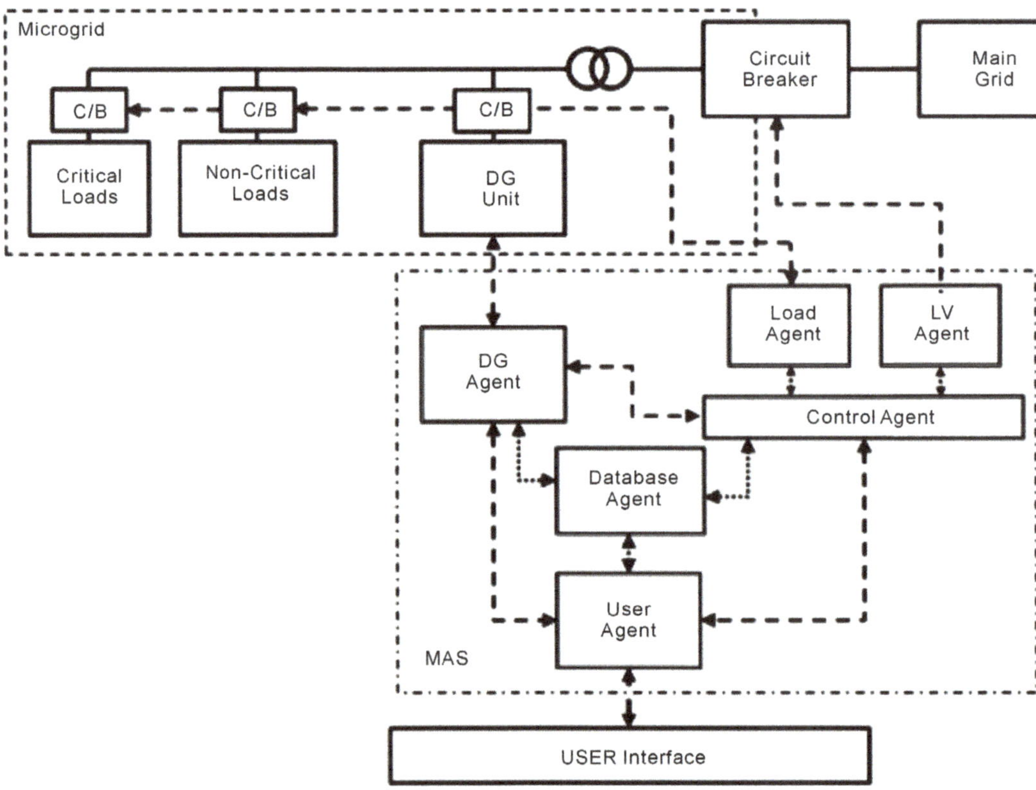

Figure 2. Dual layered control architecture.

proposed MAS system is developed using the JADE platform [16] [17] [31]. The model presents a distributed control architecture operating in two layers. The primary (strategic) layer comprises the Control agent who makes the run-time decisions. The control agent is in control of a secondary layer, overseeing the load control and microgrid connection control capabilities. The control agent is the center of the primary layer holding influence over the DG agent, controlling the LV and Load agents and interfacing with the User agent.

The command hierarchy of the proposed dual layered MAS architecture is depicted in **Figure 3**. The secondary (tactical) layer contains the DG Agent, LV Agent, Load Agent and User Agent. This layer carries out the execution of the run-time decisions of the strategic level, which mainly comprises input/output operations. The DG agent collects several information related to the DG; 1) availability, 2) connection status, 3) power rating, 4) energy source availability and 5) cost of energy.

During an upstream fault, the LV agent islands the microgrid, by tripping the circuit breaker at the point of common coupling, if the Control agent allows it. During island operation the load shedding is carried out by the Load agents to ensure that power is delivered to the most critical loads on the priority basis set by the Control agent. This is implemented by Load agents opening circuit breakers at each of the controlled local loads.

The User agent behaves as the gateway for the user to interact with the system, to obtain real-time information and to set system goals. A utility agent acting in the form of a database agent is utilized to store system information and data and messages shared between the agents. The priority list used for load management is also hosted by the database agent. The database agent acts as the access point for data for all the agents and users.

The MAS is developed for the load management during the islanded phase. As local generation capacities will most definitely be limited with respect to local loads, load shedding becomes unavoidable during islanding. Proper load management during load shedding provides better protection to the most critical loads. By assigning priority values to the local loads, the most critical loads can be identified for protection during islanded mode. By allowing priority level updates during operation the control over load shedding/management can be improved.

Both these layers are ultimately supported by a service (utility) provided by a DB (Database) agent, running in the background. This utility agent stores system information and data and messages shared between the agents and acts as a data access point for all agents.

The agents will behave as hybrid agents [32], capable of both reactive and deliberative actions. Reactive actions have the ability to respond quickly to emergencies of environment, while deliberative actions utilize the "belief, desire and intention" capability of agents [33] which allows them to adopt behaviors to achieve its objectives.

The control flow chart given in **Figure 4** describes this load management during islanded mode. The user agent is allowed to update priority levels during runtime. By revising priority levels a user can restore power to a load that was previously considered non-critical.

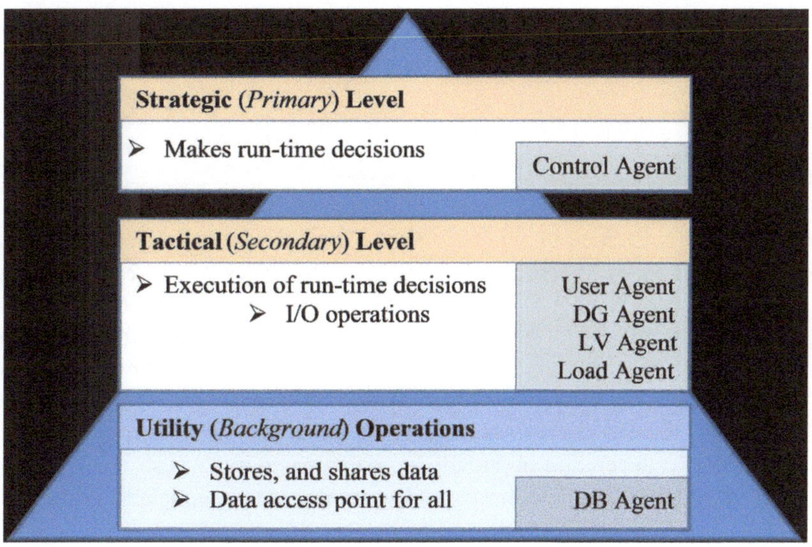

Figure 3. Command hierarchy of the proposed dual layered MAS architecture.

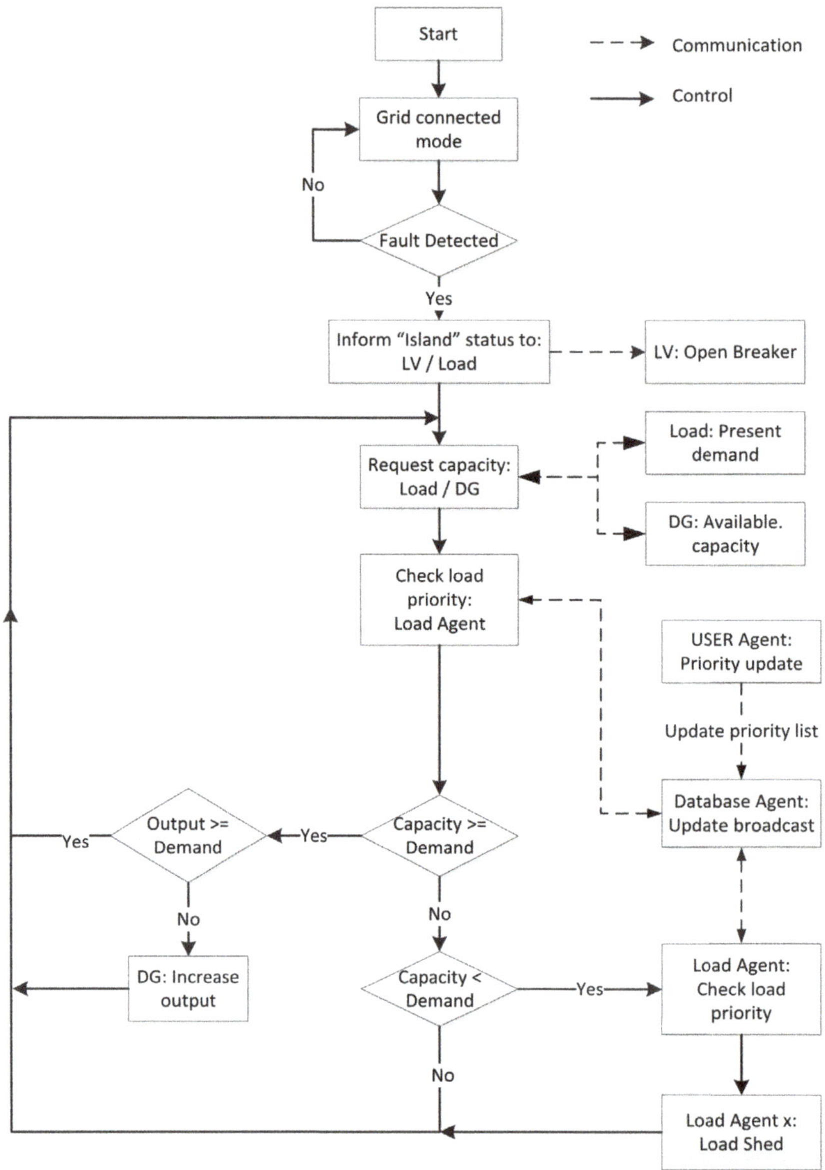

Figure 4. Control flow chart for load management.

When the local capacity is deemed unable to meet the local demand, load shedding is initialized based on the priority values of the local loads. Then the load agents with lower priorities are advised to disconnect while maintaining supply to the most critical loads. When the priority list is updated by the user agent, it is hosted by the database agent and is broadcast to all the other connected agents. This triggers a re-evaluation of capacity and demand, and thereafter any necessary load shedding is carried out based on the revised/updated priority list.

The MAS controller allows the dynamic runtime update of load priorities for load management during a main grid outage. This allows a local grid operator to increase the priority of a currently disconnected local load, and reconnect it to the local grid depending on locally available capacity. A user can change the priority of load using the USER agent via a priority update message (**Figure 3**). This is updated in the database agent, who broadcasts a load priority update to the load agents.

If another "new" local load, needs to be added during islanded operation the same routine will be followed.

The connection of the new load would depend on its priority level. If its priority precedes that of the current local loads, the new load will be connected to the load grid while the lower priority loads will be shed depending local capacity. If it has a lower priority, it will not be connected unless a user raises its priority via the USER agent.

The Proposed architecture, allows for self-organizing control. As each agent acts independently, once an agent is created or initiated it can seamlessly connect and logically interact with existing agents. This is done via the directory service provided by a utility agent acting as a database. A newly initiated agent can register in the directory and advertise its capabilities and requirements to the others. This provides self-organization capabilities to the control framework. Therefore, if one agent goes offline, the other agents can reorganize to cope with the loss. This ability increases the scalability and robustness of the controller.

4. Simulations and Results

Three simulations are carried out as case studies to verify the capability of the proposed MAS to provide control and protection for an inverter based microgrid. Within conventional power systems, frequency control is handled by rotating inertial masses of large generators. This causes a problem within smaller microgrids. Therefore, an inverter based system is selected, where the frequency control is handled by the inverter interface.

The multi agent system runs on a JAVA platform while the simulation is run on a simulating environment. Therefore, in order to establish communication between two domains, for sensing and control TCP/IP is used. A third party TCP/IP server [34] is utilized as an interface to MATLAB and it connects to another middle server which allows multiple simultaneous TCP/IP connections. Each agent requires a separate connection to the test-bed. The interface allows only a single connection at a time; therefore a separate middle server is required. This process is depicted in **Figure 5**.

The SIMULINK model for microgrid test bed designed for the simulations is shown in **Figure 6**. It consists a 480V, 15 kW embedded generator, acting as a DG unit, supplying part of the local demand, and two bulk local loads of 15 kW each. The loads are considered only at lumped load level, and connection/disconnection of part loads are not considered in this study. It is assumed that the DG units can operate at full capacity without fuel limitations during any outage.

In the inverter based microgrid is considered a grid interface unit comprising an inverter, low pass filter, Pulse Width Modulation (PWM), transformer and circuit breaker is used to connect a local DC source. The microgrid is connected to a main utility grid (11 kV) across a transformer (11 kV/400 V) and a main circuit breaker.

4.1. Intentional Islanding during Upstream Fault

The first case study is carried out to verify the effectiveness of the MAS in rapidly islanding the microgrid upon

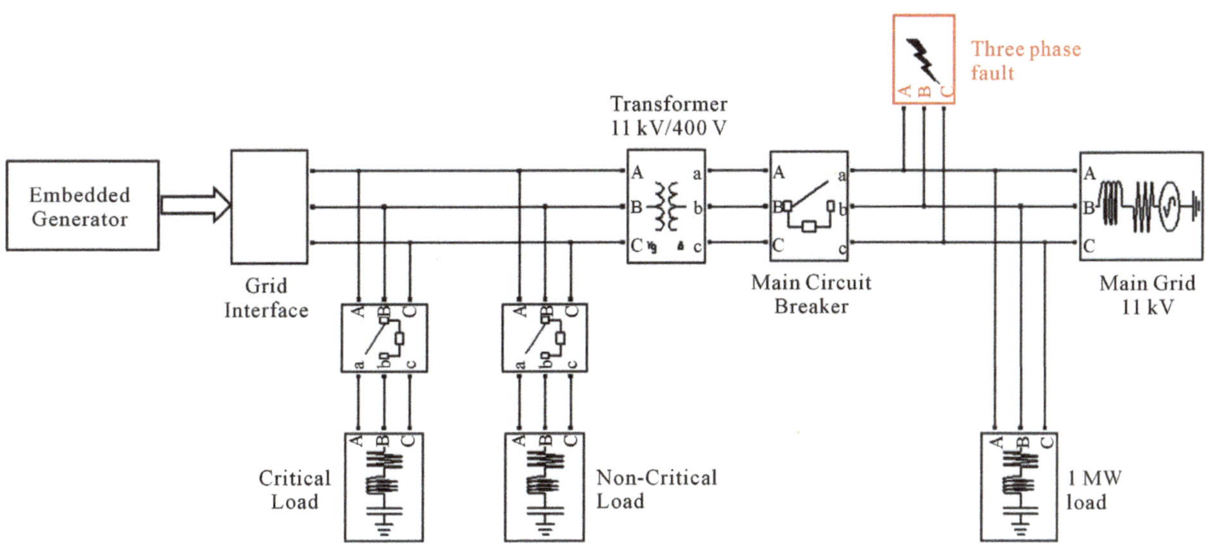

Figure 5. SIMULINK model for simulation test bed.

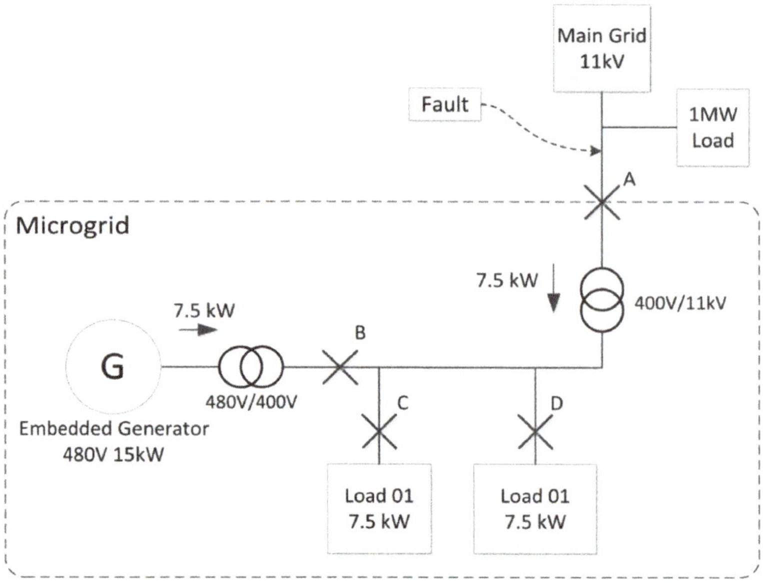

Figure 6. Simulation test-bed for case-study 1.

the detection of an upstream outage on the main grid. The test-bed used for the simulation is shown in **Figure 6**. It comprises a 15 kW DG unit producing 480 V, connected to a 400 V, 50 Hz microgrid, a transformer (480 V/400 V) connecting the DG to the microgrid, two 7.5 kW local loads, with a connection to a 11 kV main utility grid fed through another transformer (11 kV/400 V), all of which are connected across circuit breakers. Total local demand is 15 kW, 50% of which (7.5 kW) is supplied by the DG unit, while the rest is supplied by the main grid. It is assumed that the DG unit can instantly increase output using battery storage. An upstream fault is introduced to the system at t = 0.05 s into the simulation. At the beginning of the simulation breakers A, B, C and D are all closed. The test-bed shown in **Figure 5** is simulated for 0.1 s. The simulation results depicting the successful islanding process are presented in **Figure 7**.

a) Upon detection of upstream fault:

When the upstream outage is detected at t = 0.05 s the control agent informs the LV agent and the Load agents to switch to the islanded mode operation. Upon receiving the islanding order the LV agent trips the main circuit breaker A, at the PCC, isolating the microgrid from the utility [**Figure 7(a)**]. The Control agent then queries the DG agent and the load agents regarding their available capacities that the DG can provide to the microgrid and the power requirements of the connected loads.

As the available maximum capacity of 15 kW is able to meet the total demand of 15 kW, the control agent commands the DG agent to increase DG output to match the demand. Soon as the microgrid is put to island mode the supply to the loads are maintained [see **Figure 7(b)**].

b) In islanded mode:

At t = 0.05s the microgrid is separated from the main grid and the local loads are secured. After the microgrid switches to island mode the total local demand is met by the DG unit by supplying 15 kW.

All agent operations are carried out rapidly, from detecting fault, opening the main breaker, connecting the local source and shedding loads, to stabilize the microgrid within 0.02 s. Therefore, the system is able to disconnect from the main utility grid and maintain the supply to the critical loads without suffering a brownout and/or blackout.

4.2. Securing Power to Critical Loads during Intentional Islanding

The second case study addresses the capability of the MAS to secure power to the critical loads during intentional islanding. This needs to happen immediately following the transition to islanded mode. If the available supply of the local DG unit is limited and insufficient to supply the total local demand, load shedding has to be done rapidly in order to secure supply to the most critical loads. The test-bed used for the simulation is shown in **Figure 8**. The test-bed is same as the one used in case study 1 (**Figure 6**), with only the following changes. Two

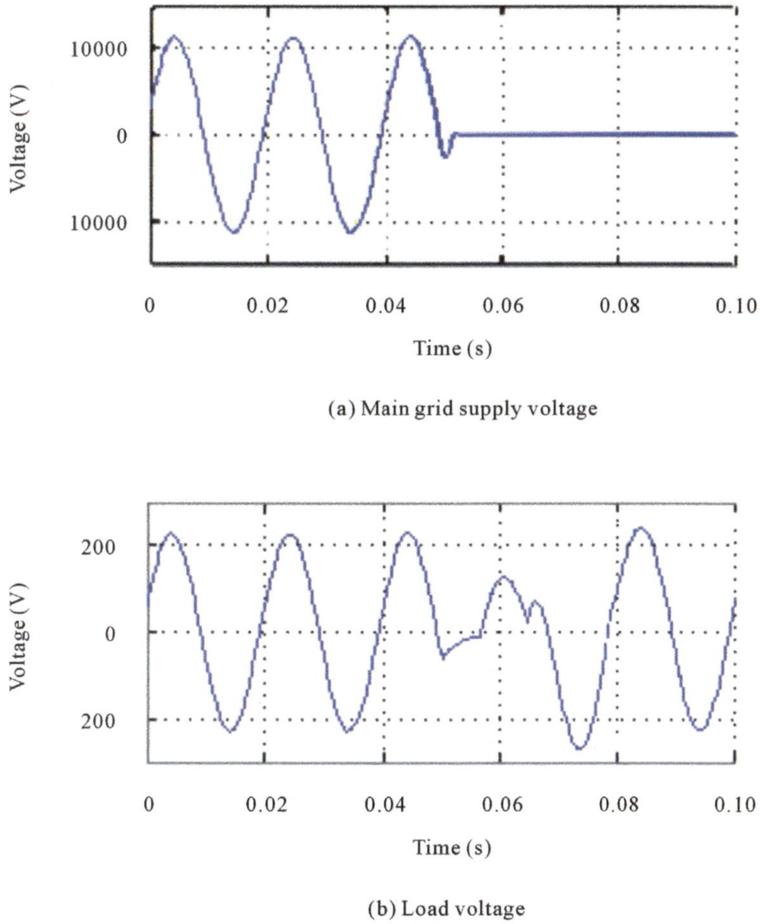

Figure 7. Voltages measured during simulations (line-to-line) at (a) Line to line voltage across the non-critical load at "A"; (b) Line to line voltage at breaker "B".

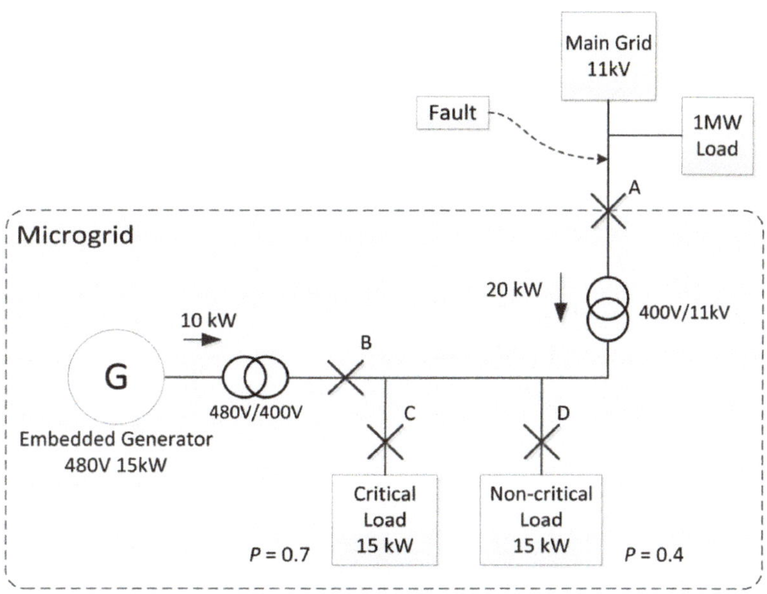

Figure 8. Simulation test-bed for case-study 2.

local loads are 15 kW each with assigned priorities (p) of p = 0.7 and p = 0.4 respectively. The total local demand is 30 kW; initially the DG unit supplies 10 kW while the other 20 kW is provided from the main grid. Similar to case study 1, an upstream fault is introduced to the system at t = 0.05 s into the simulation. At the beginning of the simulation breakers A, B, C and D are all closed.

The test-bed shown in **Figure 8** is simulated for 0.12 s. the simulation results shown in **Figure 9** depict the successful islanding and subsequent securing of power for the critical loads.

a) During grid connected mode:

While the microgrid is in grid-connected mode, the total demand is 30 kW, with 15 kW critical loads and 15 kW non-critical loads. During the grid-connected mode the DG unit provides 10 kW output and the other 20 kW is provided by the main grid. The main grid experiences a fault at t = 0.5 s.

b) Upon detection of upstream fault:

When the upstream outage is detected at t = 0.05 s the control agent informs the LV agent and the Load agents to switch to the islanded mode operation. Upon receiving the islanding order the LV agent trips the main

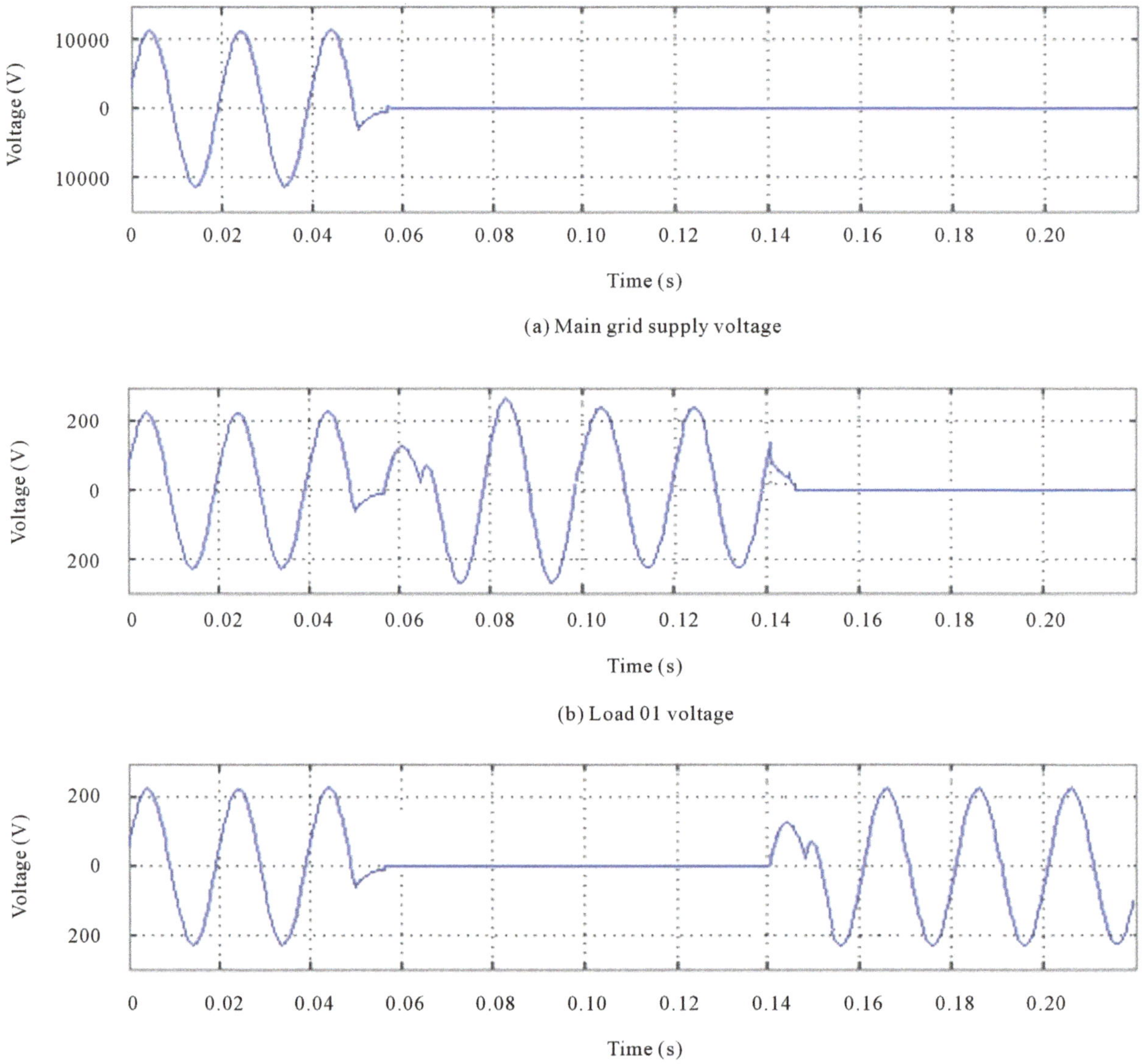

(a) Main grid supply voltage

(b) Load 01 voltage

(c) Load 02 voltage

Figure 9. Voltages measured during simulations (line-to-line) at (a) Main grid voltage at breaker "A"; (b) Load 01 measured at "C": switches from high priority to low priority; (c) Load 02 measured at "D": switches from low priority to high priority.

circuit breaker A, at the PCC, isolating the microgrid from the utility [**Figure 9(a)**]. The Control agent then queries the DG agent and the load agents regarding their available capacities the DG can provide to the microgrid and the power requirements of the connected loads.

As the total load of 30 kW exceeds the available maximum capacity of 15 kW, the control agent commands the load agents to shed the 15 kW non-critical loads to match the DG capacity and the DG unit to increase output to 15 kW. Soon as the microgrid is put to island mode the supply to the critical loads is maintained [see **Figure 9(b)**]. The load agent at the non-critical loads sheds them from the system by opening breaker D [see **Figure 9(c)**].

c) In islanded mode:

At t = 0.05s the microgrid is separated from the main grid and the load agents balance the local demand. After the microgrid switches to island mode the total local demand is met by the embedded generator supplying 15 kW.

4.3. Load Management during Islanded Mode Operation

The third case study addresses the capability of the MAS to provide load management capabilities during islanded operation. As the available local capacity is limited and insufficient to cater total local demand during islanded mode, non-critical (lower priority) loads will have to be dynamically shed in order to protect the stability of the microgrid. This simulation shows the ability of the MAS to allow for dynamic update of load priority levels. The simulation uses the test-bed shown in **Figure 8**.

The local loads are given the two same initial priority levels as shown in **Figure 8**. Local load 01 is given a priority level of p = 0.7, and load 02 is given a priority level p = 0.4. For this initial setting, their respective load agents designate load 01 as a higher priority load and load 02 as a lower priority load. At the beginning of the simulation, breakers A, B, C and D are all closed. An upstream fault is introduced to the system at t = 0.05 s into the simulation, and the microgrid switches to islanded operation. During islanded mode operation, the load 02 priority is raised to 0.9 at t = 0.14 s.

The test-bed shown in **Figure 8** is simulated for 0.2 s. the simulation results shown in **Figure 9** depict the successful islanding and subsequent load management.

a) During grid connected mode:

While the microgrid is in grid-connected mode, the total demand is 30 kW, comprising two 15 kW local loads. During the grid-connected mode the embedded generator only provides 10 kW, while the other 20 kW is supplied by the main grid to the microgrid.

b) Upon detection of upstream fault:

Similar to case-study 1 and 2, the proposed MAS is able to successfully island the microgrid during the upstream outage within 0.02 s of detecting the fault in the main utility grid [see **Figure 9(a)**]. As the local generation capacity is unable to meet the total local demand, the MAS initiates load shedding upon initially set priority levels of the loads as in case study 2. This enables the MAS to maintain supply to the most critical load, load 01, by shedding the least critical load, load 02.

c) In islanded mode:

During the island operation, the user agent can revise the initial priority assignment. This can be pre-initiated or a user can change the priorities during the islanded mode.

At t = 0.14, the user agent revises the priority value of load 02 from p = 0.4 to p = 0.9 and communicates the "revise_request" to the load agent at load 02. This update is forwarded to other agents via the directory agent, and the control agent reinitiates load management procedures. As the DG unit is still unable to meet the total demand of 30 kW, and has to shed non-critical (lower priority) loads to maintain power to the new highest priority load, load 02.

Therefore, the control agent commands the load agents to shed the load 01 of 15 kW to match the DG capacity. Thus, the supply to the new most critical load, load 01 is provided by re-closing "D" [see **Figure 9(c)**]. The load agent at the current lower priority load 01 sheds them from the system by opening breaker "C" [see **Figure 9(b)**].

When the load priorities are revised by a user via the user agent, the MAS is able to reconfigure the system, in order to provide power to the current most critical load/s. The reconfiguration is also done within 0.02 s of the revision and successfully reconnects the new critical load. The results show the capability of the MAS to safely island and maintain the supply to its critical loads, while allowing for the critical loads to be dynamically revised.

All agent operations are carried out rapidly, from detecting fault, opening the main breaker, connecting the local source and shedding loads, to stabilize the microgrid within 0.02 s. the load reconnection after the priority update is done again within 0.02 s of the revision being sent from the user agent. The rapid transitions (within one cycle) ensure a high power quality with minimum effect to the loads. Therefore, the system is able to disconnect from the main utility grid and maintain the supply to the critical loads without suffering a brownout and/or blackout.

5. Conclusions

The proposed dual layered MAS is able to successfully island the microgrid during an upstream outage within 0.02 s of detecting the fault. As the local generation capacity is unable to meet the total local demand, the MAS initiates load shedding upon pre-set priority levels of the loads. This enables the MAS to maintain the supply to the most critical load/s by shedding least critical loads until DG capacity can meet the demand. The proposed MAS controller allows for dynamic/real time load management during islanded mode of operation. When the load priorities are revised by a user via the user agent, the MAS is able to reconfigure the system, in order to provide power to the new most critical load. The reconfiguration is also done within 0.02 s of the revision and successfully reconnects the new critical load. The rapid transitions (0.02 s) ensure a high power quality with minimum effect to the loads. The results show the capability of the MAS to safely island and maintain the supply to its critical loads, even if the critical loads are revised. These results validate the effectiveness of the MAS in controlling DG units to protect and control a microgrid.

The agent behaviour can further be developed using fuzzy/neural/heuristic algorithms. The long-term extensibility of MAS can be evaluated by studying the effect of increasing number of agents on controller stability and service latency. Another important area of future study will be the developing the requirements of communication network redundancy and data network security to prevent cyber-attacks and intrusions from taking down the smartgrid.

Acknowledgements

The authors gratefully acknowledge the support provided by the Senate Research Council, University of Moratuwa (SRC/LT/2012/05).

References

[1] Li, F., Qiao, W., Sun, H., Wan, H., Wang, J., Xia, Y., Xu, Z. and Zhang, P. (2010) Smart Transmission Grid: Vision and Framework. *IEEE Transactions on Smart Grid*, **1**, 168-177. http://dx.doi.org/10.1109/TSG.2010.2053726

[2] Lasseter, R.H. and Piagi, P. (2004) Microgrid: A Conceptual Solution. *Proceedings of the Power Engineering Specialists Conference* (*PESC'*04), Aachen, 4285-4290. http://dx.doi.org/10.1109/PESC.2004.1354758

[3] Moslehi, K. and Kumar, R. (2010) A Reliability Perspective of the Smart Grid. *IEEE Transactions on Smart Grid*, **1**, 57-64. http://dx.doi.org/10.1109/TSG.2010.2046346

[4] Dou, C., Hao, D., Jin, B., Wang, W. and An, N. (2014) Multi-Agent-System-Based Decentralized Coordinated Control for Large Power Systems. *International Journal of Electrical Power & Energy Systems*, **58**, 130-139. http://dx.doi.org/10.1016/j.ijepes.2014.01.012

[5] Markus, P., Wolfgang, K., Maytal, S. and Collins, J. (2013) A Reinforcement Learning Approach to Autonomous Decision-Making in Smart Electricity Markets. *Machine Learning*, **92**, 5-39. http://dx.doi.org/10.1007/s10994-013-5340-0

[6] Salman, K., Rasheed, A.R., Leen-Kiat, S. and Sohrab, A. (2014) Multiagent Study of Smart Grid Customers with Neighborhood Electricity Trading. *Electric Power Systems Research*, **111**, 123-132. http://dx.doi.org/10.1016/j.epsr.2014.02.013

[7] Kahrobaee, S., Rajabzadeh, R.A., Leen-Kiat, S. and Asgarpoor, S. (2013) A Multiagent Modeling and Investigation of Smart Homes with Power Generation, Storage, and Trading Features. *IEEE Transactions on Smart Grid*, **4**, 659-668. http://dx.doi.org/10.1109/TSG.2012.2215349

[8] Wang, X., Zhang, P., Wang, Z., Dinavahi, V., Chang, G., Martinez, J.A., Davoudi, A., Mehrizi-Sani, A. and Abhyankar, S. (2013) Interfacing Issues in Multiagent Simulation for Smart Grid Applications. *IEEE Transactions on Power Delivery*, **28**, 1918-1927. http://dx.doi.org/10.1109/TPWRD.2013.2260566

[9] Muller, S.C., Hager, U. and Rehtanz, C. (2014) A Multiagent System for Adaptive Power Flow Control in Electrical Transmission Systems. *IEEE Transactions on Industrial Informatics*, **10**, 2290-2299. http://dx.doi.org/10.1109/TII.2014.2315499

[10] Horbani, J., Chouhan, S., Choudhry, M.A. and Feliachi A. (2013) Hybrid Multi Agent Approach for Power Distribution System Restoration. *Proceedings of the IEEE Energytech*, Cleveland, 1-5.

[11] Lim, Y., Kim, H.M. and Kinoshita, T. (2014) Distributed Load-Shedding System for Agent-Based Autonomous Microgrid Operations. *Energies*, **7**, 385-401. http://dx.doi.org/10.3390/en7010385

[12] Sanjari, M.J. and Gharehpetian, G.B. (2014) Game Theoretic Approach to Cooperative Control of Distributed Energy Resources in Islanded Microgrid Considering Voltage and Frequency Stability. *Neural Computing and Applications*, **25**, 343-351. http://dx.doi.org/10.1007/s00521-013-1497-5

[13] Sanjari, M.J. and Gharehpetian, G.B. (2014) Incomplete Information Based Decentralized Cooperative Control Strategy for Distributed Energy Resources of VSI-Based Microgrids. *Neural Computing and Applications*, **25**, 189-197. http://dx.doi.org/10.1007/s00521-013-1473-0

[14] Niannian, C. and Mitra, J. (2014) A Multi-Level Control Architecture for Master-Slave Organized Microgrids with Power Electronic Interfaces. *Electric Power Systems Research*, **109**, 8-19. http://dx.doi.org/10.1016/j.epsr.2013.11.027

[15] Ling, W. and Liu, D. (2014) A Distributed Fault Localization, Isolation and Supply Restoration Algorithm Based on Local Topology. *International Transactions on Electrical Energy Systems*, in Press. http://dx.doi.org/10.1002/etep.1859

[16] JADE Agent Development Toolkit (2014) http://jade.tilab.com

[17] Bellifemine, F.L., Caire, G., Poggi, A. and Rimassa, G. (2003) A White Paper EXP in Search of Innovation. http://jade.tilab.com/papers/2003/WhitePaperJADEEXP.pdf

[18] Mcarthur, S.D.J., Davidson, E.M., Catterson, V.M., Dimeas, A.L., Hatziargyriou, N.D., Ponci, F. and Funabashi, T. (2007) Multi-Agent Systems for Power Engineering Application Part I: Concepts, Approaches, and Technical Challenges. *IEEE Transaction on Power Systems*, **22**, 1743-1752. http://dx.doi.org/10.1109/TPWRS.2007.908471

[19] Mcarthur, S.D.J, Davidson, E.M., Catterson, V.M., Dimeas, A.L., Hatziargyriou, N.D., Ponci, F. and Funabashi, T. (2007) Multi-Agent Systems for Power Engineering Applications—Part II: Technologies, Standards, and Tools for Building Multi-Agent Systems. *IEEE Transaction on Power System*, **22**, 1753-1759. http://dx.doi.org/10.1109/TPWRS.2007.908472

[20] Kulasekera, A.L., Gopura, R.A.R.C., Hemapala, K.T.M.U. and Perera, N. (2012) A Review on Multi-Agent Systems in Microgrid Applications. *Proceeding of the Innovative Smart Grid Technologies*, India, 173-177.

[21] Kumar, N.H.S.V.S. and Doolla, S. (2013) Multiagent-Based Distributed-Energy-Resource Management for Intelligent Microgrids. *IEEE Transactions on Industrial Electronic*, **60**, 1678-1687. http://dx.doi.org/10.1109/TIE.2012.2193857

[22] Wooldridge, M. and Weiss, G. (1999) Intelligent Agents, Multi-agent Systems. Cambridge, 3-51.

[23] Franklin, S. and Graesser, A. (1996) Is It an Agent or Just a Program? *Proceeding of the 3rd International Workshop Agent Theories, Architectures, and Languages*, New York, 21-35.

[24] Cai, N. and Mitra, J. (2010) A Decentralized Control Architecture for a Microgrid with Power Electronic Interfaces. *Proceeding of the North American Power Symposium*, USA, 1-8.

[25] Foo Eddy, Y.S., Gooi, H.B. and Chen, S.X. (2015) Multi-Agent System for Distributed Management of Microgrids. *IEEE Transactions on Power System*, **30**, 24-34. http://dx.doi.org/10.1109/TPWRS.2014.2322622

[26] Xu, Y. and Liu, W. (2011) Novel Multiagent Based Load Restoration Algorithm for Microgrids. *IEEE Transactions on Smart Grid*, **2**, 152-161. http://dx.doi.org/10.1109/TSG.2010.2099675

[27] Zhang, J., Ai, Q., Jiang, C.W., Wang, X.G., Zheng, Z.H. and Gu, C.H. (2009) The Application of Multi Agent System in Microgrid Coordination Control. *Proceeding of the International Conference on Sustainable Power Generation and Supply*, Nanjing, 1-6.

[28] Zheng, W. and Cai, J. (2010) A Multi-Agent System for Distributed Energy Resources Control in Microgrid. *Proceeding of the 5th International Conference on Critical Infrastructure (CRIS)*, Beijing, 1-5.

[29] Aquino-Lugo, A.A., Klump, R. and Overbye, T.J. (2011) A Control Framework for the Smart Grid for Voltage Support Using Agent-Based Technologies. *IEEE Transactions on Smart Grid*, **2**, 173-180. http://dx.doi.org/10.1109/TSG.2010.2096238

[30] Kulasekera, A.L., Gopura, R.A.R.C., Hemapala, K.T.M.U., Perera, N. and Pallegedara, A. (2012) Dual Layered Multi Agent System for Intentional Islanding Operation of Microgrids. *Proceeding of the 17th International Symposium on Artificial Life and Robotics*, Oita, 994-997.

[31] Bellifemine, F., Caire, G. and Greenwood, D. (2007) Developing Multi-Agent Systems with JADE. John Wiley &

Sons Ltd., 51-111. http://dx.doi.org/10.1002/9780470058411

[32] Dou, X.C. and Liu, B. (2013) Multi-Agent Based Hierarchical Hybrid Control for Smart Microgrid. *IEEE Transactions on Smart Grid*, **4**, 771-778. http://dx.doi.org/10.1109/TSG.2012.2230197

[33] Georgeff, M., Pell, B., Pollack, M., Tambe, M., Wooldridge, M., Müller, J.P., Rao, A.S. and Singh, M.P. (1999) The Belief-Desire-Intention Model of Agency. *Proceeding of the Intelligent Agents*, 1-10. http://dx.doi.org/10.1007/3-540-49057-4_1

[34] Zimmermann, W. (2011) IOlib-Hardware Input/Output Library Module for Matlab and Simulink. http://www.it.hs-esslingen.de/~zimmerma/software/index_uk.html

5

Unsteady Double Wake Model for the Simulation of Stalled Airfoils

Néstor Ramos-García, Antoine Cayron, Jens Nørkær Sørensen

Department of Wind Energy, Fluid Mechanics Section, Technical University of Denmark, Lyngby, Denmark
Email: nerga@dtu.dk

Abstract

In the present work, the recent developed Unsteady Double Wake Model, USDWM, is used to simulate separated flows past a wind turbine airfoil at high angles of attack. The solver is basically an unsteady two-dimensional panel method which uses the unsteady double wake technique to model flow separation and its dynamics. In this paper, the calculated integral forces have been successfully validated against wind tunnel measurements for the FFA-W3-211 airfoil. Furthermore, the computed highly unsteady flow field is analyzed in detail for a set of angles of attack ranging from light to deep stall conditions.

Keywords

Wind Energy, Airfoil, Separated Flow, Stall, Panel Method

1. Introduction

Airfoil aerodynamics is one of the key subjects in the wind energy field. With the fast growing wind turbine dimensions new challenges have arisen for scientists and engineers in the sector; thick airfoils, high Reynolds numbers and sub-sonic flow conditions are some of the new obstacles that the wind energy community is facing. However, besides the new challenges, flow separation, which has been one of the main difficulties faced by aerodynamicist in the last decades, remains a hot-topic nowadays. An all-times challenge in aerodynamics has been to predict the airfoil characteristics at high angles of attack, in the stalled region, where the flow partially or completely leaves the airfoil forming a separated flow region. It is well-known that even highly demanding Navier-Stokes codes with sophisticated turbulence models fail in accurately predicting the correct behavior of airfoils in the deep stall region.

With the aim of capturing the overall physics of separated flows over airfoils using vortex theory, Maskew and Dvorak [1] developed a simplified model based in an inviscid flow solver, or panel method, which could accurately simulate steady flows around airfoils at high angles of attack. Following this idea, Marion *et al.* [2] developed a similar model focusing in the deep stall region. An unsteady version of the model, capable of taking into account the dynamic effects of the separated wake was first developed by Vezza and Galbraith [3] three decades ago. This approach has been recently revisited by two different research groups [4] [5] which included an integral boundary solver to compute the separation location. All three approaches modelled the separated region with a set of time-updated point vortices, while different approaches were used to represent the airfoil contour as well as the handling of the first released wake panels.

In the present work, a combination of the different approaches has been implemented, making possible to carry out the flow simulations with a small time step and a large number of vortex singularities. Moreover, the separation point is calculated externally with a viscous-inviscid strong coupling procedure, improving the accuracy of its prediction. In what follows the main characteristics of the model will be outlined. For more detailed information about the USDWM the reader is kindly referred to the work carried by Antoine [6] during his stay at DTU Wind Energy.

2. Numerical Model

2.1. Potential Flow Background

If the flow around a body with a free stream velocity U_∞ is considered, at a point in the domain surrounding the solid body the velocity can be defined as the sum of the free stream velocity and the induced velocity u,

$$U = U_\infty + u \tag{1}$$

If the flow is considered to be incompressible, inviscid and irrotational, u can be expressed as the gradient of the potential, $\nabla\Phi$, with Φ satisfying the Laplace equation. The general solution to the Laplace equation can be obtained through a source and vorticity distribution around the body contour. Moreover, the shedding wake can be modelled with downstream converging point vortices. The velocity induced at any point in the domain can therefore be written as the superposition of the body and the wake singularities as follows,

$$u = u_\sigma + u_\gamma + u_{\gamma w\,TE} + u_{\gamma w\,SEP} + u_{\Gamma\,TE} + u_{\Gamma\,SEP} \tag{2}$$

2.2. Problem Formulation

In practice, as sketched in **Figure 1**, the airfoil contour is discretized in N panels, with each of the panels represented by a linear distribution of vorticity, $\gamma_1, \ldots, \gamma_{N+1}$. Where γ_1 is the trailing vorticity of the lower trailing edge panel and γ_{N+1} can be seen as the leading vorticity of the upper trailing edge panel. The vorticity released at the trailing edge and the separation points is first taken into account as two uniform panel distributions, $\gamma_{w\,SEP}$ and $\gamma w_{\,TE}$, which later are convected into the separated wake region as point vortices. Finally a constant source distribution is applied over the airfoil, σ.

Hence a total number of N + 4 unknowns have to be determined. The equations necessary in order to solve the system are presented in what follows,

Equation (1) to N: Neumann non-penetration condition applied at the center of each one of the panels.

$$U_i \cdot n_i = 0 \tag{3}$$

Equation N + 1: Kutta condition [7].

$$\gamma_1 + \gamma_{N+1} = \gamma_{W\,TE} \tag{4}$$

Equation N + 2: vortex strength at the separation panel.

$$\gamma_{W\,SEP} = \gamma_{SEP} \tag{5}$$

Equation N + 3: Kelvin's theorem implementation.

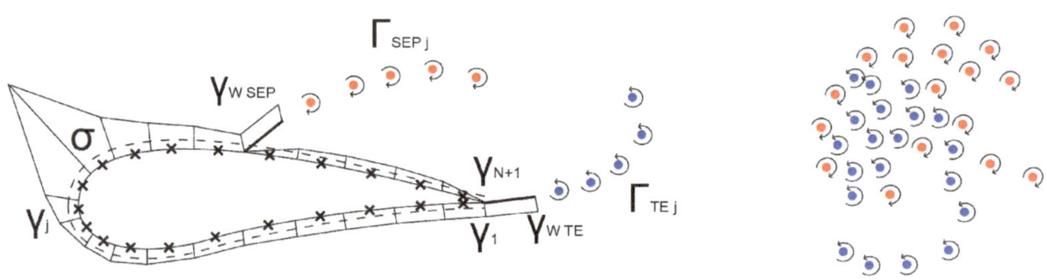

Figure 1. Sketch of the singularity distributions in the USDWM model.

$$\left(\gamma_{W\,TE}\Delta_{W\,TE}\right)^t + \left(\gamma_{W\,SEP}\Delta_{W\,SEP}\right)^t = \Gamma_B^{\ t} - \Gamma_B^{\ t-1} \quad (6)$$

Equation N + 4: zero vortex strength at N + 1

$$\gamma_{N+1} = 0 \quad (7)$$

The system of equations is closed and the unknowns can be determined. It is important to note that the influence of the point vortices in the wake is included in the right hand side of the system, and their strength remains constant in time until they are merged by coalescence or enter the far wake model. Once the system of equations has been solved, the length and angle of the trailing edge and the separation panels is updated as follows,

$$l_{TE} = \frac{|\gamma_{W\,TE}|}{2} \quad (8)$$

$$l_{SEP} = \frac{|\gamma_{W\,SEP}|}{2} \quad (9)$$

$$\theta_{TE} = \begin{cases} \alpha_1 & if \quad \gamma_{W\,TE} < 0 \\ \alpha_{N+1} & if \quad \gamma_{W\,TE} > 0 \end{cases} \quad (10)$$

$$\theta_{SEP} = \tan^{-1}\frac{w_{W\,SEP}}{u_{W\,SEP}} \quad (11)$$

Convergence is obtained once the residual value of the above parameters l_{TE}, l_{SEP}, θ_{TE}, θ_{SEP} as well the total airfoil circulation, Γ_B, is lower than a threshold, in this case set to 10^{-4}.

Finally, the surface pressure coefficient, C_p is calculated using the unsteady Bernoulli equation and accounting for the increase in total pressure over that at infinity in the separated region,

$$C_p = 1 - \frac{u^2}{U_\infty^{\ 2}} - \frac{2}{U_\infty^{\ 2}}\frac{\partial \phi}{\partial t} - \frac{2}{U_\infty^{\ 2}}\Delta h \quad (12)$$

where Δh is the total pressure jump at the interface of the two wake sheets, which only adopts a non-zero value inside the wake region. The head jump can be expressed as the difference between the heads on both sides of the wake sheets,

$$\Delta h = h_1 - h_N = \frac{\partial}{\partial t}\left[\phi_1 - \phi_N\right] + \left[U_1^{\ 2} - U_N^{\ 2}\right] \quad (13)$$

3. Simulations

To first validate the implemented model, fixed angle of attack simulations have been carried out. In what follows we will analyze both, the average aerodynamic coefficients as well as the instantaneous values driven by the vortex shedding. In all the simulations presented here 80 panels have been used to represent the airfoil surface, the time step is set to 0.004 s, the free-stream velocity is 1 m/s and the chord length is fixed to 1m. Moreover, the maximum number of point vortices in the wake has been fixed to 6000 and a coalescence criterion has been used in order to reduce the number of point vortices in the far wake. The coalescence radius between particles is fixed to 0.04 and all vortices located more than four chords downstream from the trailing edge are candidates for the merging. The flow past the airfoil has been simulated a total time of 50 seconds, therefore a total of 25,000 vortices have been released in the wake.

The FFA-W3-211 airfoil has been chosen for the study. The 21% thick airfoil was tested in the low speed wind tunnel L2000 [8], in the Royal Institute of Technology, Stockholm. Measurements were performed at a Reynolds number of 1.8×10^6 and a turbulence intensity of 0.15%. Simulations with the USDWM have been carried out for flows past the FFA airfoil at four different angles of attack: 12, 15, 19 and 23 degrees. The separation location, which is a necessary input to the USDWM is computed using the VI solver Q^3UIC [9]. Q^3UIC solves the integral form of the boundary layer equations using a strong viscous inviscid coupling procedure. All boundary layer calculations in Q^3UIC have been carried out with natural transition at a Reynolds number of 1.8 $\times 10^6$ and a turbulence intensity of 0.15%.

The instantaneous flow visualization at the final simulated time of 50 seconds is shown in the left side of **Figures 2-5**. The lift coefficient variation with time is presented on the right side of the figures. The length scales of the vortex shedding are directly related with the angle of attack, α, and therefore with the size of the separated region. It is seen how the deviations of the instantaneous C_l from its average value are closely related to α as well.

Figure 6 shows the calculated lift and drag coefficients in comparison with Q^3UIC and the experimental data. Generally a good agreement is obtained in terms of average values of C_l and C_d. Low angles of attack computations present a small standard deviation and vice versa, clearly relating the size of the shedding structures to the instantaneous variations in C_l. There is a tendency to slightly over predict the average C_l at high angles of attack.

The energy spectra of the vertical fluctuating velocity component calculated at a point in the airfoils wake is shown in **Figure 7** and briefly analyzed for two different angles of attack. The point is situated at the airfoils

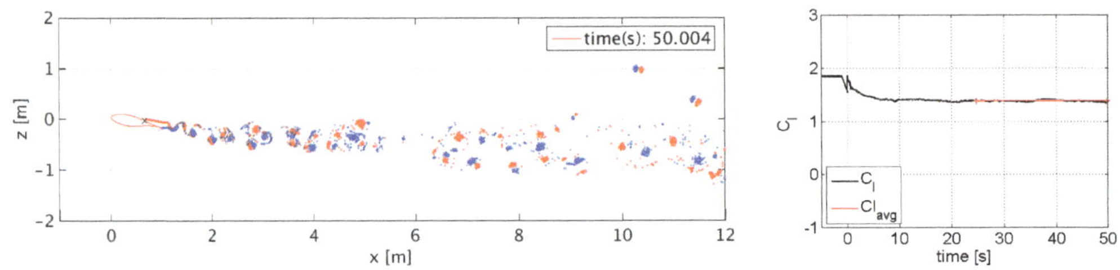

Figure 2. USDWM simulation of the flow past the FFA-W3-211 airfoil at an angle of attack of 12 deg and a Reynolds number of 1.8×10^6. (left) Instantaneous flow visualization at t = 50 s. (right) Instantaneous and average value of C_l.

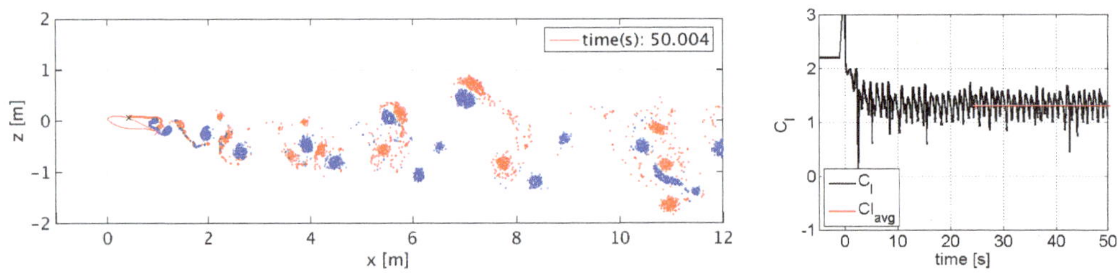

Figure 3. USDWM simulation of the flow past the FFA-W3-211 airfoil at an angle of attack of 15 deg and a Reynolds number of 1.8×10^6. (left) Instantaneous flow visualization at t = 50 s. (right) Instantaneous and average value of C_l.

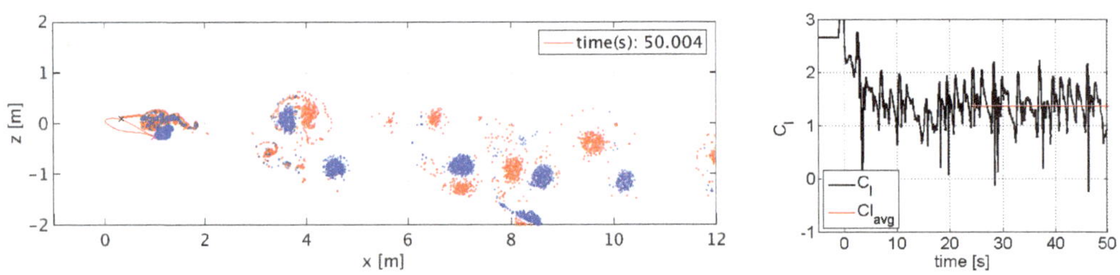

Figure 4. USDWM simulation of the flow past the FFA-W3-211 airfoil at an angle of attack of 19 deg and a Reynolds number of 1.8×10^6. (left) Instantaneous flow visualization at t = 50 s. (right) Instantaneous and average value of C_l.

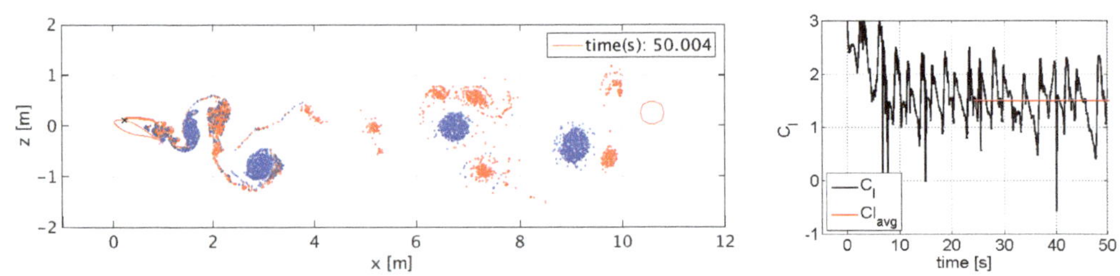

Figure 5. USDWM simulation of the flow past the FFA-W3-211 airfoil at an angle of attack of 23 deg and a Reynolds number of 1.8×10^6. (left) Instantaneous flow visualization at t = 50 s. (right) Instantaneous and average value of C_l.

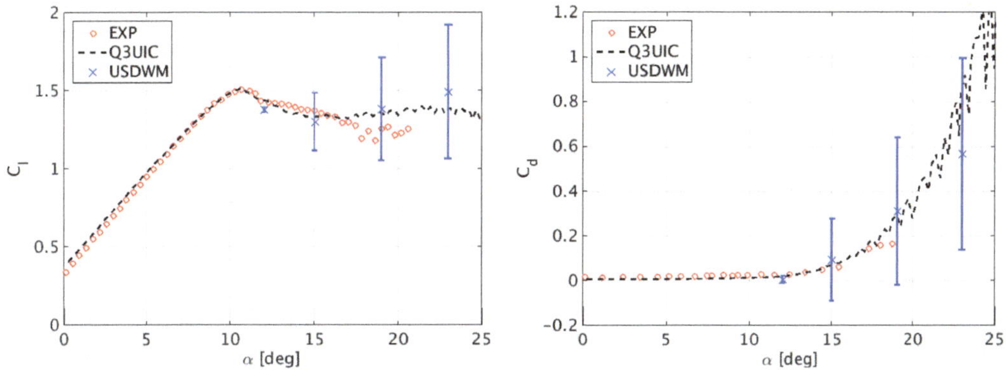

Figure 6. (left) Lift force coefficients and (right) drag force coefficients for flows past the FFA-W3-211 airfoil at a Reynolds number of 1.8×10^6.

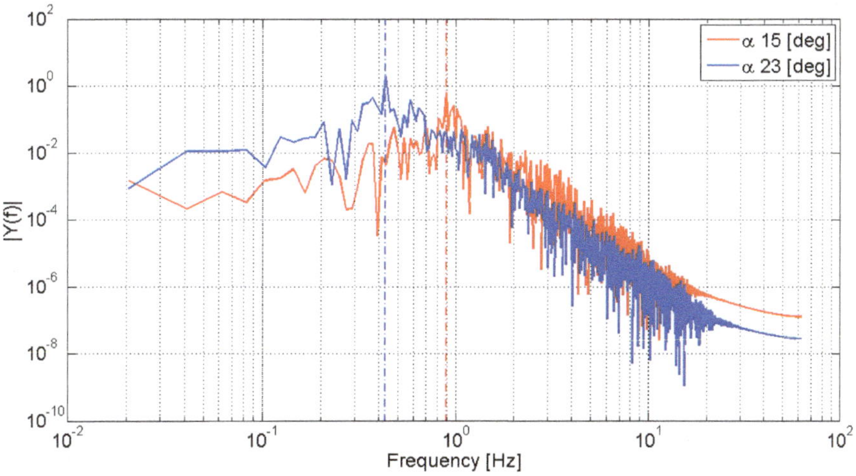

Figure 7. Energy spectra of the vertical fluctuating velocity at a location 2 chords dowstream the leading edge of the FFA-W3-211 airfoil and at its c/4 height.

quarter-chord height and a downstream distance of two chords from the leading edge. The USDWM is capable of capturing the variation in the shedding frequency with the change in angle of attack. Simulations at high angles of attack, where the airfoil presents large separated flow regions, exhibit low shedding frequencies. In massive separated cases, the distance between the two shedding layers is larger, making the interaction slower. At lower angles of attack, with small separated bubbles, the interaction between both shedding layers is much faster, and therefore the shedding frequency is larger.

4. Conclusion

The implemented Unsteady Double Wake Model is capable of capturing accurately the average aerodynamic coefficients of a typical wind turbine airfoil under stall conditions. Moreover the model is able to predict the change in the shedding frequency as well as in the size of the shed vortex structures as a function of the angle of attack. Further work has to be done to validate the accuracy of the predicted unsteady aerodynamic quantities as well as the vortex shedding characteristics.

Acknowledgements

The authors would like to acknowledge the support from the Danish Council for Strategic Research for the project Off Wind China (Sagsnr. 0603-00506B) and the European Union's Seventh Programme for research, technological development and demonstration for the project "AVATAR: AdVanced Aerodynamic Tools for large Rotors" (FP7-ENERGY-2013-1/no. 608396).

References

[1] Maskew, B. and Dvorak, F.A. (1978) The Prediction of C_{lmax} Using a Separated Flow Model. *Journal of the American Helicopter Society.*

[2] Marion, L., Ramos-García, N. and Sørensen, J.N. (2014) Inviscid Double Wake Model for Stalled Airfoils. The Science of Making Torque from Wind. IOP Publishing. *Journal of Physics*: Conference Series. http://dx.doi.org/10.1088/1742-6596/524/1/012132

[3] Vezza, M. and Galbraith, R.A.McD. (1985) An Inviscid Model of Unsteady Aerofoil Flow with Fixed Upper Surface Separation. *International Journal for Numerical Methods in Fluids*, **5**, 577-592. http://dx.doi.org/10.1002/fld.1650050607

[4] Zanon, A., Giannattasio, P. and Simão Ferreira, C.J. (2013) A Vortex Panel Method for the Simulation of the Wake Flow past a Vertical Axis Wind Turbine in Dynamic Stall. *Wind Energy*, **16**, 661-680. http://dx.doi.org/10.1002/we.1515

[5] Riziotis, V.A. and Voutsinas, S.G. (2008) Dynamic Stall Modelling on Airfoils Based on Strong Viscous Inviscid Interaction Coupling. *International Journal for Numerical Methods in Fluids*, **56**, 185-208. http://dx.doi.org/10.1002/fld.1525

[6] Cayron, A. (2015) Unsteady Inviscid Double-Wake Model for Two-Dimensional Stalled Airfoils. Msc. Project Report. Technical University of Denmark.

[7] Wu, F., Zeng, N., Zhang, L. and Wu, D. (2004) Interaction of Two-Dimensional Impulsively Started Airfoils. *Journal of Marine Science and Application*, **3**. http://dx.doi.org/10.1007/BF02918638

[8] Björk, A. (1990) A Guide to Data Files from Wind Tunnel Test of a FFA-W3-211 Airfoil at FFA. FFA-V-019, Flygtekniska Försöksanstalten, Sweden.

[9] Ramos-García, N., Sørensen, J.N. and Shen, W.Z. A Strong Viscous-Inviscid Interaction Model for Rotating Airfoils. *Wind Energy*, **17**, 1957-1984. http://dx.doi.org/10.1002/we.1677

6

Validation of the Actuator Line Model for Simulating Flows past Yawed Wind Turbine Rotors

Wen Zhong Shen[1], Wei Jun Zhu[1], Hua Yang[2]

[1]Department of Wind Energy, Technical University of Denmark, Lyngby, Denmark
[2]School of Hydraulic, Energy and Power Engineering, Yangzhou University, Yangzhou, China
Email: wzsh@dtu.dk

Abstract

The Actuator Line/Navier-Stokes model is validated against wind tunnel measurements for flows past the yawed MEXICO rotor and past the yawed NREL Phase VI rotor. The MEXICO rotor is operated at a rotational speed of 424 rpm, a pitch angle of −2.3°, wind speeds of 10, 15, 24 m/s and yaw angles of 15°, 30° and 45°. The computed loads as well as the velocity field behind the yawed MEXICO rotor are compared to the detailed pressure and PIV measurements which were carried out in the EU funded MEXICO project. For the NREL Phase VI rotor, computations were carried out at a rotational speed of 90.2 rpm, a pitch angle of 3°, a wind speed of 5 m/s and yaw angles of 10° and 30°. The computed loads are compared to the loads measured from pressure measurement.

Keywords

Actuator Line Model, MEXICO Rotor, NREL Phase VI Rotor, Yawed Wind Turbine Rotor

1. Introduction

In order to precisely predict the performance of wind turbines in wind farms, it is required to have a detailed knowledge on the inflow conditions in front of the turbines which are usually created by the atmospheric turbulence, the ground and the upstream turbines. To compute such complex flows, using a standard CFD code, which solves both the boundary layer on the blades and the wake, is too expensive. In order to reduce computational costs, the Actuator Line/Navier-Stokes (AL/NS) technique [1] was developed at the Technical University of Denmark (DTU) such that the procedure of solving the boundary layer is skipped and the loading on the rotor blades is represented by using rotating body forces. This enables more mesh points to be used in the wake region or in the flow regions between the turbines. The Actuator Line (AL) model has been further developed by including the atmospheric turbulence and the wind shear by using the immersed boundary technique for flows past two wind turbines [2] [3]. In order to include the influence of landscape, a hybrid Actuator Line/Navier-Stokes model has been developed where the rotor is solved with the Actuator Line model and the ground is solved with a standard Navier-Stokes solver for Very Large Eddy Simulation (VLES). The flow past a wind turbine on a Gaussian hill was simulated in [4].

An important check is whether the AL/NS technique can predict the loads on the blades and the wake behind the turbines correctly. In the first part of the MexNext project [5], the AL/NS technique has been validated against detailed measurements for the MEXICO rotor in standard axial conditions [6] [7]. Results showed that the model can predict the loading and the wake with good accuracy.

In this paper, the AL/NS technique is validated against wind tunnel measurements for wind turbine rotors in yaw. Both measurements for the MEXICO and NREL Phase VI rotors in yaw are used.

2. Numerical Methods

The numerical method used in the paper is the Actuator Line/Navier-Stokes model which was developed in [1]-[4] [6] [7]. The principle of the model is summarized here.

The EllipSys3D code [8] [9] is used as the Navier-Stokes solver of the AL/NS model. The EllipSys3D code was developed at the Department of Wind Energy, Technical University of Denmark (DTU). The code is based on a multi-block/cell-centered finite volume discretization of the steady/unsteady incompressible Navier-Stokes equations in primitive variables (pressure-velocity). The predictor-corrector method is used. In the predictor step, the momentum equations are discretized using a second-order backward differentiation scheme in time and second-order central differences in space, except for the convective terms that are discretized by the QUICK upwind scheme. In the corrector step, the improved Rhie-Chow interpolation [10] is used in order to avoid numerical oscillations from velocity-pressure decoupling. In order to make it more consistent, the improved SIMPLEC scheme for collocated grids [11] is used. Using the scheme the solution is in-dependent of relaxation parameter and time-step. The obtained Poisson pressure equation is solved by a five-level multi-grid technique. Since the EllipSys3D code is programmed using a multi-block topology, it can be parallelized relatively easily using Message Passing Interface (MPI). The turbulence model used for Large Eddy Simulation is the mixed scale turbulence model developed at LIMSI [12] [13].

To determine the body forces on the rotor blades, we use a blade-element approach combined with airfoil characteristics. The computational domain is chosen to be fixed with the ground and three rotating blades are represented with a rotating body force. At each time step, the flow solver gives a Cartesian velocity field. The velocity at a given blade position is calculated by identifying the index of blade position and performing a tri-linear interpolation. In order to find the loading, the obtained Cartesian velocity (u, v, w) is transformed into the local velocity of the blades,

$$V_n = w\cos\gamma + u\sin\gamma \tag{1}$$

$$V_t = v\cos\theta - \left(u\cos\gamma - w\sin\gamma\right)\sin\theta \tag{2}$$

where (V_n, V_t) are the velocity in the normal and tangential directions of the rotor, respective, γ is the yaw angle, θ is the azimuth angle. Since the rotor is rotating with an angular velocity Ω, the flow angle is determined as

$$\phi = \tan^{-1}\left(\frac{V_n}{\Omega r - V_t}\right) \tag{3}$$

The angle of attack at each cross section is defined as $\alpha = \phi - \beta$ where β is the sum of local twist and pitch angles on the rotor. The relative velocity is

$$V_{rel} = \sqrt{V_n^2 + \left(\Omega r - V_t\right)^2} \tag{4}$$

The force per spanwise unit length is

$$\vec{f} = \frac{d\vec{F}}{dr} = \frac{1}{2}\rho V_{rel}^2 c\left(C_l\vec{e}_L + C_d\vec{e}_D\right) \tag{5}$$

where $C_l = C_l\left(\alpha, \mathrm{Re}\right)$ and $C_d = C_d\left(\alpha, \mathrm{Re}\right)$ are the lift and drag coefficients, respectively. The airfoil data obtained directly from 2D measurements or computations need to be corrected for rotational effects caused by Coriolis and centrifugal forces, especially for cross-sections near the root. At the same time, airfoil cross-sections near the blade tip are influenced from the fact of pressure equalization from the pressure and suction sides at the tip such that the tip flow is different from the corresponding 2D flow at the same angle of attack. To take into account tip effects, a function F_1 is applied on the 2D airfoil data [14]-[16]. The function is

$$F_1 = \frac{2}{\pi}\cos^{-1}\left[\exp(-g\frac{B(R-r)^n}{2r^n \sin\phi})\right] \quad (7)$$

where B is number of blades, the function g is

$$g = \exp\left(\frac{-0.125(B\Omega R/U_\infty - 21)}{1-\min\left(dc/dr\right)}\right) + 0.1 \quad (8)$$

$$n = 1 + 0.5\min(dc/dr) \quad (9)$$

and c is blade chord.

The g function adjusts the influence of the tip vortices on the pressure distribution in the blade tip region. The number of blades, B, determines the distance between the tip vortices at a fixed tip speed ratio whereas the tip speed ratio, $\Omega R/U_\infty$, determines both the distance between the tip vortices at a fixed number of blades, and the pitch of the vortex structure. Their influences on the pressure distribution on the blade are similar and thus are considered together in the g function. Finally, the obtained 2D force is put into a Cartesian regularized volume force.

To take into account the dynamic effects for yawed rotors, the Beddoes-Leishman type dynamic stall model [17] has been implemented in the code.

3. Results

AL/NS computations are carried for both yawed MEXICO and NREL Phase VI rotors. The obtained results are presented in the following two subsections.

3.1. MEXICO Rotor in Yaw

Computations for flows past the MEXICO rotor at a rotational speed of 424 rpm, a pitch angle of −2.3°, wind speeds of 10, 15, 24 m/s and yaw angle of 15°, 30° and 45° were carried out by employing the AL/NS model on a Cartesian mesh consisting of 11.8 M mesh points in a domain of [−16R, 16R] × [−16R, 16R] × [−16R, 16R], with the finest mesh size of R/30 where R is the rotor radius. Two sets of airfoil data: original airfoil data (OAD) and modified airfoil data (MAD) are used for flows past the yawed MEXICO rotor. For more information about the airfoil data, the reader is referred to [6]. To validate the computations (with MAD), the force distributions during one revolution are compared to the DNW measurements in **Figure 1** and **Figure 2**. **Figure 1** shows the loading at a wind speed of 10 m/s and a yaw angle of 30°. The normal force at radial positions of 35%, 60%, 82% r/R follows quite well the experiments while at 92% r/R a phase shift occurs between the computations and the experiments. The variation of the computed tangential force agrees in general with the measurements with slight over-predictions of the mean tangential force. In **Figure 2**, the force distribution at a wind speed of 15 m/s and a

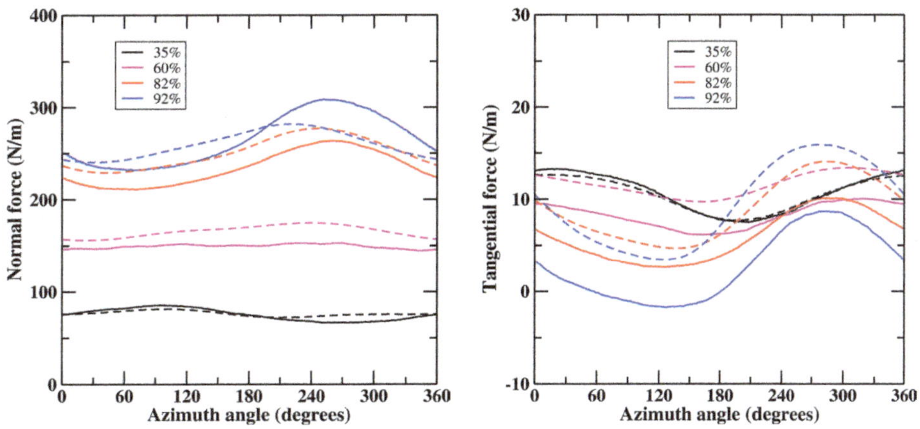

Figure 1. Comparison between computations and experimental data for flows past the MEXICO 25 kW rotor at a wind speed of 10 m/s and a yaw angle of 30° (solid line: experiment; dashed line: computation with MAD).

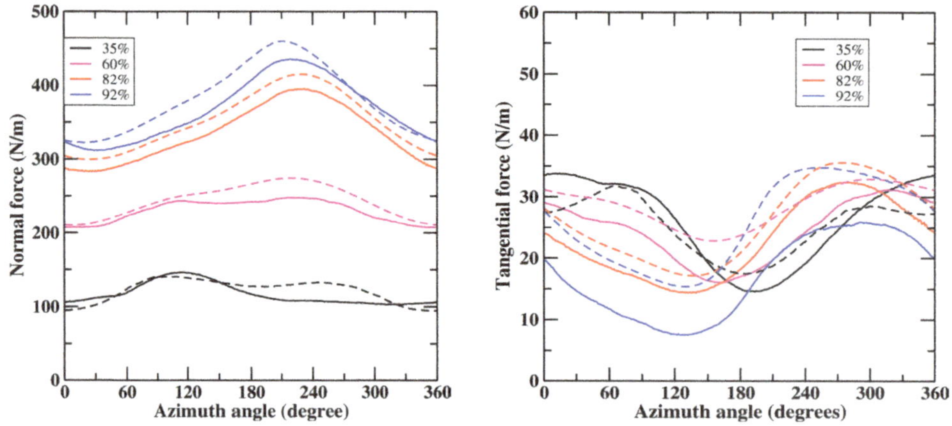

Figure 2. Comparison between computations and experimental data for flows past the MEXICO 25 kW rotor at a wind speed of 15 m/s and a yaw angle of 45° (solid line: experiment; dashed line: computation with MAD).

yaw angle of 45° is plotted. From the figure, good agreements for both normal and tangential force are seen between the computations and measurements.

The radial, tangential and axial velocity components in the rotor coordinates are also plotted. In **Figure 3**, the velocity at 0.15 m before the rotor is compared to measurement. From the figure, good agreements are seen. The velocity at 0.15 m after the rotor is also plotted and compared to experiment in **Figure 4**. Excellent agreements are seen. It is worth noting that in the tip region big velocity variations are seen in the experiments where a smooth transition occurs in the computations. This is probably caused by the too coarse mesh used in the computations.

In order to show the structure behind a yawed rotor, iso-vorticity is plotted in **Figure 5**. The structure shows that the wake develops first in the axial direction, and then bends to a direction with an angle which is slightly negative (below the wind direction).

3.2. NREL Phase VI in Yaw

Computations for flows past the NREL Phase VI rotor [18] at a rotational speed of 90.2 rpm, a pitch angle of 3°, a wind speed of 5 m/s and yaw angle of 10° and 30° were carried out by employing the AL/NS model on a Cartesian mesh consisting of 11.8 M mesh points in a domain of [−16R, 16R] × [−16R, 16R] × [−16R, 16R], with the finest mesh size of R/30 where R is the rotor radius. The airfoil data used in the computations are that described in [15]. To validate the computations, the force distributions during one revolution are compared to the measurements in **Figure 6** and **Figure 7**. **Figure 6** shows the loading at a wind speed of 5 m/s and a yaw angle of 10°. The normal force at radial positions of 30%, 47%, 63%, 80% r/R follows quite well the experiments while at 95% r/R computation over-predicts the normal force which has something to do about tip loss correction. The variation of the computed tangential force follows in general with the measurements but the values are under-predicted. In **Figure 7**, the force distribution at a wind speed of 5 m/s and a yaw angle of 30° is plotted. From the figure, good agreements for both normal and tangential force are seen between the computations and measurements.

4. Conclusion

The Actuator Line/Navier-Stokes model has been validated for both loading and velocity field against measurements for flows past the yawed MEXICO rotor, and for loading for the yawed NREL Phase VI rotor. In general, the model can predict correctly the force changes during a revolution. The velocity behind the rotor can also be predicted correctly in the near field. From the study we can conclude that the AL/NS model is ready to simulate flows past a wind turbine in yaw and to simulate complex flows in wind farms.

Acknowledgements

This work was supported by the Energy-Technological Development and Demonstration Program under the

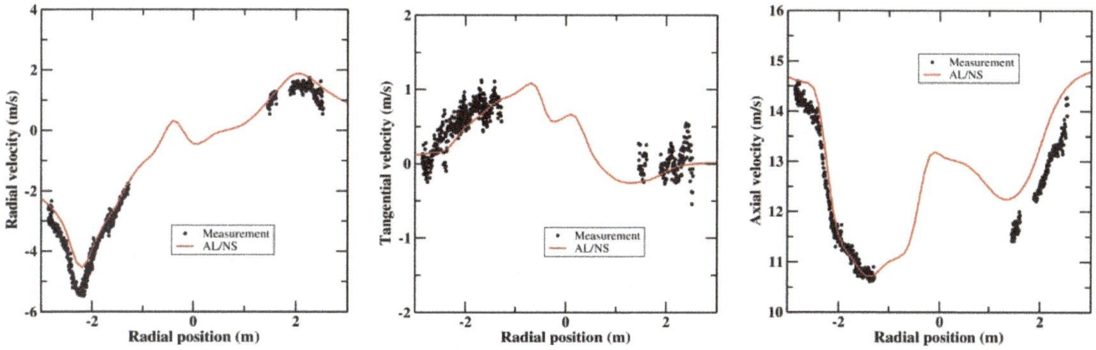

Figure 3. Radial, tangential and axial velocity at 0.15 m before the rotor for flows past the MEXICO rotor at a wind speed of 15 m/s and a yaw angle of 30°.

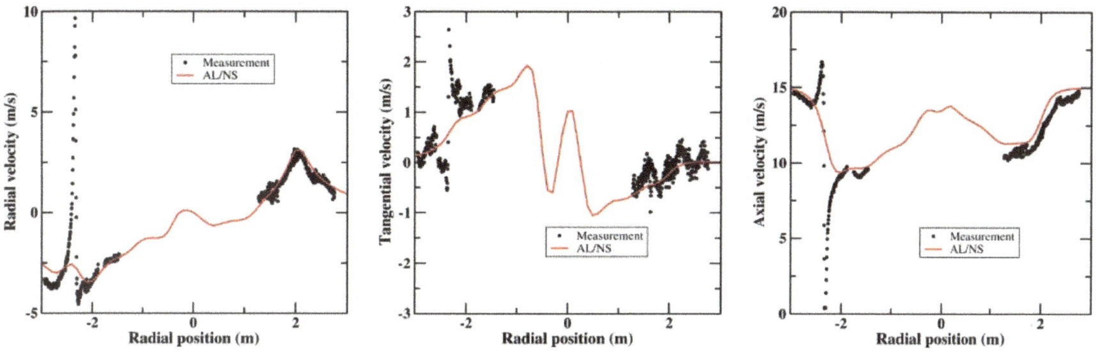

Figure 4. Radial, tangential and axial velocity at 0.15 m after the rotor for flows past the MEXICO rotor at a wind speed of 15 m/s and a yaw angle of 30°.

Figure 5. Iso-vorticity surface plot for flows past the MEXICO rotor at a wind speed of 15 m/s and a yaw angle of 30°.

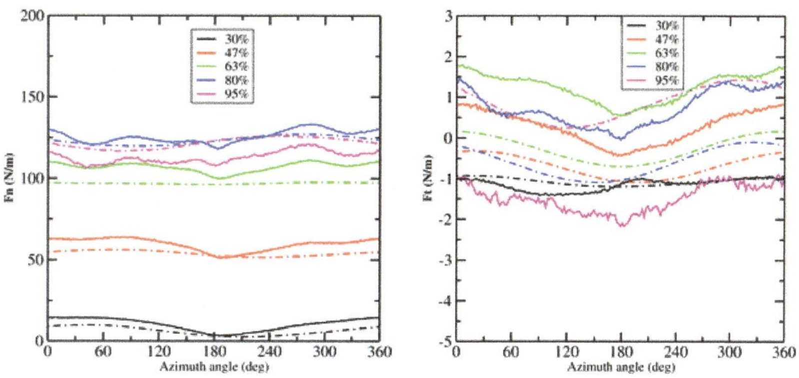

Figure 6. Comparison between computations and experimental data for flows past the NREL Phase VI rotor at a wind speed of 5 m/s and a yaw angle of 10° (solid line: experiment; dashed line: computation).

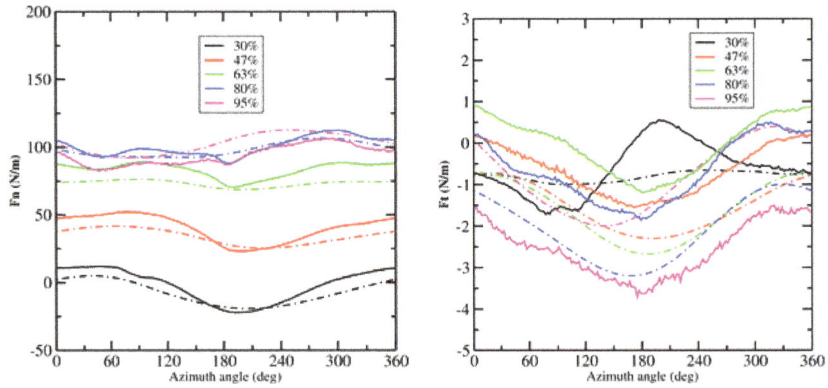

Figure 7. Comparison between computations and experimental data for flows past the NREL Phase VI rotor at a wind speed of 5 m/s and a yaw angle of 30° (solid line: experiment; dashed line: computation).

project (J.nr. 64012-0146) and the Danish Council for Strategic Research under the project (Sags nr. 0603-00506B). The authors wish to give thanks to the international partners for their collaborations in the Mex Next II project within the framework of IEA Wind Annex 29.

References

[1] Sørensen, J.N. and Shen, W.Z. (2002) Numerical Modeling of Wind Turbine Wakes. *Journal of Fluids Engineering*, **124**, 393-399. http://dx.doi.org/10.1115/1.1471361

[2] Troldborg, N., Sørensen, J.N. and Mikkelsen, R. (2010) Numerical Simulations of Wake Characteristics of a Wind Turbine in Uniform Inflow. *Wind Energy*, **13**, 86-99. http://dx.doi.org/10.1002/we.345

[3] Troldborg, N., Larsen, G.C., Madsen, H.A., Hansen, K.S., Sørensen, J.N. and Mikkelsen, R. (2011) Numerical Simulations of Wake Interaction between Two Wind Turbines at Various Inflow Conditions. *Wind Energy*, **14**, 859-876. http://dx.doi.org/10.1002/we.433

[4] Shen, W.Z., Sørensen, J.N. and Mikkelsen, R. (2009) Simulation of Flow past Wind Turbines Located on a Hill by a Hybrid Actuator/Navier-Stokes Method. *Proceedings of EWEC* 2009, 10 pages, Marseille, France.

[5] Schepers, J.G., Boorsma, K., Cho, T., Gomez-Iradi, S., Schaffarczyk, P., Jeromin, A., Shen, W.Z., Lutz, T., Meister, K., Stoevesandt, B., Schreck, S., Micallef, D., Pereira, R., Sant, T., Madsen, H. and Sørensen, N. (2012) Final Report of IEA Task 29, Mexnext (Phase 1): Analysis of Mexico Wind Tunnel Measurements. **2012**, Published by ECN Wind Energy.

[6] Shen, W.Z., Zhu, W.J. and Sørensen, J.N. (2012) Actuator Line/Navier-Stokes Computations for the MEXICO Rotor: Comparison with Detailed Measurements. *Wind Energy*, **15**, 811-825. http://dx.doi.org/10.1002/we.510

[7] Nilsson, K., Shen, W.Z., Sørensen, J.N., Breton, S.P. and Ivanell, S. (2015) Validation of the Actuator Line Method Using Near Wake Measurements of W.Z. the MEXICO Rotor. *Wind Energy*, **18**, 449-467. http://dx.doi.org/10.1002/we.1714

[8] Michelsen, J.A. (1992) Basis 3D—A Platform for Development of Multiblock PDE Solvers. *Technical Report*, AFM 92-05, Technical University of Denmark, Denmark.

[9] Sørensen, N.N. (1995) General Purpose Flow Solver Applied to Flow over Hills. Risø-R-827-(EN), Risø National Laboratory, Roskilde, Denmark.

[10] Shen, W.Z., Michelsen, J.A. and Sørensen, J.N. (2001) An Improved Rhie-Chow Interpolation for Unsteady Flow Computations. *AIAA Journal*, **39**, 2406-2409. http://dx.doi.org/10.2514/2.1252

[11] Shen, W.Z., Michelsen, J.A., Sørensen, N.N. and Sørensen, J.N. (2003) An Improved SIMPLEC Method on Collocated Grids for Steady and Unsteady Flow Computations. *Numerical Heat Transfer*, *Part B*, **43**, 221-239. http://dx.doi.org/10.1080/713836202

[12] Ta Phuoc, L. (1994) Modèles de Sous Maille Appliqués aux Ecoulements Instationnaires Décollés. *Proceedings of the DRET Conference*: *Aérodynamique Instationnaire Turbulente-Aspects Numériques et Expérimentaux*, Paris, France, DGA/DRET editors.

[13] Shen, W.Z., Zhu, W.J. and Sørensen, J.N. (2009) Aero-Acoustic Computations for Turbulent Airfoil Flows. *AIAA Journal*, **47**, 1518-1527. http://dx.doi.org/10.2514/1.40399

[14] Shen, W.Z., Sørensen, J.N. and Mikkelsen, R. (2005) Tip Loss Corrections for Actuator/Navier-Stokes Computations.

Journal of Solar Energy Engineering, **127**, 209-213. http://dx.doi.org/10.1115/1.1850488

[15] Shen, W.Z., Mikkelsen, R., Sørensen, J.N. and Bak, C. (2005) Tip Loss Correction for Wind Turbine Computations. *Wind Energy*, **8**, 457-475. http://dx.doi.org/10.1002/we.153

[16] Shen, W.Z., Zhu, W.J. and Sørensen, J.N. (2014) Study of Tip Loss Corrections Using CFD Rotor Computations. *Journal of Physics: Conference Series*, **555**, 012094. http://dx.doi.org/10.1088/1742-6596/555/1/012094

[17] Leishman, J.G. and Beddoes, T.S. (1989) A Semi-Empirical Model for Dynamic Stall. *Journal of American Helicopter Society*, **34**, 3-17.

[18] Schreck, S. (2001) The NREL Full-Scale Wind Tunnel Experiment-Introduction to the Special Issue. *Wind Energy*, **5**, 77-84. http://dx.doi.org/10.1002/we.72

7

Cost Effective Operating Strategy for Unit Commitment and Economic Dispatch of Thermal Power Plants with Cubic Cost Functions Using TLBO Algorithm

E. B. Elanchezhian, S. Subramanian, S. Ganesan

Department of Electrical Engineering, Annamalai University, Chidambaram, India
Email: elanchezhian.eb@gmail.com

Abstract

This paper deals with a Unit Commitment (UC) problem of a power plant aimed to find the optimal scheduling of the generating units involving cubic cost functions. The problem has non convex generator characteristics, which makes it very hard to handle the corresponding mathematical models. However, Teaching Learning Based Optimization (TLBO) has reached a high efficiency, in terms of solution accuracy and computing time for such non convex problems. Hence, TLBO is applied for scheduling of generators with higher order cost characteristics, and turns out to be computationally solvable. In particular, we represent a model that takes into account the accurate higher order generator cost functions along with ramp limits, and turns to be more general and efficient than those available in the literature. The behavior of the model is analyzed through proposed technique on modified IEEE-24 bus system.

Keywords

Cubic Cost Functions, Ramp Rate, Teaching Learning Based Optimization, Unit Commitment

1. Introduction

The non-storable nature of electrical energy calls for permanent adjustment of production to consumption. Since consumption varies constantly throughout a day, any utility should forecast these variations, and each day, draw up a schedule for its production facilities for the next day. These processing functions are an essential stage in preparation of "real time" control of an electrical network. One of these processing functions is the determination of thermal Unit Commitment (UC). The UC problem is a large-scale, non-convex, non-linear, and mixed integer optimization problem. The optimal solution of the problem can be obtained by complete enumeration, which is prohibitive in practice owing to its excessive computational resource requirements. So, attempts are being continuously made to solve this problem by reliable iterative and heuristic methods.

A bibliographical survey on UC reveals that a good amount of numerical optimization techniques [1]-[4] and meta-heuristic [5]-[13] methods have been applied to achieve efficient and near optimal solutions. Traditionally to solve the UC effectively, conventional techniques need the incremental fuel cost curves to be featured of monotonically increasing and continuous. But practically, the generating units actually have non-monotonic incremental fuel cost curves. For simplicity and easy solving purposes, the input-output characteristics of thermal generators are usually approximated by quadratic functions or piecewise linear functions [1]-[13]. However, many approaches have been addressed that the Economic Dispatch (ED) solution accuracy can be improved or the exact cost can be obtained with higher order functions or Cubic Cost Functions (CCF) [14]-[17]. Earlier papers in the literature have not considered the third order cost functions because they focus on their algorithms and lost the accuracy by approximating the realistic cost curve. In our previous work, we proposed an economic dispatch model using CCF [17]. Even though the UC research is solved for past years, less work is carried out for UC with Cubic Cost Functions (UC-CCF).

The evolutionary and population based techniques in the literature survey are probabilistic techniques and necessitate some common controlling parameters like population size, number of generations and elite size. Moreover, the algorithms also require finding of some of their own control parameters. Genetic Algorithm (GA) needs crossover rate and mutation rate, Particle Swarm Optimization (PSO) has its own parameters like inertia weight, and social and cognitive parameters. The global solution of any function is only achieved with the proper tuning of these algorithm-specific parameters. Improper tuning may lead to local optimal solution or increase in convergence time.

In this paper, a recently developed heuristic algorithm named teaching learning based optimization (TLBO) algorithm based on the effect of the influence of a teacher on the output of learners in a class, introduced by Rao *et al.* [18], is utilized for the solution of UC problem. This TLBO algorithm has been implemented in various problem domains of engineering and technology. Most recently, it has been utilized in solving a few areas in power system [17] [19]-[21]. Unlike other population based techniques, TLBO requires only determination of common controlling parameters like population size and number of generations for its functionality. In the present work, TLBO based UC (TLBO-UC) with CCF (TLBO-UC-CCF) is carried on 26-unit test system considering ramp rate constraints for a time horizon of 24 hours to prove the scalability of the algorithm.

2. Unit Commitment Problem Formulation

Mathematically, the unit commitment problem can be formulated as a mixed integer non-linear problem. The objective function and constraints are formulated as given below.

2.1. Objective Function

The objective function of UC problem is the minimization of the total cost (F_{Tuc}) (Equation (1)) which is the sum of the fuel cost (F) and the start up cost (SUC) of individual units (N) for the given period (T) subject to various constraints. Mathematically, the UC problem model can be formulated as

$$\text{Minimize } F_{Tuc} = \sum_{t=1}^{T}\sum_{i=1}^{N} U_{it}\left[F_i\left(P_i(t)\right) + SUC_i\left(1 - U_{i(t-1)}\right)\right] \tag{1}$$

where F_{Tuc} is the total operating cost in \$, $F_i\ (P_i\ (t))$ is the fuel cost of unit i at hour t, $P_i\ (t)$ is the output power of ith unit at hour t, U_{it} is the on/off status of ith unit at hour t.

The major component of the operating cost, for thermal units, is the power production cost of the committed units that is modelled as third order cubic function as shown in Equation (2)

$$F_i\left(P_i\right) = a_i + b_i P_i + c_i p_i^2 + d_i P_i^3 \ (\$/\text{h}) \tag{2}$$

where a_i, b_i, c_i, d_i are cost function coefficients of unit i in \$, \$/MW, \$/MW^2, \$/MW^3 respectively.

2.2. Constraints

2.2.1 System Power Balance

$$\sum_{i=1}^{N} U_i P_i(t) = P_d(t), \; t = 1,2,\cdots,T \tag{3}$$

where $P_d(t)$ is the power demand at t^{th} interval.

2.2.2. Spinning Reserve Constraints

The sum of the maximum power generating capacities of all the committed units at a time instant should be at least equal to the sum of the known power demand and minimum spinning reserve requirement at that time instant, *i.e.*

$$\sum_{i=1}^{N} U_{it} P_{i(\max)} \geq P_d(t) + R_t, \; t = 1,2,\cdots,T \tag{4}$$

where $P_{i\,(\max)}$ is the known maximum power that can be generated by unit i at any time instant and R_t is the minimum spinning reserve requirement at time t.

2.2.3. Generation Limit Constraints

$$P_{i(\min)} \leq P(t)_i \leq P_{i(\max)} \quad \text{when } U_{it} = 1$$

$$P_i(t) = 0 \qquad \text{when } U_{it} = 0 \tag{5}$$

where $P_{i\,(\min)}$ and $P_{i\,(\max)}$ represents the minimum and maximum generation limits of thermal units.

2.2.4. Unit Minimum up/down Time Constraints

$$\left[X_{i(t-1)}^{on} - T_i^{on}\right] * \left[U_{i(t-1)} - U_{i(t)}\right] \geq 0 \tag{6}$$

$$\left[X_{i(t-1)}^{off} - T_i^{off}\right] * \left[U_{i(t)} - U_{i(t-1)}\right] \geq 0 \tag{7}$$

where $X_{i(t)}^{on}$ and $X_{i(t)}^{off}$ is the time duration for which unit i has been on and off respectively at hour t.

2.2.5. Ramp Rate Limits

$$P_i(t) - P_i(t-1) \leq UR_i \quad \text{As generation increases} \tag{8}$$

$$P_i(t-1) - P_i(t) \leq DR_i \quad \text{As generation decreases} \tag{9}$$

where UR and DR represents the generator ramp up and ramp down limits

3. Teaching Learning Based Optimization

By inspiring a teaching-learning process, Rao *et al.* proposed relatively a new population based meta-heuristic optimization algorithm called, Teaching Learning Based Optimization (TLBO). It works on the effect of influence of a teacher on the output of learners in a class. The key advantage of this algorithm is it has no user-defined parameter which makes it better than other optimization algorithms. The TLBO method works on the philosophy of teaching and learning. Teacher and learners are the two vital components of the algorithm. The working of TLBO is divided into two parts, "teacher phase" and "learner phase". In this algorithm, a group of learners is considered as population and design variables are considered as different subjects offered to the learners and the learner's result is analogous to the "fitness" value of the optimization problem. In the entire population, the best solution is considered as the teacher. The UC problem is a highly combinatorial optimization problem can be solved in two folds: optimal unit selection and dispatch among online generating units. Usually, unit selection is expressed as binary variables and real numbers for dispatch. Thus, the problem contains both types of variables. The applicability of TLBO for solving such mixed integer problems is explored.

The steps involved in the search procedure of the TLBO algorithm for the proposed UC-CCF problem is illustrated by flowchart in **Figure 1** and are summarized as follows.

Initialization of TLBO-UC-CCF problem

Step 1: Define the *UC-CCF* optimization problem as minimization problem.

Step 2: Population size (*Ps*), number of design variables (*Nd*) which represents number of generating units, minimum up/down time, initial status, maximum and minimum generation limits (limits of design variables) and stopping criteria (maximum number of iterations) are defined in this step.

Teacher phase

Step 3: Evaluate the difference between existing mean and best mean result by utilizing Teaching factor (T_f).

Learner phase

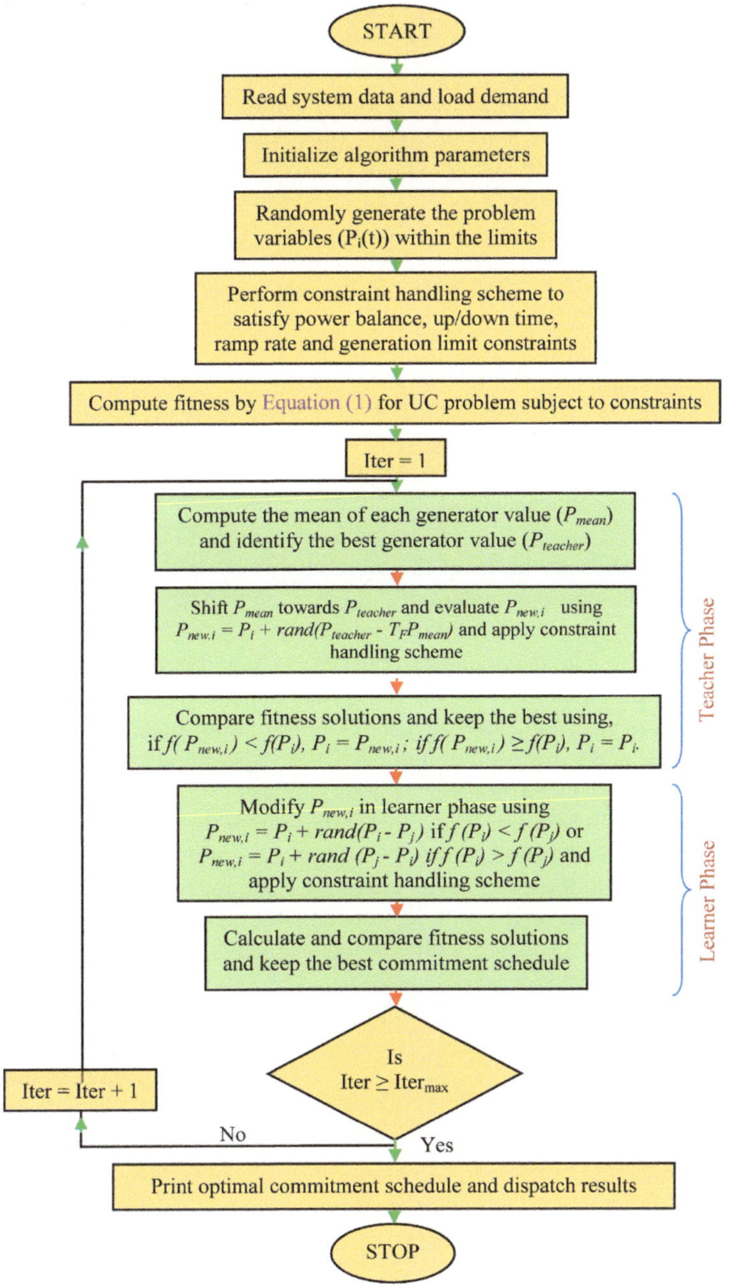

Figure 1. Flowchart of TLBO based UC-CCF.

Step 4: Update the learner's generation solution with the help of teacher's generation.
Step 5: Update the learner's generating solution by utilizing the generating solution of some other learner.
Termination criteria
Step 6: Repeat the procedure from Step 2 to 5 till the maximum number of iterations is met.

Constraint Handling Techniques

A key factor in the application of heuristic algorithms to UC problem is how the algorithm handles the constraints relating to the problem. In the proposed solution method, the constraints of the UC are implemented using a combination of preservation and penalty function methods. The generating capacity limits and ramp up/down rate constraints are also handled by the preservation method. The production of the initial population and in the two phases of TLBO, the output of all committed units minus the reference unit are chosen arbitrary within their respective generating capacity limits, whereas the output of the reference unit is constrained by the system power balance constraint. So, for the dispatchable units other than reference unit, the generating capacity limit constraint is automatically satisfied in both initial population and two phases of TLBO. Therefore, the violation of the capacity limit constraint should be only considered as the penalty term for the reference unit. The minimum up-time/down-time duration constraints of the UC problem are checked for each unit over the scheduling horizon in each interval. If there is any violation in the minimum up or minimum down time constraint then repair mechanism [11] is used to overcome the violation.

4. Numerical Simulation Results and Discussions

In this section, the numerical results aimed at showing the efficiency and the effectiveness of the proposed approach is presented. The proposed TLBO-UC method is tested on modified IEEE-24 bus system having 26 units over a 24-hour scheduling horizon. In order to avoid misleading results due to the stochastic nature of the TLBO, 100 runs were averaged with each run starting with random initial populations. The performance of the proposed problem is implemented using MATLAB-7.9 and realized in 2.40 GHz, i3 processor with 4 GB RAM. The obtained results are compared with other methods in the literature.

This test system consists of 26 units involving cubic cost equations. System particulars involving system data, load demand, ramp rate and capacity limits of the aforementioned units are obtained from literature [2] [3] [14] [15]. In order to determine the best parameters of the proposed TLBO, a number of simulations are carried out. After a number of careful experimentation, following optimal values of TLBO-UC parameters have finally been settled and most appropriate to adopt: Population size (Ps) = 30 and Number of iterations = 100. These parameters are obtained from fine tuning process which is problem dependant task. Simulations are carried out for a 24 hour horizon. The generation cost, % reserve and start-up cost for 24 hour horizon is shown in **Table 1**. The optimal commitment schedule and generation output derived by the proposed algorithm considering the ramp rate and cubic cost equation is shown in **Table 2** for 24 hour horizon.

In this combination for unit scheduling and generation output, the total generation cost obtained by the proposed approach is $795488.869. **Table 3** shows the comparison of the total generation cost and execution time achieved by TLBO method with the popular methods, e.g. Piecewise Linear Iterative (PLI) [15] and Dynamic Programming-Sequential and Truncated Combination (DP-STC) [15]. The corresponding generation cost using the PLI and DP-STC is $795698.33 and $795489.09 respectively. The total operation cost considering start-up cost obtained by TLBO is $800158.869.

For the 26 unit system, the number of units committed for the entire scheduling period is shown in **Figure 2**. **Figure 3** illustrates variation of load demand and the summation of the maximum capacity of the committed units for each hour in the entire time horizon. **Figure 4** exemplify the generating units ON/OFF status of the 26 unit system for 24 hour horizon. Execution time complexity of each optimization method is very important for its application to real systems. **Table 3** shows the execution time of 0.32 seconds which is less than the other methods in literature. The performance analysis of the test system involving CCF as considered by the authors of the present work being totally a new one, no comparison could be done with regard to the performance analysis results. The best, average and worst fuel costs among the 100 runs of algorithm satisfying the system constraints are listed in **Table 4**.

In this case, two criteria of goodness are considered. First, the popular "*Mean*" and "*Standard Deviation*" that has been used approximately in all of optimization algorithms are computed and listed in **Table 4**. The small

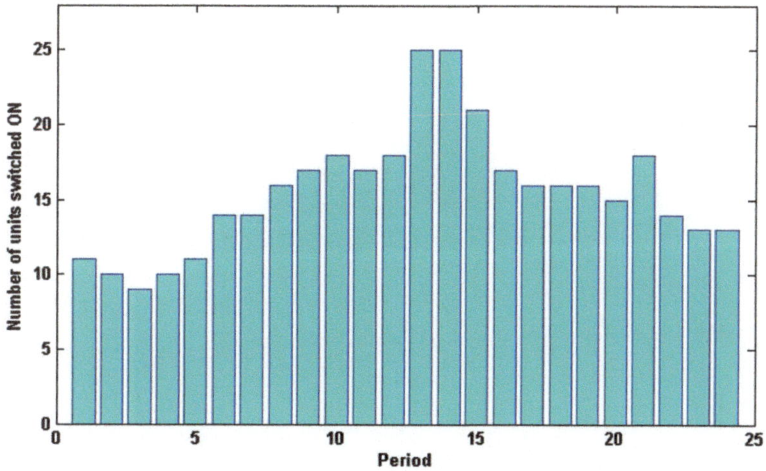

Figure 2. Number of units switched ON for 24 hours.

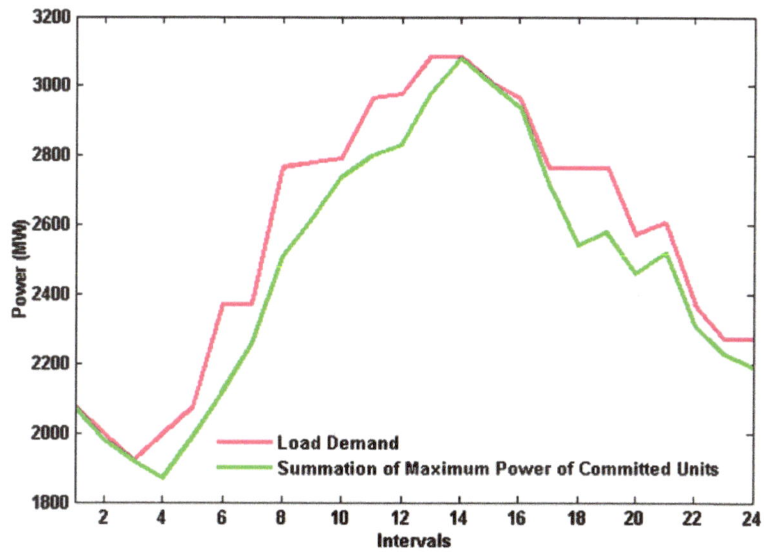

Figure 3. Power demand and sum of maximum capacity of committed units for 24 hours.

values of the standard deviation confirm the capability and reliability of the TLBO to find the best compromise solution. The success rate is defined as ($Run_{Success}$/$\mathrm{Run}_{\mathrm{Tot}}$) x 100, where $Run_{Success}$ is the number of successful experiments which converge to the best solution within the range and $\mathrm{Run}_{\mathrm{Tot}}$ is the total number of runs performed. Results of the success rate for 26-unit with cubic cost functions is 63% and is listed in **Table 4** which demonstrates that TLBO has satisfactory success rate. From the results, it is clear that the proposed method is robust and it is applicable to practical systems.

Like other evolutionary algorithms, TLBO uses the stochastic techniques, thus randomness is an intrinsic feature of these techniques. Several runs with different initial learners have been conducted to test the performance and consistency of the proposed algorithm. The spread of best costs for 100 runs are calculated and graphically displayed in **Figure 5** to illustrate the robustness of the proposed algorithm. To gain further insights into the solution quality of the test case, the obtained statistical results are reported in **Table 4** by means of best, worst and average values. From the statistical results it is evident that the generation costs obtained by different trails are close to the best solution, thereby validating that the proposed method has the higher probability of attaining reliable and quality solution.

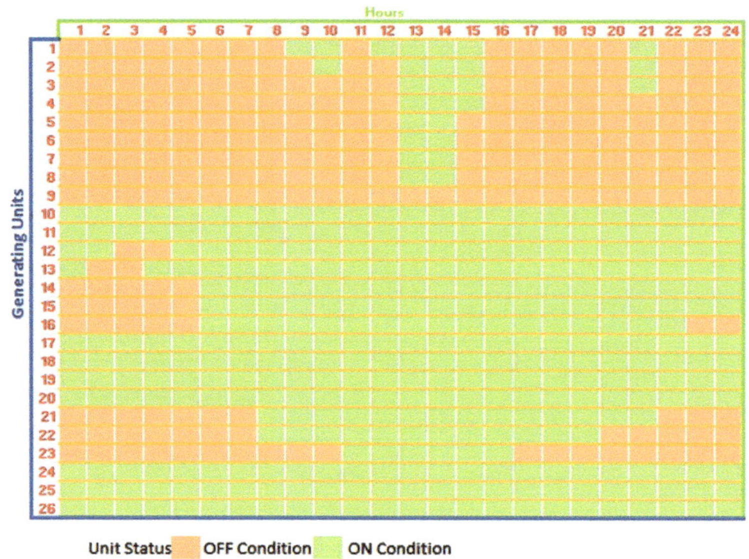

Figure 4. Units ON/OFF representation of 26 unit test system for 24 hours.

Table 1. Operation costs and available reserve for 24 hours.

Hour (h)	Demand (MW)	Reserve %	Generation cost ($)	Start-up cost ($)
1	2070	0.19	23627.42	3220
2	1980	0.90	22268.16	0
3	1920	0.10	21351.82	0
4	1870	6.41	20701.82	80
5	1990	4.05	22482.68	80
6	2120	10.7	25286.92	300
7	2260	4.80	27778.05	0
8	2510	9.32	33640.50	600
9	2620	5.76	36251.28	0
10	2740	1.86	39154.75	0
11	2800	5.56	40740.86	300
12	2830	4.94	41474.64	0
13	2980	3.40	45727.71	90
14	3080	0.16	48818.60	0
15	3010	0.10	45959.10	0
16	2940	0.84	44085.14	0
17	2720	1.73	38603.78	0
18	2540	8.24	34335.21	0
19	2580	6.79	35276.52	0
20	2460	4.32	32211.16	0
21	2520	3.34	33715.07	0
22	2310	2.70	28726.08	0
23	2230	1.93	27013.60	0
24	2190	3.69	26257.99	0
		Total	**795488.8690**	**4670.0**

Table 2. Optimal scheduling of modified IEEE-24 bus system.

Units → / Hours ↓	1	2	3	4	5	6	7	8	9	10	11	12	13
1	0.00	0.00	0.00	0.00	0.00	0.00	0.00	0.00	0.00	76.0	76.00	75.75	72.25
2	0.00	0.00	0.00	0.00	0.00	0.00	0.00	0.00	0.00	72.91	69.90	67.20	DN
3	0.00	0.00	0.00	0.00	0.00	0.00	0.00	0.00	0.00	76.00	74.00	DN	UP
4	0.00	0.00	0.00	0.00	0.00	0.00	0.00	0.00	0.00	43.71	41.10	UP	15.19
5	0.00	0.00	0.00	0.00	0.00	0.00	0.00	0.00	0.00	76.00	75.10	15.20	53.65
6	0.00	0.00	0.00	0.00	0.00	0.00	0.00	0.00	0.00	76.00	75.62	53.70	69.68
7	0.00	0.00	0.00	0.00	0.00	0.00	0.00	0.00	0.00	76.00	76.00	76.00	76.00
8	0.00	0.00	0.00	0.00	0.00	0.00	0.00	0.00	0.00	76.00	76.00	76.00	76.00
9	2.40	0.00	0.00	0.00	0.00	0.00	0.00	0.00	0.00	76.00	76.00	76.00	76.00
10	10.0	2.40	0.00	0.00	0.00	0.00	0.00	0.00	0.00	76.00	76.00	76.00	76.00
11	0.00	0.00	0.00	0.00	0.00	0.00	0.00	0.00	0.00	76.00	76.00	76.00	76.00
12	2.40	0.00	0.00	0.00	0.00	0.00	0.00	0.00	0.00	76.00	76.00	76.00	76.00
13	11.5	2.40	2.40	2.40	2.40	4.00	4.00	4.00	0.00	76.00	76.00	76.00	76.00
14	12.0	12.0	12.0	12.0	12.0	20.0	20.0	15.0	0.00	76.00	76.00	76.00	76.00
15	12.0	12.0	11.7	9.30	0.00	0.00	0.00	0.00	0.00	76.00	76.00	76.00	76.00
16	0.00	0.00	0.00	0.00	0.00	0.00	0.00	0.00	0.00	76.00	76.00	76.00	76.00
17	0.00	0.00	0.00	0.00	0.00	0.00	0.00	0.00	0.00	76.00	76.00	76.00	76.00
18	0.00	0.00	0.00	0.00	0.00	0.00	0.00	0.00	0.00	76.00	76.00	76.00	76.00
19	0.00	0.00	0.00	0.00	0.00	0.00	0.00	0.00	0.00	76.00	76.00	76.00	76.00
20	0.00	0.00	0.00	0.00	0.00	0.00	0.00	0.00	0.00	76.00	76.00	76.00	76.00
21	2.40	2.40	2.40	0.00	0.00	0.00	0.00	0.00	0.00	76.00	76.00	76.00	76.00
22	0.00	0.00	0.00	0.00	0.00	0.00	0.00	0.00	0.00	76.00	76.00	76.00	76.00
23	0.00	0.00	0.00	0.00	0.00	0.00	0.00	0.00	0.00	76.00	76.00	76.00	76.00
24	0.00	0.00	0.00	0.00	0.00	0.00	0.00	0.00	0.00	76.00	76.00	76.00	76.00

Units → / Hours ↓	14	15	16	17	18	19	20	21	22	23	24	25	26
1	0.00	0.00	0.00	155	155	155	155	0.00	0.00	0.00	350	400	400
2	0.00	0.00	0.00	155	155	155	155	0.00	0.00	0.00	350	400	400
3	0.00	0.00	0.00	155	155	155	155	0.00	0.00	0.00	350	400	400
4	0.00	0.00	0.00	155	155	155	155	0.00	0.00	0.00	350	400	400
5	UP	UP	UP	155	155	155	155	0.00	0.00	0.00	350	400	400
6	25.0	25.0	25.0	155	155	155	155	UP	UP	0.00	350	400	400
7	68.9	62.0	55.1	155	155	155	155	UP	UP	0.00	350	400	400

Continued

8	100	100	98.0	155	155	155	155	69.0	69.0	0.00	350	400	400
9	100	100	100	155	155	155	155	124	119.6	UP	350	400	400
10	100	100	100	155	155	155	155	179	174.6	UP	350	400	400
11	100	100	100	155	155	155	155	188.50	168.5	69	350	400	400
12	100	100	100	155	155	155	155	174.8	154.8	124	350	400	400
13	100	100	100	155	155	155	155	197	197	178.9	350	400	400
14	100	100	100	155	155	155	155	197	197	197	350	400	400
15	100	100	100	155	155	155	155	197	197	197	350	400	400
16	100	100	100	155	155	155	155	197	195.1	173.9	350	400	400
17	100	100	100	155	155	155	155	183	163	DN	350	400	400
18	100	100	100	155	155	155	155	92.8	73.2	DN	350	400	400
19	100	100	100	155	155	155	155	112.8	93.2	0.00	350	400	400
20	100	100	100	155	155	155	155	86	DN	0.00	350	400	400
21	100	100	100	155	155	155	155	138.8	DN	0.00	350	400	400
22	85.30	78.6	72.1	155	155	155	155	DN	0.00	0.00	350	400	400
23	81.40	74.6	DN	155	155	155	155	DN	0.00	0.00	350	400	400
24	61.50	54.5	0.00	155	155	155	155	0.00	0.00	0.00	350	400	400

Table 3. Comparison of best results obtained by different methods.

Method	Total start-up cost ($)	Total generation cost ($)	Total operation cost ($)	Execution time (s)
PLI	-	795698.3300	-	0.33
DP-STC	-	795489.0900	-	0.34
TLBO	4670	795488.8690	800158.869	0.32

Table 4. Performance analysis of best feasible solution for 26 unit test system by TLBO.

Test system	26-unit system
Population size	30
Best cost ($)	795488.8690
Average cost ($)	795509.9318
Worst cost ($)	795553.5564
Standard deviation	22.515
Success rate 795,488 ($) - 795,510 ($)	63%
Average CPU time (s)	0.32

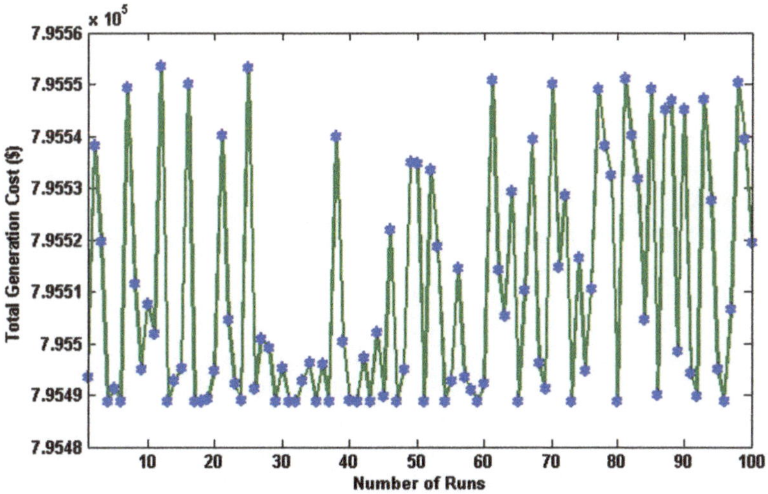

Figure 5. Distribution of the best generation cost of each simulation run for 24 hours.

5. Conclusions

In this paper, a nature inspired TLBO technique has been proposed for solving unit commitment problem with ramp constraints on the thermal generating units involving cubic cost functions. The keys of the effectiveness of the approach are the efficient algorithm for ED problems with cubic cost equations recently proposed that exactly solves the UC problem involving non-convex higher order cost functions without any form of approximation. The sophisticated heuristics for producing a ramp feasible and demand feasible solution is developed. Solving ramp constrained UC problem turns out to be efficient with high provable accuracy on large scale realistic instances in reasonable computational time. The proposed TLBO-UC-CCF method remarkably reduces the generation cost and computation time, and yields more accurate generation scheduling, which shows its adaptability to any higher order generation cost functions.

Incorporating reserve constraints in the UC problem is usually difficult; indeed, they are often used in literature. However, extending our approach to reserve constrained UC with higher order cost functions is not straightforward, and it will be the subject of a future work.

Acknowledgements

The authors gratefully acknowledge the authorities of Annamalai University, Annamalai Nagar, Tamilnadu, India for the facilities provided to carry out this research work.

References

[1] Wood, A.J. and Woolenberg, B.F. (1996) Power Generation, Operation and Control. Wiley, New York.

[2] Tong, S.K., Shahidehpour, S.M. and Ouyang, Z. (1991) A Heuristic Short-Term Unit Commitment. *IEEE Transactions on Power Systems*, **6**, 1210-1216. http://dx.doi.org/10.1109/59.119268

[3] Wang, C. and Shahidehpour, S.M. (1993) Effects of Ramp Rate Limits on Unit Commitment and Economic Dispatch. *IEEE Transactions on Power Systems*, **8**, 1341-1350. http://dx.doi.org/10.1109/59.260859

[4] Wang, S.J., Shahidehpour, S.M., Kirschen, D.S., Mokhtari, S. and Irisarri, G.D. (1995) Short Term Generation Scheduling with Transmission and Environmental Constraints Using an Augmented Lagrangian Relaxation. *IEEE Transactions on Power Systems*, **10**, 1294-1301. http://dx.doi.org/10.1109/59.466524

[5] Dudek, G. (2004) Unit Commitment by Genetic Algorithm with Specialized Search Operators. *Electric Power Systems Research*, **72**, 299-308. http://dx.doi.org/10.1016/j.epsr.2004.04.014

[6] Wang, L.F. and Singh, C. (2009) Unit Commitment Considering Generator Outages through a Mixed-Integer Particle Swarm Optimization. *Applied Soft Computing*, **9**, 947-953. http://dx.doi.org/10.1016/j.asoc.2008.11.010

[7] Eslamian, M., Hosseinian, S.H. and Vahidi, B. (2009) Bacterial Foraging-Based Solution to the Unit Commitment

Problem. *IEEE Transactions on Power Systems*, **24**, 1478-1488. http://dx.doi.org/10.1109/TPWRS.2009.2021216

[8] Ebrahimi, J., Hosseinian, S.H. and Gharehpetian, G.B. (2011) Unit Commitment Problem Solution using Shuffled Frog Leaping Algorithm. *IEEE Transactions on Power Systems*, **26**, 573-581. http://dx.doi.org/10.1109/TPWRS.2010.2052639

[9] Abookazemi, K., Ahmad, H., Tavakolpour, A. and Hassan, M.Y. (2011) Unit Commitment Solution Using an Optimized Genetic System. *International Journal of Electrical Power & Energy Systems*, **33**, 969-975. http://dx.doi.org/10.1016/j.ijepes.2011.01.009

[10] Vaisakh, K. and Srinivas, L.R. (2011) Evolving Ant Colony Optimization Based Unit Commitment. *Applied Soft Computing*, **11**, 2863-2870. http://dx.doi.org/10.1016/j.asoc.2010.11.019

[11] Chandrasekaran, K., Hemamalini, S., Simon, S.P. and Padhy, N.P. (2012) Thermal Unit Commitment Using Binary/Real Coded Artificial Bee Colony Algorithm. *Electric Power Systems Research*, **84**, 109-119. http://dx.doi.org/10.1016/j.epsr.2011.09.022

[12] Datta, D. and Dutta, S. (2012) A Binary-Real-Coded Differential Evolution for Unit Commitment Problem. *International Journal of Electrical Power & Energy Systems*, **42**, 517-524. http://dx.doi.org/10.1016/j.ijepes.2012.04.048

[13] Roy, P.K. and Sarkar, R. (2014) Solution of Unit Commitment Problem Using Quasi-Oppositional Teaching Learning Based Algorithm. *International Journal of Electrical Power & Energy Systems*, **60**, 96-106. http://dx.doi.org/10.1016/j.ijepes.2014.02.008

[14] Moon, Y.H., Park, J.K., Kook, H.J. and Lee, Y.H. (2001) A New Economic Dispatch Algorithm Considering Any Higher Order Generation Cost Functions. *International Journal of Electrical Power & Energy Systems*, **23**, 113-118. http://dx.doi.org/10.1016/S0142-0615(00)00043-0

[15] Lin, W.-M., Gow, H.-J. and Tsay, M.-T. (2007) A Partition Approach Algorithm for Non-Convex Economic Dispatch. *International Journal of Electrical Power & Energy Systems*, **29**, 432-438. http://dx.doi.org/10.1016/j.ijepes.2006.11.002

[16] Saber, A.Y., Chakraborthy, S., Abdur Razzak, S.M. and Senjyu, T. (2009) Optimization of Economic Load Dispatch of Higher Order General Cost Polynomials and Its Sensitivity Using Modified Particle Swarm Optimization. *Electric Power Systems Research*, **79**, 98-106. http://dx.doi.org/10.1016/j.epsr.2008.05.017

[17] Elanchezhian, E.B., Subramanian, S. and Ganesan, S. (2014) Economic Power Dispatch with Cubic Cost Models Using Teaching Learning Algorithm. *IET Generation, Transmission & Distribution*, **8**, 1187-1202. http://dx.doi.org/10.1049/iet-gtd.2013.0603

[18] Rao, R.V., Savsani, V.J. and Vakharia, D.P. (2012) Teaching-Learning-Based Optimization: An Optimization Method for Continuous Non-Linear Large Scale Problems. *Information Sciences*, **183**, 1-15. http://dx.doi.org/10.1016/j.ins.2011.08.006

[19] Sultana, S. and Roy, P.K. (2014) Optimal Capacitor Placement in Radial Distribution Systems Using Teaching Learning Based Optimization. *International Journal of Electrical Power & Energy Systems*, **54**, 387-398. http://dx.doi.org/10.1016/j.ijepes.2013.07.011

[20] Singh, M., Panigrahi, B.K. and Abhyankar, A.R. (2013) Optimal Coordination of Directional Over-Current Relays Using Teaching Learning-Based Optimization (TLBO) Algorithm. *International Journal of Electrical Power & Energy Systems*, **50**, 33-41. http://dx.doi.org/10.1016/j.ijepes.2013.02.011

[21] Patel, S.J., Panchal, A.K. and Kheraj, V. (2014) Extraction of Solar Cell Parameters from a Single Current-Voltage Characteristic Using Teaching Learning Based Optimization Algorithm. *Applied Energy*, **119**, 384-393. http://dx.doi.org/10.1016/j.apenergy.2014.01.027

8

Hybrid Dissolved Gas-in-Oil Analysis Methods

Leonidha Londo[1], Rajmonda Bualoti[2], Marialis Çelo[2], Nako Hobdari[2]

[1]Transmission System Operator, Tirana, Albania
[2]Electric Engineering Faculty, Polytechnic University of Tirana, Tirana, Albania
Email: bualoti@icc-al.org

Abstract

Dissolved gas analysis is the most widely used diagnostic test in power transformers. The aim of this paper is to introduce the dissolved gas analysis (DGA) methods able to diagnose the transformer conditions. The faults cause the transformer oil, pressboard, and other insulating materials to decompose and generate gases, some of which dissolve in the oil. The results of DGA must be accurate if faults are to be diagnosed reliably. There are different established methods used in industry for interpreting DGA results. We will compare the result of IEEE Key Gas Methods and Rogers' Ratios. The transformer conditions are evaluated by the Key Gas Method with total combustible gas method (TCGM) and then verified by the Rogers' Ratios. As result, the aging pattern and trend of the power transformer deterioration can be determined. The 30 sample data from IEEE with known faults and dissolved gas concentrations were used as the basis of comparison.

Keywords

Dissolved Gas Analysis, Key Gas Methods, Rogers' Ratios, Transformer

1. Introduction

A transformer is the most important equipment for power supply to consumers. Failure of one power transformer may cause long interruptions in supply, costly repairs and loss of revenue. For uninterrupted power supply to consumers, proper maintenance, particularly preventive maintenance, is very necessary. The failure in magnetic, electric and dielectric circuits as well as structural failure may cause extensive damage to the equipment and surroundings. Proper operation and maintenance procedure may help to prevent failure and extend life of operation of the transformer.

2. Risk of Failure of Transformers

The mineral oil uses in transformers, acts as a dielectric medium and also as a heat transfer agent. The breakdown of electrical insulating materials and related components inside the transformer liberates gases within the unit. The distribution of these gases can be related to the type of electrical fault, and the rate of gas generation can indicate the severity of the fault. The identity of the gases being generated by a particular unit can be very useful information in any preventive maintenance program [1]-[3]. There are several techniques in detecting those fault gases and DGA was recognized as the most informative method. This method involves sampling of the oil to measure the concentration of the dissolved gases. The risk of failure of transformers in service, based on dissolved gas analysis (DGA), depends on three main parameters: the type of fault involved, the location of the fault (in oil or in paper), and the amount of gases formed (concentrations and rates).

- The most dangerous faults are: high-energy arcing faults in oil and paper (D2), low-energy arcing faults in paper (D1), and hot spots in paper of high temperatures (>700°C T3 and 300°C - 700°C T2).
- Less dangerous faults are: low-energy arcing faults in oil (D1), hot-spots in oil (>700°C T3 and 300°C - 700°C T2), and hot spots in paper of low temperature (<300°C T1).
- Non-dangerous faults are: hot spots in oil (<300°C T1), producing only "stray gassing" of oil, corona partial discharges (PD) (unless very high levels of hydrogen are formed), catalytic reactions with water, and aging of paper.

A risk model for transformer failures [4]-[6], based on aging, can be represented by Perks formula:

$$f(t) = \frac{A + \propto e^{\beta t}}{1 + \mu e^{\beta t}}. \tag{1}$$

Figure 1 is the corresponding exponential curve for 50% failure rate at the age of 50 year. To include the random events (lighting, vandalism, etc.) separate from the aging component, the constant "A" is set at 0.005. In **Figure 2** is represented the age of the main transformers of Albanian Transmission System. We can see that for some transformers the failure rate is higher than 10%. They are expensive equipments; that's why there has been a growing interest in the technique to diagnose, determine and decide the condition assessment of transformer insulation [7] [8].

3. Identification of Type of Faults

The general type of fault (PD, D1, D2, T1, T2, and T3) can be identified by several methods: e.g., Rogers, Key Gas, IEC ratios, and Duval Triangle [9] [10]. Electrical and thermal stresses such as arching, partial discharges and overheating cause degradation of dielectric oil and solid dielectric cellulose materials.

The degradation of insulation produces different gases. Different degradation mechanisms generate different gases thus making it possible to determine the degrading part of the transformer. Power transformer gas-in-oil analysis (DGA) can be used for effective diagnostics and monitoring. The IEEE and IEC guides [11] [12] offer a variety of ratio-based tools to diagnose DGA data. A quick summary of the different tools found in the current IEEE guide and the IEC guide is in **Table 1**.

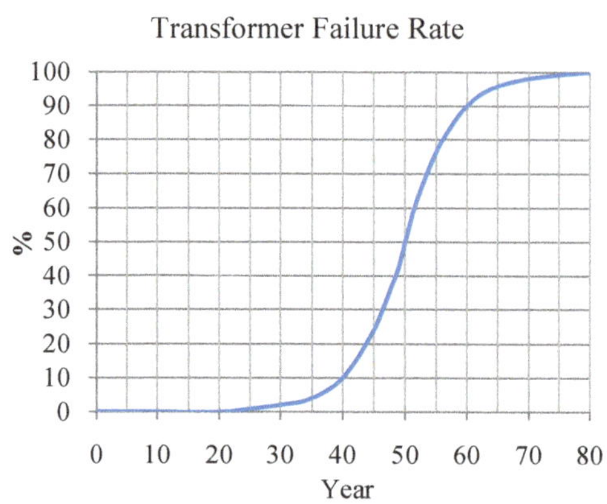

Figure 1. The curve for transformer failure rate.

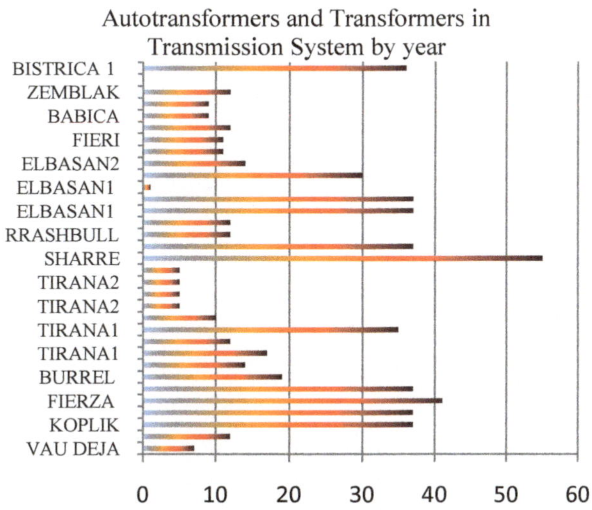

Figure 2. Autotransformers and Transformers ages in Albanian Transmission System.

Table 1. A summary of the different tools.

Tool	Reference Standard	
	IEEE C57. 104-1991	IEC 60599-1999
Key Gas Procedure	X	
Rogers Ratios	X	
TDCG Procedure	X	
Basic Gas Ratios		X
Duval Triangle		X
Doernenburg Ratios	X	

DGA diagnostic tools vary in their complexity and accuracy. **Table 2** shows the comparison among the most widely used DGA diagnostic methods.

The Albanian Power Companies started to invest to DGA laboratory tests from two years, so there is no previous dissolved gas history data. In IEC Standards 60599 is suggested a four-level criterion to classify risks to transformers evaluating individual gas and TDCG concentrations when there is no previous dissolved gas history. That is why among different methods for interpreting DGA results, we will use hybrid method based in the Key Gas Method with total dissolved combustible gas method to be evaluated and then verified by the Rogers' Ratios.

4. Gases Detected and Their Relevance

Typical gases generated from mineral oil/cellulose (paper and pressboard) insulated transformers include: Hydrogen, H_2; Methane, CH4; Ethane, C_2H_6; Ethylene, C_2H_4; Acetylene, C_2H_2; Carbon Monoxide, CO; Carbon Dioxide, CO_2. Additionally, oxygen and nitrogen are always present; their concentrations vary with the type of preservation system used on the transformer. In addition, gases such as propane, butane and others can be formed as well, but their use for diagnostic purposes is not widespread. The concentration of the different gases provides information about the type of incipient-fault condition present as well as the severity. In **Table 3** and **Table 4** are shown fault interpretations from dissolved gases according [13].

5. Key Gas Method and TDCG method

The key gas method is mainly depends on the quantity of fault gases release in mineral oil when fault occur.

Table 2. Comparison of DGA diagnostic methods.

Features	Rogers Ratio	Key Gas	TDCG
Data can be quickly and easily interpreted		□□	□
Early detection of fault	□□	□□	□□
Fault types can be identify accurately	□□	□□	
Able to identify in detail a specific fault type	□□		
Can interpret based on individual or incomplete fault gases		□	
Oil volume independent	□□		
Transformer type independent	□□		□

Table 3. Fault interpretation from dissolved according.

Gas	**Normal** ppm	**Abnormal** ppm	Interpretation
H_2	150	1000	Arcing corona
CH_4	25	80	Sparking
C_2H_6	10	35	Local overheating
C_2H_4	20	100	Severe overheating
C_2H_2	15	70	Arcing
CO	500	1000	Severe overloading
CO_2	10,000	15,000	Severe overloading
TDCG	720	2285	

Table 4. Fault interpretations from gases detected in the insulation oil.

No.	Gases Detected in Oil Sample	Fault Interpretations
1	Nitrogen plus 5% or less oxygen	Normal operation
2	Nitrogen, carbon monoxide, and carbon dioxide	Transformer winding insulation overheated; Key gas is carbon monoxide **CO**
3	Nitrogen, ethylene, and methane—some hydrogen and ethane	Transformer oil is overheated; minor fault causing oil breakdown. Key gas is ethylene $\mathbf{C_2H_4}$
4	Nitrogen, hydrogen, small quantities of ethane and ethylene	Corona discharge in oil; key gas is hydrogen $\mathbf{H_2}$
5	Same as 4 with carbon dioxide and carbon monoxide	Corona involving paper insulation; key gas is hydrogen $\mathbf{H_2}$
6	Nitrogen, high hydrogen and acetylene; minor quantities of methane and ethylene	High-energy arcing; key gas is acetylene $\mathbf{C_2H_2}$
7	Same as 6 with carbon dioxide and carbon monoxide	High-energy arcing involves paper insulation of winding; key gas is acetylene $\mathbf{C_2H_2}$

Fault gases are caused by corona (partial discharge), thermal heating and arcing. The Key Gas Method considers the following four general fault types:

1) Thermal fault due to overheated oil;
2) Thermal fault due to overheated cellulose;
3) Electrical fault due to corona;
4) Electrical fault due to arcing.

The standard of IEEE Std C57.104-1991 indicates the key gases and their relative proportions for four fault types (**Table 5**). Generally, the thermal decomposition of oil produced more than 60% of ethylene (C_2H_4) and thermal decomposition of cellulose produce key gas carbon monoxide (CO) is 90%. In case of corona in oil mainly produce large amount principal gas hydrogen nearly 80% and due to arcing key gas acetylene produced 30% with trace quantity of hydrogen.

The total dissolved combustible gas (TDCG) is the definition of the sum of the combustible gas concentrations as follows:

$$TDCG = H_2 + CH_4 + C_2H_6 + C_2H_4 + C_2H_2 + CO \quad (2)$$

InIEC Standards 60599 it is suggested a four-level criterion (**Table 6**) to classify risks to transformers when there is no previous dissolved gas history as follow:

Condition 1: TDGC below this level indicates the transformer is operating satisfactorily;

Condition 2: TDGC within this range indicates greater than normal combustible level. Action should be taken to establish a trend;

Condition 3: TDGC within this range indicates a high level of decomposition. Immediate action should be taken to establish a trend;

Condition 4: TDGC within this range indicates excessive decomposition. Continued operation could result in failure of the transformer.

6. Roger's Ratio Method

This method uses five gases (H_2, C_2H_6, CH4, C_2H_2 and C_2H_4) to generate codes based on their composition ratio [14]. The codes are then used to categorize a range of ratios used to diagnose the fault as shown in **Table 7**.

The method utilizes the ratio of gas concentration to indicate fault types. Rogers Ratio method can be used when any of individual gases exceeds its normal limit and it does not depend on specific gas concentrations. The Rogers Ratio method utilizes four ratios; C_2H_6/CH_4, C_2H_2/C_2H_4, CH_4/H_2, and C_2H_4/C_2H_6 that leads to twenty-two proposed diagnosis as shown in **Table 8** [15]. Rogers Ratio method diagnosis provides more interpretation details in terms of temperature range of decomposition.

In [16], the Rogers Ratio method diagnosis was revised, as shown in **Table 9**, to include only six diagnosis interpretations; normal, low-energy density arcing-PD, arcing-high-energy discharge (D2), low temperature thermal (T1), thermal 300˚C - 700˚C (T2) and thermal >700˚C (T3).

This method does not consider dissolved gases below normal concentration limits; also certain ratio values are inconsistent with the diagnostic assigned and lead to invalid codes [17].

7. A Hybrid DGA Analysis Methods

The hybrid method proposed in this paper is a step-by-step procedure (flow chart **Figure 3**) to diagnose faults

Table 5. The key gases and their relative proportions for four fault types.

Faults/Gases	CO	H_2	CH_4	C_2H_6	C_2H_4	C_2H_2
Overheated oil	0	2	16	19	63	0
Overheated cellulose	92	0	0	0	0	0
Corona in oil	0	85	13	1	1	0
Arcing in oil	0	60	5	2	3	30

Table 6. Dissolved gas concentrations (ppm).

Status	H_2	CH_4	C_2H_2	C_2H_4	C_2H_6	CO	CO_2	TDCG
Condition 1	<100	<120	<35	<50	<65	<350	<2500	<720
Condition 2	101 - 700	121 - 400	36 - 50	51 - 100	66 - 100	351 - 570	2500 - 4000	721 - 1920
Condition 3	701 - 1800	401 - 1000	51 - 80	101 - 200	101 - 150	571 - 1400	4001 - 10,000	1921 - 4630
Condition 4	>1800	>1000	>80	>200	>150	>1400	>10,000	>4630

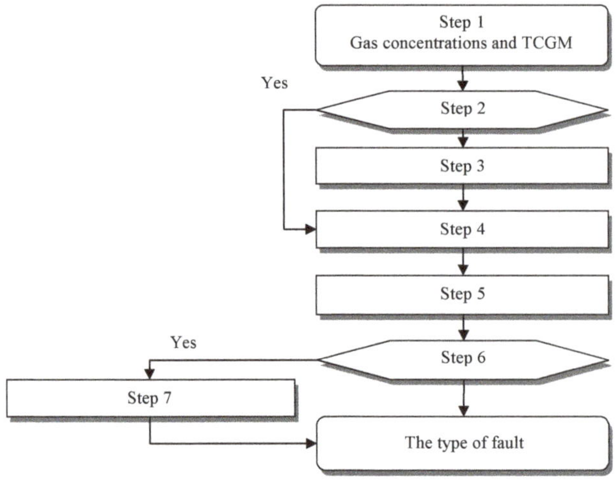

Figure 3. The flow chart of the hybrid Methods.

Table 7. Code definition of Rogers refined ratio method.

Gas Ratio	Range	Code
CH_4/H_2 (R1)	Not greater than 0.1	5
	Between 0.1 and 1.0	0
	Between 1.0 and 3.0	1
	Not less than 3.0	2
C_2H_6/CH_4 (R_4)	Less than 1.0	0
	Not less than 1.0	1
C_2H_4/C_2H_6 (R5)	Less than 1.0	0
	Between 1.0 and 3.0	1
	Not less than 3.0	2
C_2H_2/C_2H_4 (R_2)	Less than 0.5	0
	Between 0.5 and 3.0	1
	Not less than 3.0	2

using the Key Gas Method with total combustible gas method (TCGM) and then verified by the Rogers' Ratios.

Step 1. Gas concentrations are obtained by DGA and TCGM is evaluated;

Step 2. If at least one of the gas concentrations (in ppm) for H_2, CH_4, C_2H_2, C_2H_4, C_2H_6 and TCGM exceeds the values for abnormal limit (see **Table 3**), the unit is considered faulty; proceed to Step 4;

Step 3. If at least one of the gas concentrations (in ppm) for H_2, CH_4, C_2H_2, C_2H_4, C_2H_6 and TCGM exceeds the values for limit C_2 (see **Table 6**), the unit is considered faulty; precede to Step 4, otherwise "No fault condition";

Step 4. Determining the values of ratio obtained from **Table 7** in the order of ratio CH_4/H_2, $C2H_2/CH_4$, C_2H_4/C_2H_6 and C_2H6/CH_4 procedure. Define the codes from **Table 7**;

Step 5. Each successive code is compared to the values obtained from **Table 8** to define the fault;

Step 6. If all succeeding ratios for a specific fault type fall within the values (column) given in **Table 9**, the type of fault is defined;

Step 7. The gas concentrations for H_2, CH_4, C_2H_2, C_2H_4, C_2H_6 in % of TCGM is evaluated; If relative proportions of the keys gases for a specific fault type fall within the values given in **Table 5**, the suggested diagnosis is valid. The report of fault is defined.

Table 8. Fault diagnosis table for Rogers ratio method.

No.	Codes	Fault Type
1	5000	B: Partial discharges of low energy density or hydrolysis
2	5001	E: Partial discharges of increasing energy density
3	5010	D: Coincidental partial discharges and conductor overheating
4	5100	C: Partial discharges of high energy density, possibly with tracking
5	0000	A: No fault: normal deterioration
6	0001	N: Thermal fault of temperature range 100˚C - 200˚C
7	0010	I: Insulated conductor overheating
8	0011	J: Complex thermal hotspot and conductor overheating
9	0100	F: Low energy discharge: flashover without power follow through
10	0101	K: Coincidental thermal hotspot and low energy discharge
11	0110	G: Low energy discharge: continuous sparking to floating potential
12	0120	H: High energy discharge: arc with power follow through
13	0200	F: Low energy discharge: flashover without power follow through
14	0210	G: Low energy discharge: continuous sparking to floating potential
15	0220	H: High energy discharge: arc with power follow through
16	1000	M: Thermal fault of low temperature range < 150˚C
17	1001	N: Thermal fault of temperature range 100˚C - 200˚C
18	1010	O: Thermal fault of temperature range temperature range 150˚C - 300˚C overheating of copper due to eddy currents
19	1020	P: Thermal fault of high temperature range 300˚C - 700˚C: bad contacts/joints: core and tank circulating currents
20	1100	K: Coincidental thermal hotspot and low energy discharge
21	2001	N: Thermal fault of temperature range 100˚C - 200˚C
22	2020	P: Thermal fault of high temperature range 300˚C - 700˚C: bad contacts/ joints: core and tank circulating currents

Table 9. Latest version of RRM diagnosis.

Case	R_2 C_2H_2/C_2H_4	R1 CH_4/H_2	R5 C_2H_4/C_2H_6	Suggested Fault Diagnosis
0	<0.1	>0.1 to <1	<1	Unit normal
1	<0.1	<0.1	<1	Low-energy density arcing-PD
2	0.1 to 3	0.1 to 1	>3	Arcing–High-energy discharge
3	<0.1	>0.1 to <1	1 to 3	Low temperature thermal
4	<0.1	>1	1 to 3	Thermal less than 700˚C
5	<0.1	>1	>3	Thermal exceeding 700˚C

8. Results

The method was tested against all the 30 cases in the data set [18]. In **Table 10**, we show a set of data that illustrate the method as well as the evaluation of the key gases concentrations according **Table 6** (Step 1).

In **Table 11** are represented the result of application of step 2 and 3 of above methods and the first step evaluation of faults. In **Table 12** are represented the Rogers' Ratios and the fault according to the Rogers Code Method.

Table 10. Set of data used in analysis and the evaluation the key gases concentrations.

No.	H_2 (ppm)	CH_4 (ppm)	C_2H_2 (ppm)	C_2H_4 (ppm)	C_2H_6 (ppm)	CO (ppm)	TDGC (ppm)	Actual Fault	H_2	CH_4	C_2H_2	C_2H_4	C_2H_6	CO	TDGC
1	33	79	5	215	30	56	418	DHE	C1	C1	C1	C4	C1	C1	C1
2	266	584	1	862	328	230	2271	TF	C2	C3	C1	C4	C4	C1	C3
3	9474	4345	12,752	6517	353	504	33,945	TF-H	C4	C4	C4	C4	C4	C2	C4
4	507	1053	17	1440	297	3034	6348	DHE	C2	C4	C1	C4	C4	C4	C4
5	441	207	261	224	43	234	1410	TF-M	C2	C2	C4	C4	C1	C1	C2

Table 11. Set of data used in evaluation of fault according the key gases concentrations.

No.	H_2 %	CH_4 %	C_2H_2 %	C_2H_4 %	C_2H_6 %	CO %	Fault
1	Normal	Normal	Normal	Severe Overheating	Normal	Normal	TF
2	Normal	Sparking	Normal	Severe Overheating	Local Overheating	Normal	TF
3	Arcing, Corona	Sparking	Arcing	Severe Overheating	Local Overheating	Normal	TF-H
4	Normal	Sparking	Normal	Severe Overheating	Local Overheating	Severe Overloading	DHE
5	Normal	Sparking	Arcing	Severe Overheating	Local Overheating	Normal	TF

Table 12 represented the result of application of step 4 and 5 of above methods based in the following reasoning:

1) Since total combustible gases (TDCG) are LESS the 720 ppm, the Rogers' Ratios result indicated below is considered less significant. Since C_2H_4 is ABNORMAL a Core & Tank circulating currents overheated joints;

2) Since total combustible gases (TDCG) are GREATER than 720 ppm, the Rogers' Ratios result indicated below should be considered more significant. Winding circulating currents;

3) Since total combustible gases (TDCG) are GREATER than 720 ppm, the Rogers' Ratios result indicated below should be considered more significant. Arc, with power follow through;

4) Since total combustible gases (TDCG) are GREATER than 720 ppm, the Rogers' Ratios result indicated below should be considered more significant. Core & Tank circulating currents overheated joints;

5) Since total combustible gases (TDCG) are GREATER than 720 ppm, the Rogers' Ratios result indicated below should be considered more significant. Arc, with power follow through.

From **Tables 10-12**, we can see the evaluation of the results for 5 cases has given a successful prediction with the Key Gas Method with total combustible gas method (TCGM) for cases 2 - 5 and a successful prediction with the Rogers' Ratios for case 2.

The evaluation of the results for 30 cases in the data set give a successful prediction of the Key Gas Method with total combustible gas method (TCGM) of about 78%, a successful prediction of the Rogers' Ratios of about 45% and the successful prediction of the Hybrid Method of about 87%.

9. Case Study Albanian Power Companies

The Albanian Power Companies started to invest in the last years to DGA laboratory tests to monitor main HPPs transformer units [19]. The data consists of numerical measurement of each individual key gas in ppm. In **Table 13** are represented the data measured in transformer of HEC Koman, one of the main power plant in Albania in two consecutive months.

By looking at the relative proportions of gases in the DGA results and the monthly increase it could be possible to identify the type of fault occurring in the transformer. It is evident an increase of Carbon Dioxide, Oxygen and humidity, while TDCG are LESS the 720 ppm. So, the rehabilitation of oil through degasification and dehumidification was necessary. After this process, as presented in **Table 14**, the relative proportions of gases in the DGA results at normal value.

Table 12. Set of data used in evaluation of fault according the Rogers Code Method.

No.	H_2 (ppm)	CH_4 (ppm)	C_2H_2 (ppm)	C_2H_4 (ppm)	C_2H_6 (ppm)	CO (ppm)	TDGC (ppm)	Actual Fault	R1 CH_4/H_2	R4 C_2H_6/CH_4	R5 C_2H_4/C_2H_6	R2 C_2H_2/C_2H_4	Roger Codes	Fault
1	33	79	5	215	30	56	418	DHE	2.39	0.38	7.17	0.02	1020	TF
2	266	584	1	862	328	230	2271	TF	2.20	0.56	2.63	0.00	1010	TF-M
3	9474	4345	12,752	6517	353	504	33,945	TF-H	0.46	0.08	18.46	1.96	0021	DHE
4	507	1053	17	1440	297	3034	6348	DHE	2.08	0.28	4.85	0.01	1020	TF-H
5	441	207	261	224	43	234	1410	TF-M	0.47	0.21	5.21	1.17	0021	DHE

Table 13. The data measured in transformer of HEC Koman.

Unit	Sample Date	Hydrogen (ppm)	Methan (ppm)	Acetylen (ppm)	Ethylene (ppm)	Ethan (ppm)	Carbon Monoxid (ppm)	**Carbon Dioxid (ppm)**	**Oxygen (ppm)**	TDCG (ppm)	**Water (ppm)**
TR4 HEC KOMAN	5/1/2015 18:00	13.7	16.3	0	17.8	64	314.8	**3565**	**19937.4**	427	**34.7**
	3/12/2014 23:00	12.6	14.1	0	8	74.8	238.2	**2744**	**17353.7**	347	**39.9**
	Monthly increase	1.1	2.2	0	9.8	−10.8	76.6	**821**	**2583.7**	80	**-5.2**

Table 14. The data measured in transformer of HEC Koman after the rehabilitation of oil.

Unit	Sample Date	Hydrogen (ppm)	Methane (ppm)	Acetylene (ppm)	Ethylene (ppm)	Ethan (ppm)	Carbon Monoxide (ppm)	Carbon Dioxide (ppm)	Oxygen (ppm)	TDCG (ppm)	Water (ppm)
TR4	16/02/2015 18:00	0	1	0	1	3.5	0	100	500	5.5	9

10. Conclusions

For uninterrupted power supply to consumers, proper maintenance of transformer, particularly preventive maintenance, is very necessary. The IEC standards 60599 and IEEE standards C57.104-1991 help in interpretation of DGA results, which together with further field inspections and experienced personnel judgment can give an answer to what is going on inside a suspected unhealthy transformer. The Key Gas Method in principal gives no answer in multiple fault condition.

To increase the accuracy of the interpretation and the certainty of the transformer condition, a hybrid method based on the Key Gas Method with total combustible gas method (TCGM) verified by the Rogers' Ratios is used to improve the accuracy of interpretation of DGA results.

The result of the prediction done to test the accuracy of the method shows the improvement of accuracy with 9% compared with the simple Key Gas Method.

References

[1] Griffin, P.J. (1998) Criteria for the Interpretation of Data for Dissolved Gases in Oil from Transformers (A Review). Special Technical Publication 998, ASTM, Philadelphia, 89-107.

[2] Suleiman, A.A., *et al.* (2012) Improving Accuracy of DGA Interpretation of Oil-Filled Power Transformers Needed for Effective Condition Monitoring. *IEEE International Conference on Condition Monitoring and Diagnosis*, Bali, 23-27 September 2012, 374-378.

[3] Lewand, L.R. and Griffin, P.J. (2000) The Effective Use of Laboratory Analysis of Insulating Oil as a Maintenance

Tool. *Proceedings of the Sixty-Seventh Annual International Conference of Doble Clients*, Watertown, 27-31 March 2000, Sec 5-8.

[4] Pradhan, M.K. and Ramu, T.S. (2004) Criteria for Estimation of End of Life of Power and Station Transformers in Service, Electrical Insulation and Dielectric Phenomena, CEIDP. *Annual Report Conference*, Boulder, 17-20 October 2004, 220-223.

[5] William H. Bartley, P.E. (2004) An Analysis of Transformer Failures. Part 2. Hartford Steam Boiler Inspection and Insurance Company. https://www.hsb.com/TheLocomotive/Article.aspx?id=169&terms=Bartley

[6] Chen, Q.M. and Egan, D.M. (2006) A Bayesian Method for Transformer Life Estimation Using Perks' Hazard Function. *IEEE Transactions on Power Systems*, **21**, 1954-1965. http://dx.doi.org/10.1109/TPWRS.2006.881129

[7] Ding, X.Q. and Cai, H. (2001) On-Line Transformer Winding's Fault Monitoring and Condition Assessment. *Proceedings of* 2001 *International Symposium on Electrical Insulating Materials* (*ISEIM* 2001), Himeji, 19-22 November 2001, 801-804.

[8] McGrail, T. and Wilson, A. (1997) On-Line Gas Sensors. IEE Colloquium on Condition Monitoring of Large Machines and Power Transformers, London, 19 June 1997, 1/1-1/4.

[9] Sun, H.-C., Huang, Y.-C. and Huang, C.-M. (2012) A Review of Dissolved Gas Analysis in Power Transformers. *Energy Procedia*, **14**, 1220-1225. http://dx.doi.org/10.1016/j.egypro.2011.12.1079

[10] Stebbins, R.D., Kelly, J.J. and Myers, S.D. (1997) Power Transformer Fault Diagnosis. *IEEE PES WM*, *Panel Session*, New York, 6 February 1997, 463-468.

[11] IEEE Standards C57.104-1991. IEEE Guide for the Interpretation of Gases Generated in Oil-Immersed Transformers, 1991.

[12] IEC Standards 60599 (1999) Mineral Oil-Impregnated Electrical Equipment in Service—Guide to the Interpretation of Dissolved and Free Gases Analysis. Second Edition.

[13] Franzén, A. and Karlsson, S. (2007) Failure Modes and Effects Analysis of Transformers. Technical Report TRITA-EE 2007:040, KTH School of Electrical Engineering.

[14] Rogers, R.R. (1975) UK Experiences in the Interpretation of Incipient Faults in Power Transformers by Dissolved Gas-in-Oil Chromatography Analysis (A Progress Report). *Minutes of Forty-Second International Conference of Doble Clients*, Section 10-201

[15] Rogers, R.R. (1978) IEEE and IEC Codes to Interpret Incipient Faults in Transformers, Using Gas in Oil Analysis. *IEEE Transactions on Electrical Insulation*, **EI-13**, 349-354. http://dx.doi.org/10.1109/TEI.1978.298141

[16] IEEE Guide for the Interpretation of Gases Generated in Oil-Immersed Transformers—Redline. IEEE Std C57.104-2008 (Revision of IEEE Std C57.104-1991)—Redline, pp. 1-45, 2009.

[17] Abu-Siada, A. and Islam, S. (2012) A New Approach to Identify Power Transformer Criticality and Asset Management Decision Based on Dissolved Gas-in-Oil Analysis. *IEEE Transactions on Dielectrics and Electrical Insulation*, **19**, 1007-1012. http://dx.doi.org/10.1109/TDEI.2012.6215106

[18] Beykverdi, M., Faghihi, F. and Pour, A.M. (2014) A New Approach for Transformer Incipient Fault Diagnosis Based on Dissolved Gas Analysis (DGA). *Nova Journal of Engineering and Applied Sciences*, **3**, 1-7. www.novaexplore.com

[19] Londo, L., Çelo, M. and Bualoti, R. (2015) Assessment of Transformer Condition Using the Improve Key Gas Methods. *International Journal of Engineering Research & Technology* (*IJERT*), **4**, 48-55.

9

Optimization of Liquid Ammonia Treatment for Enzymatic Hydrolysis of *Miscanthus sinensis* Anderss

Liping Zhang, Hehuan Peng, Qinling Yu, Ying Zhang, Zhijun Wang, Xianwei Tang

School of Engineering, Zhejiang A&F University, Linan, China
Email: 253455672@qq.com

Abstract

Miscanthus biomass yield high, high photosynthetic efficiency, fast growth, easy breeding, widely distributed, is suitable as raw materials for the production of fuel ethanol. In this paper, the method of the pretreatment of liquid ammonia for Miscanthus, using the pretreatment of liquid ammonia to overcome biomass recalcitrance, adding cellulase enzyme, and the monosaccharide content is determined by using high performance liquid chromatography (HPLC). Effect of pretreatment process on the enzyme solution effect of awn, conducts the research to the water processing rate, temperature, residence time, amount of ammonia, the amount of H_2O_2 parameters. Data shows that, in the moisture content of 80%, temperature 130˚C, loading amount of ammonia 2:1, residence time 10 min, the glucan and xylan enzyme conversion rate are 73.23% and 73.28% respectively. After a mass balance, dry base mans per 100 g can get glucose 28.96 g and xylan 17.25 g.

Keywords

Miscanthus, Liquid Ammonia Pretreatment, Enzymatic Hydrolysis, Fuel Ethanol

1. Introduction

With the energy crisis becomes more prominent, the development and utilization of new energy have becoming imminent. The bio-energy has become a hot energy in today's world. Renewable energy fuel ethanol is much cleaner compared to fossil fuel.

Miscanthus (*Miscanthus sinensis* Anderss) belongs to angiosperms Gramineae. It is herbaceous perennial and tall plant, composing stable community [1]. Miscanthus plants blossom after heading, leaf stem ratio increasing, high cellulose content, especially the stalk, crude cellulose content in 50% above, and to the Miscanthus as the most. According to the determination, the total content of cellulose in Miscanthus is 80.75% [2], it is very suitable as biological energy materials.

The main components of the plant cell wall are cellulose, hemicellulose, lignin. Lignin interior besides have strong hydrogen bonding connection, but also with the semi cellulose by covalent bond to form stable LCCs.

These structure characteristics make lignocellulose hydrolysis recalcitrant exist strong [3] [4]. Although many microbes can decompose separate existence of the cellulose, but because in cell wall cellulose by lignin protection, so it is not susceptible to microbial degradation of [5] [6]. The most critical step in the production of fuel ethanol by lignocellulosic materials is the pre-processing of material [7], it can decide the preprocessing step of cellulose and hemicellulose conversion rate and cost [8]. The pretreatment methods mainly include: acid pretreatment, alkali pretreatment, high temperature hot water method, acid catalysis method, high temperature hot water steam blasting method and ammonia fiber explosion method, each treatment effect have advantages and disadvantages [9]-[12]. Through the cycle of ammonia explosion treatment of corn straw can improve the enzymatic hydrolysis rate, maximum glucan yield reaches 85.7% [13]. Presented in this paper is put forward on the basis of the AFEX liquid ammonia pretreatment of this new technology, experiment for influence in different conditions of enzymatic hydrolysis rate, experimental results obtained the highest saccharification rate of fuel ethanol, resulting in lower production costs and higher yield [14]. The total conversion of *Erianthus enzymolysis* of glucan and xylan rates were 69.34% and 82.60% with liquid ammonia pretreatment method, than not pretreatment of raw materials were increased by 573% and 1056%, the yield is 8 times of monosaccharide untreated [15].

2. Materials and Methods

2.1. Experimental Materials

Miscanthus used in this study was grown in Lin'an County (30.11, 119.38), Zhejiang Province, China, harvested in June, 2012. After harvesting the whole plant awn of the ground part is cut into 1 - 2 cm long strip, The Miscanthus samples were dried to <10% moisture (total weight basis) using a 40˚C oven, then the micro grinder (FZ102) crushed into 50 - 100 mesh powder. After mixing evenly, using instrument using water (MA35, Sartorius) to measure the moisture content, labeled with a plastic bag sealed, placed in the refrigerator and temperature is set to −20˚C in storing spare.

For high performance liquid chromatography (HPLC) analysis of glucose, xylan standard are purchasing Sigma-Aldr ICH company, the purity was more than 99%.

The addition of hydrolytic enzymes for cellulose (NovozymeNS50013), β-glucosidase (Novozyme NS50010) and xylanase (Shandong chanson Biotechnology Co. Ltd.).

2.2. Experimental Methods

2.2.1. Acid Solution on the Material Composition Analysis

Experimental determination of determining carbohydrate awn material and insoluble lignin (Klason lignin) components using LAP002 and LAP003 methods. Using acid hydrolysis method aims to determine biomass carbohydrates and insoluble lignin (Klason lignin) content of component in raw material. Concrete steps are first weighing awn materials stem based 0.3 g, adding 72% sulfuric acid 3 + 0.01 mL, keep 2 h in 30˚C ± 1˚C water bath, adding deionized water 84 mL (diluted to 4% sulfuric acid), in the 121˚C autoclave in 1h, after vacuum filtration, the filter paper on solids drying to constant weight, using high performance liquid chromatography (high performance liquid chromatography, HPLC) determine sugar content, and the calculation of glucan, xylan, Arabia sugar percentage. Solid quality calcined to remove ash quality after the acid insoluble lignin content, namely Klason lignin content.

2.2.2. Liquid Ammonia Pretreatment Method (LAT)

Pretreatment with LAT: take quantitative raw materials according to a dry basis, according to the set water content by adding quantitative water, place in a reaction kettle full mix, and set the heating temperature. Quantitative liquid ammonia injection liquid cylinders, heat to about 50˚C and then ammonia injected into the reaction kettle, rapidly heat to the set temperature and keep a certain predetermined time. Subsequent release of ammonia, remove the processing of raw materials and take pre drying in a ventilated cabinet, after 24 h, remove it to the oven (lower than 45˚C) drying about 12 h (moisture content below 10%), weighed and measured to determine the water content, determine the biomass yield after the pretreatment of raw materials, labeled and placed in the ice box refrigeration spare −20˚C. The choice of technological parameters of LAT optimization respectively: temperature, moisture content, residence time, ammonia dosage.

2.2.3. Enzymatic Hydrolysis

Enzymatic hydrolysis with the relevant rules of LAP009, the reaction volume increased to 15 mL (the original 10 mL). Weigh 0.15 g dextran equivalent biomass use electronic scale into enzymolysis bottle, by calculation, sequentially add 750 ul buffer, 150 ul antibiotics, after adding, take the enzyme solution bottle into table, and set the temperature of 50˚C, the speed of 150 r/min pre heating 1h. Then remove the enzyme solution bottle, add cellulase 38 ul, β-glucosidase 48 ul, xylanase 15 ul. Add the amount of cellulase was 15 FPU/(g equivalent of dextran), β-glucosidase 64 pNPGU/g. Added into the table started to enzymatic hydrolysis and to record the start time of enzymatic hydrolysis. Enzymatic hydrolysis of the sampling interval time were 24 h, 72 h and 144 h.

The enzyme solution sampling: with 1000 uL pipetting Gun take 1mL enzyme samples, inject into centrifuge tube, take into the heating constant temperature heater temperature is set as 100˚C in 20 min, and then put it into the speed for centrifugal 15,000 r/min centrifuge disposable needle in 5 min, and filter, then take into HPLC sample bottle, put into cold storage of the spare −20˚C.

2.2.4. The Quantitative Analysis of Sugar

The quantitative determination of monosaccharides by HPLC method, the determination of glucose, xylan and Arabia sugar content, and then converted into the glucan, xylan or Arabia in raw materials. The experiment used for Agilent Technologies 1200 Series (with auto injector), using differential refractive index detector, the column was BioRad's HPX-87H, the mobile phase was 0.005 mol/L dilute sulfuric acid, the flow rate is 0.5 mL/min, the column temperature was 50˚C, sample size 15 μL. For high performance liquid chromatography (HPLC) analysis of the control group's (sample) glucose, xylan, Arabia sugar were purchased from Sigma-Aldrich company, the purity was more than 99%.

2.2.5. The Calculation of the Hydrolysis Rates and Total Sugar Yield

Enzymatic hydrolysis rate was actually obtained similar monosaccharide content of monosaccharide content and raw material ratio in (or a similar ratio of dextran, chitosan) by enzymatic hydrolysis rate as an example, the calculation formula is:

$$d = \frac{g * 15/1000 * 162/180}{0.375 * c/100} * 100$$

where: g is the concentration of glucose in the enzyme solution.

c is the raw material of cellulose (dextran) percentage content, obtained by acid hydrolysis method.

15 express that the enzyme solution is the total 15 mL, 162/180 express that mass conversion of dextran after polymerization lose water, 0.375 express that dry basis of biological (g).

3. The Results and Analysis

3.1. Miscanthus Component Analysis

By **Table 1** can see, Miscanthus of dextran (cellulose) content of 39.54%, the content of xylan content is 23.56%, the two together 63.10%.

3.2. Effects of Moisture Content and Components on the Enzymatic Hydrolysis

Effect of moisture content on the enzymatic hydrolysis rate as shown in **Figure 1**. The experiment to control the temperature of 130˚C, pretreatment of residence time is 10min, liquid ammonia and raw material ratio of 2:1. The results show that, when the moisture content is 80% and 150%, the conversion rate of glucan enzyme is high, 72 h sampling the enzymolysis rate were 67.07% and 68.02%, with the untreated 24.13% respectively improved 178% and 182%. Xylan enzymatic conversion rate is high in water content is 80% and 100%, respectively 73.28% and 73.77%, with the untreated 14.07% respectively improved 421% and 424%. The experiment-

Table 1. Compositional analysis of Miscanthus.

Component (%)	Glucan	Xylan	Klason lignin	Ash
Miscanthus	39.54 ± 0.74	23.56 ± 1.51	22.85 ± 0.78	4.5 ± 0.02

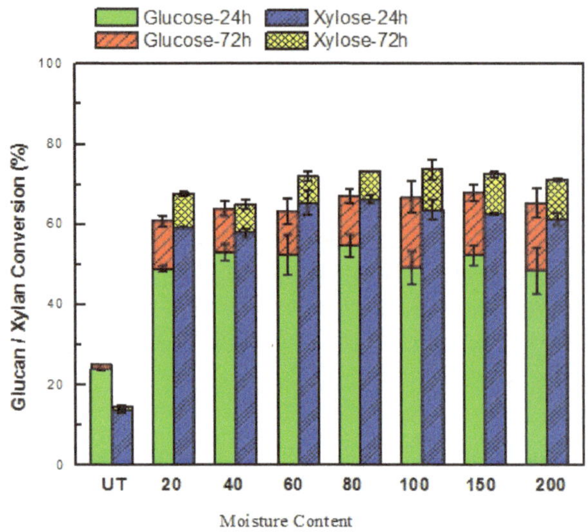

Figure 1. Effects of moisture content on glucan/xylan conversion of LAT-treated *Miscanthus sinensis* Anderss.

tal results show that the moisture content is 100%, the enzyme solution 144 h, dextran and xylanase hydrolysis rates were 74.54% and 77.66%. As can be seen from the graph, in the range of parameters, variation of water content for Miscanthus conversion rate influence is not very big.

3.3. Effects of Temperature and Components on the Enzymatic Hydrolysis

The experiment will control the moisture content in 80%, pretreatment of residence time is 10 min, liquid ammonia and raw material ratio of 2:1, temperature of solution for the influence of the rate of enzyme as shown in **Figure 2**. The experimental results show that, dextran and xylanase hydrolysis rate improved with the increase of temperature basically, and the conversion rate increased with the temperature rise is small. Research has shown that, glucose production needs higher reaction temperature [16]. When the temperature is 170˚C, conversion rate of glucan hydrolysis is the highest, enzymatic hydrolysis of 72 h sampling rate is 74.84%, with the untreated 24.13% increased 2.1 times. Xylan enzyme hydrolysis conversion rate at the temperature of 170˚C is the highest, 75.17%, increased by 434% with the untreated 14.07%. The experimental results show that, the temperature of 170˚C, enzyme 144 h, dextran and xylanase hydrolysis rates were 81.49% and 79.15%.

3.4. Effects of Ammonia Loading and Components on the Enzymatic Hydrolysis

Ammonia is used to increase the removal of grass lignin content [17], so ammonia loading is an important parameter of liquid ammonia pretreatment. Many scholars show that Rice straw and starch rich corn green is the best carrier for 1:1 [18]. And the best loading amount of ammonia of bamboo is 1:2 [14]. Results from the study can be seen, for different biomass materials, the optimal amount of load of ammonia is not the same. So it is very important for the loading amount of ammonia Miscanthus optimization. The amount of ammonia loading refers to the use of volume and biomass feedstock ammonia dry based mass ratio. The experiment conditions were: 80%moisture content, temperature fixed at 130˚C, residence time of 10min. As you can see from **Figure 3**, with loading amount of ammonia increase from 0.5:1 to 3:1, the conversion rate of glucan and xylan increase slowly, and to increase the proportion to 5:1, there has a slight decline. When the load of ammonia concentration of 3:1, 72 h sampling of glucan and xylan hydrolysis rates were 78.19% and 76.03%, 144 h sampling of glucan and xylan hydrolysis rates were 84.26% and 79.83%. So the loading amount of ammonia is not the bigger the better, but from a certain extent, increase the content of ammonia will promote the enzymolysis effect.

3.5. Effects of Residence Time and Components on the Enzymatic Hydrolysis

The experimental temperature fixed at 130˚C, 80% moisture content, ammonia and raw material ratio of 2:1, the residence time solution for the influence of the rate of enzyme as shown in **Figure 4**. When the residence time is

5 min, glucan and xylan conversion rate is the lowest, 72 h sampling were 58.86% and 66.19%, and When the residence time is 10 min, glucan and xylan conversion rate reach the highest point, glucan and xylan conversion rates were 73.23% and 72.20%, compare with the conditions of residence time is 5min increase by 24% and 9%. So the effect of increasing residence time on conversion rate of dextran is large and xylan is small, the residence time of 10 min for the glucan and xylan conversion rates are 73.23% and 72.20%, the conversion rate is relatively high. Considering energy consumption and cost, choose 10 min residence time as the optimum pretreatment conditions.

3.6. Mass Balance

Pretreatment and enzymatic hydrolysis mass balance process as shown in **Figure 5**. In the preprocessing optimal conditions (80% moisture content, temperature fixed at 130˚C, load the amount of ammonia 2:1, residence time is 10 min), cellulase dosage of 15 FPU/ (g equivalent dextran), 100 g dry Miscanthus enzyme 45.97 g can be obtained after 72 h of monosaccharide, without pretreatment materials available sugars as 11.57 g, transformation rate increase 2.97 times. Miscanthus enzyme the highest glucose and xylan yield were 73.23% and 73.20% after pretreatment of LAT.

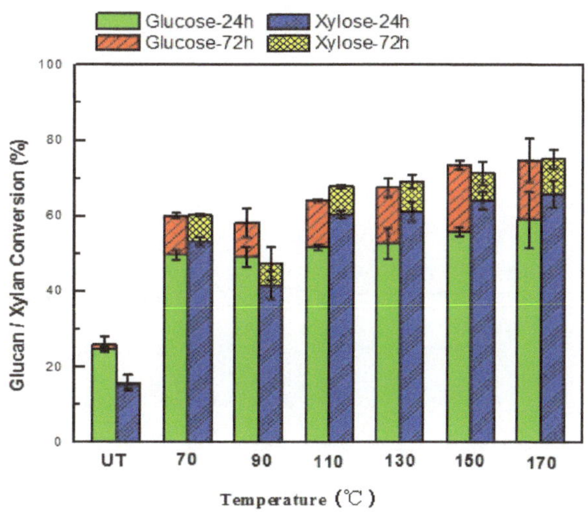

Figure 2. Effects of temperature on glucan/xylan conv-ersionof LAT-treated *Miscanthus sinensis* Anderss.

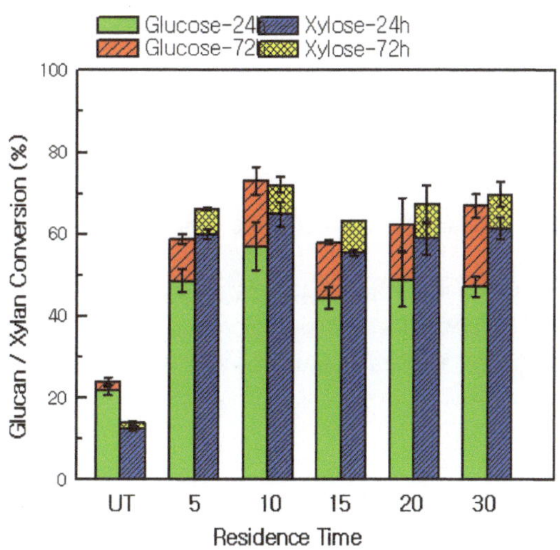

Figure 3. Effects of ammonia loading on glucan/xylan conversion of LAT-treated *Miscanthus sinensis* Anderss.

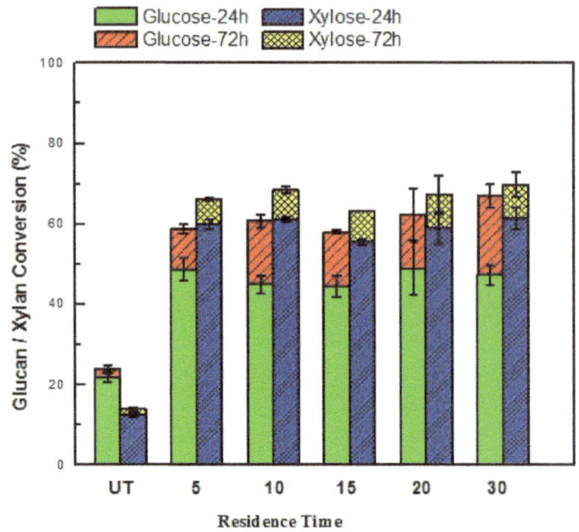

Figure 4. Effects of time residence on glucan/xylan conversion of LAT-treated *Miscanthus sinensis* Anderss.

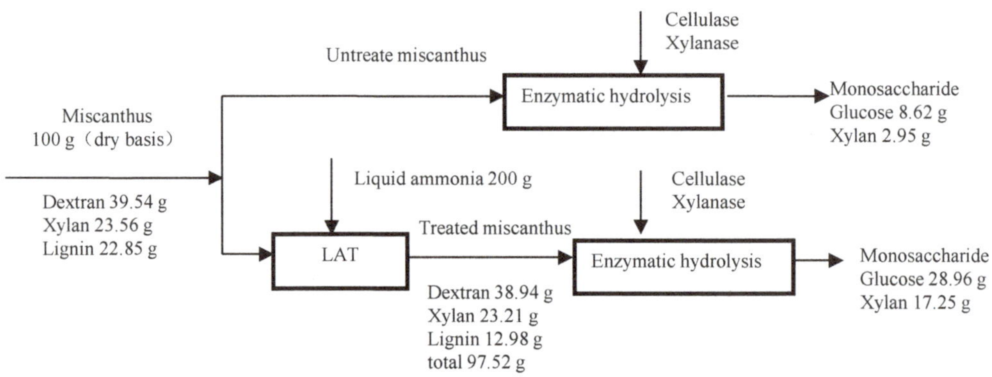

Figure 5. Flow chart of mass balance during pretreatment and hydrolysis process for Miscanthus.

4. Conclusion

On the condition of Miscanthus LAT processing optimize moisture content, temperature, loading amount of ammonia and residence time, equilibrium of raw materials of enzymatic hydrolysis rate and consumption, draw the optimum conditions is 80% moisture content, temperature fixed at 130˚C, loading amount of ammonia 2:1, residence time 10 min, at this time, glucan and xylan enzymatic conversion 73.23% and 73.20%, after the mass balance, per 100 g dry Miscanthus can obtain glucose 28.96 g, xylanase 17.25 g.

References

[1] Yao, X.J. (2006) Agricultural Resources and Sustainable Development of Agriculture. *Journal of Chengdu University: Social Sciences Edition*, **5**, 48-49.

[2] Zhao, N.X. and Xiao, Y.F. (1990) Anhui Province Miscanthus Plant Resources and Its Exploitation and Utilization. *Wuhan Botany Research*, **4**, 375-379.

[3] Liu, C. and Wyman, C. (2003) The Effect of Flow Rate of Compressed Hot Water on Xylan, Lignin, and Total Mass Removal from Corn Stover. *Industrial Engineering Chemistry Research*, **42**, 5409-5416. http://dx.doi.org/10.1021/ie030458k

[4] Anvar, U. and Buranov, G.M. (2008) Lignin in Straw of Herbaceous Crops. *Industrial Crops and Products*, **28**, 237-259. http://dx.doi.org/10.1016/j.indcrop.2008.03.008

[5] Li, G.X. and Zhang, F.S. (2000) Solid Waste Composting and Organic Fertilizer Production. Chemical Industry Press, Beijing.

[6] Xi, B.D., Liu, H.L., *et al.* (2002) Composting of Cellulose and Lignin Biodegradation Research Status. *Techniques and*

Equipment for Environmental Pollution Control, **3**, 19-23.

[7] Sánchez, Ó.J. and Cardona, C.A. (2007) Trends in Biotechnological Production of Fuel Ethanol from Different Feedstocks. *Bioresource Technology*, **11**, 1-25.

[8] Balat, M., Balat, H. and Ŏz, C. (2008) Progress in Bioethanol Processing. *Progress in Energy and Combustion Science*, **34**, 551-573. http://dx.doi.org/10.1016/j.pecs.2007.11.001

[9] Yu, Z., Li, G.Y., *et al.* (2008) Comparison of Different Pretreatment for *Arundo donax* Saccharification on the Production of Fuel Ethanol Effect. *Guizhou Agricultural Sciences*, **36**, 117-119.

[10] Mosier, N., Wyman, C., Dale, B., *et al.* (2005) Features of Promising Technologies for Pretreatment of Lignocellulosic Biomass. *Bioresource Technology*, **96**, 673-686. http://dx.doi.org/10.1016/j.biortech.2004.06.025

[11] Geddes, C.C., Nieves, I.U. and Ingram, L.O. (2011) Advances in Ethanol Production. *Current Opinion in Biotechnology*, **22**, 312-319. http://dx.doi.org/10.1016/j.copbio.2011.04.012

[12] Kumar, P., Barrett, D.M., Delwiche, M.J., *et al.* (2009) Methods for Pretreatment of Lignocellulosic Biomass for Efficient Hydrolysis and Biofuel Production. *Industrial Engineering Chemistry Research*, **48**, 3713-3729. http://dx.doi.org/10.1021/ie801542g

[13] Zhang, C., *et al.* (2013) Recycled Aqueous Ammonia Expansion (RAAE) Pretreatment to Improve Enzymatic Digestibility of Corn Stalks. *Bioresource Technology*, **138**, 314-320. http://dx.doi.org/10.1016/j.biortech.2013.03.091

[14] Shao, Q.J., Chundawat, S.P., Krishnan, C., *et al.* (2010) Enzymatic Digestibility and Ethanol Fermentability of AFEX-Treated Starch-Rich Lignocellulosics Such as Corn Silage and Whole Corn Plant. *Biotechnology for Biofuels*, **3**, 12. http://dx.doi.org/10.1186/1754-6834-3-12

[15] Liu, J.J., Peng, H.H., *et al.* (2013) Erianthus Enzymatic Conversion of Fermentable Sugars Liquid Ammonia Pretreatment and Parameters Optimization. *Journal of Biological Engineering*, **29**, 233-241.

[16] Boonmanumsin, P., Treeboobpha, S., Jeamjumnunja, K., *et al.* (2012) Release of Monomeric Sugars from *Miscanthus sinensis* by Microwave-Assisted Ammonia and Phosphoric Acid Treatments. *Bioresource Technology*, **103**, 425-431. http://dx.doi.org/10.1016/j.biortech.2011.09.136

[17] Rodríguez, H., Padmanabhan, S., *et al.* (2011) Addition of Ammonia and/or Oxygen to an Ionic Liquid for Delignification of Miscanthus. *Bioresource Technology*, **102**, 7946-7952. http://dx.doi.org/10.1016/j.biortech.2011.05.039

[18] Zhong, C., Lau, M.W., Balan, V, *et al.* (2009) Optimization of Enzymatic Hydrolysis and Ethanol Fermentation from AFEX-Treated Rice Straw. *Applied Microbiology and Biotechnology*, **84**, 667-676. http://dx.doi.org/10.1007/s00253-009-2001-0

10

Airfoil Trailing Edge Noise Generation and Its Surface Pressure Fluctuation

Wei Jun Zhu, Wen Zhong Shen

Department of Wind Energy, Technical University of Denmark, Lyngby, Denmark
Email: wjzh@dtu.dk

Abstract

In the present work, Large Eddy Simulation (LES) of turbulent flows over a NACA 0015 airfoil is performed. The purpose of such numerical study is to relate the aerodynamic surface pressure with the noise generation. The results from LES are validated against detailed surface pressure measurements where the time history pressure data are recorded by the surface pressure microphones. After the flow-field is stabilized, the generated noise from the airfoil Trailing Edge (TE) is predicted using the acoustic analogy solver, where the results from LES are the input. It is found that there is a strong relation between TE noise and the aerodynamic pressure. The results of power spectrum density show that the fluctuation of aerodynamic pressure is responsible for noise generation.

Keywords

Large Eddy Simulation, Trailing Edge Noise, Surface Pressure

1. Introduction

As one of the important renewable energy source, wind energy development is growing very fast in recent years. To achieve the ambition of 100% renewable energy in 2050, wind energy is considered as the most important renewable energy source in Denmark. Continuously improving wind energy efficiency has become the major task to keep wind power as an ever growing portion of the world renewable energy. However, among some other problems, public acceptance due to noise from wind turbines becomes a significant barrier for future wind energy development. The issue of wind turbine noise becomes a large factor of uncertainty during wind turbine design and wind farm planning. There is a need of developing wind turbine noise prediction tool that can be used to design low noise wind turbines and silent wind farms.

So far the methods that can be used for wind turbine noise prediction range from the rule of thumb models to advanced Computational Aero-Acoustic methods (CAA). From the simple models to the CAA methods, the computational time grows exponentially. Some of the most widely used methods are summarized below:

- The rule of thumb models.
- Semi-empirical engineering models: BPM [1] [2], NAF Noise [3].
- More theoretically based models: Amiet [4], Lowson [5], TNO [6]-[8], Howe [9] [10].

- CAA methods: Acoustic analogy [11] [12], flow/acoustic splitting technique [13]-[15], Direct Numerical Simulation (DNS).

The engineering tools are convenient to provide reasonable good predictions. The CAA methods are able to give physical understanding of the flow induced noise mechanisms. Unless DNS is applied, modelling of turbulence cannot be avoided where LES is so far a reliable model for turbulent flow and noise investigations. The aim of the present study is to focus on turbulent flows over an airfoil and its noise generation. By understanding more about the flow characteristics with respect to noise generation, it is possible to design low noise airfoils with special treatment at trailing edge.

2. Numerical Approach

The numerical approach solving the Navier-Stokes equations and the TE noise is presented in this section. Based on the results from LES, the acoustic analogy is performed to calculate noise level at receiver.

2.1. Flow Solver

The Reynolds number of the current study is around 3 million. Although sound propagation is based on compressible air, the flow can still be considered as incompressible because the flow and acoustic solver can be split [13]. The filtered incompressible Navier-Stokes equations, momentum, turbulent stresses and eddy viscosity equations are applied to obtain the turbulent flow data. The filtered incompressible equations are solved by the in-house EllipSys3D code [16] [17]. The code is based on a multi-block/cell-centered finite volume discretization of the steady/unsteady incompressible Navier-Stokes equations in primitive variables (pressure and velocity). The EllipSys3D code is programmed using a multi-block topology, and therefore it is parallelized using a message-passing interface.

$$\frac{\partial \bar{U}_i}{\partial t}+\frac{\partial\left(\bar{U}_i\bar{U}_j\right)}{\partial x_j}=-\frac{1}{\rho}\frac{\partial \bar{P}}{\partial x_i}+\nu\frac{\partial^2 \bar{U}_i}{\partial x_j^2}+\frac{\partial \tau_{ij}}{\partial x_j} \tag{1}$$

$$\tau_{ij}=\nu_t\left(\frac{\partial \bar{U}_i}{\partial x_j}+\frac{\partial \bar{U}_j}{\partial x_i}\right)-\frac{2}{3}k\delta_{ij} \tag{2}$$

$$\nu_t=C\left|\bar{\omega}\right|^{\alpha}k^{(1-\alpha)/2}\Delta^{(1+\alpha)} \tag{3}$$

$$k=\frac{1}{2}\sum_{j=1}^{3}(U_j-\bar{U}_j)^2\approx\frac{1}{2}\sum_{j=1}^{3}(U_j-\tilde{\bar{U}}_j)^2 \tag{4}$$

The turbulent stresses in Equation (1) is based on the mix model [18] [19] through Equations (2)-(4) where the first filter is identified with an over bar "-" which relates to the finest mesh. The second filter is denoted with a tilde "~" above. The second filter is also denoted as the test filter, it is computed on a mesh twice as coarse as the finest mesh. In Equation (3), ω is vorticity, $\Delta = (\Delta_x\Delta_y\Delta_z)^{1/3}$ is the reference filter size based on the grid volume. The model constants are $C = 0.04$ and $\alpha = 0.5$.

2.2. Acoustic Solver

In this work, the formulation proposed by Farassat [20] is applied. The formulation is the solution of the Lighthill's acoustic analogy [11] with surface sources only when the surface moves at subsonic speed. This formulation has been successfully used for helicopter rotor and propeller noise predictions. At the retarded or emission time, the thickness and loading noise equations are written as

$$4\pi p_T'(\mathbf{x},t)=\frac{\partial}{\partial t}\int_{f=0}\left[\frac{\rho_0 v_n}{r(1-M_r)}\right]_{ret}dS \tag{5}$$

$$4\pi p_L'(\mathbf{x},t)=\frac{1}{c}\frac{\partial}{\partial t}\int_{f=0}\left[\frac{p\cos\cos\theta}{r(1-M_r)}\right]_{ret}dS+\int_{f=0}\left[\frac{p\cos\cos\theta}{r^2(1-M_r)}\right]_{ret}dS \tag{6}$$

The right hand sides of Equations (5) and (6) are the integrations of time history variables obtained from flow calculations. The variables include wall normal velocity v_n, the Mach number of the source in the radiation direction M_r, the pressure on the solid wall surface p, the distance between source and receiver r, the angle between the radiation direction and the local wall normal direction θ. **Figure 1** shows a sketch of an airfoil where dS indicates one of the typical wall element that is integrated over the entire airfoil surface. It is obvious that the the angle θ will contribute to the noise directivity. The acoustic solver may run in parallel with flow model, in practice the acoustic solver starts when the flow-field is fully established.

3. Results and Discussions

3.1. Brief Description of Surface Pressure Measurements

The wind tunnel used for the surface pressure measurements is optimized for aerodynamic performance. Since TE noise signal is difficult to obtain directly, the surface pressure measurements are performed to evaluate the noise source mechanisms [21]. The measurements are performed for a NACA 0015 airfoil, at a flow speed of 50 m/s, and the airfoil chord is 900 mm. The surface pressure microphones are identified with different numbers and mounted on the airfoil surface. The locations of the microphones are depicted in **Figure 2**. The microphones are placed along the chord direction, e.g., from leading edge to the trailing edge. They are also spread in the spanwise direction to minimize the flow disturbance due to the upstream microphones. Pressure signals from some of these measurement points (Marked with circles in **Figure 2**) will be compared with LES results.

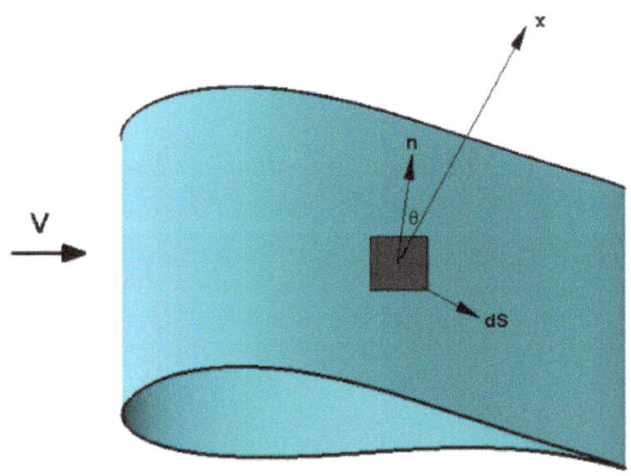

Figure 1. Example of an airfoil surface element and the vectors.

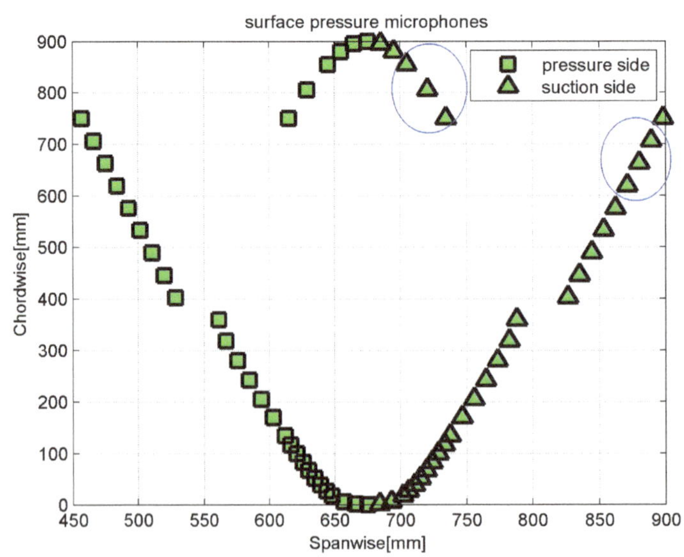

Figure 2. Locations of the surface pressure microphones.

3.2. LES Results

A C-type mesh is created for numerical simulation. The mesh is block-structured with 48^3 grid points in each block. The total number of blocks is 216 which contains a total number of grid points around 24 million. The first wall cell size is in the order of 10^{-5} chord and the ratio of $\Delta x/\Delta y$ is around 25 along the airfoil wall surface. An iso-vorticity plot is shown in **Figure 3**. At the angle of attack of 4°, the flow separation is seen on the suction side surface. In **Figure 4**, the spectral density of the incompressible pressure is compared with measurement data. As seen from the figure, the agreements are found at frequencies below 1000 Hz. The deviation at higher frequencies is due to two reasons: the limited grid size used for LES filter; the wind tunnel fan noise causes high frequency noise, especially the tonal components.

An additional computation is carried out for the same airfoil with a serrated TE. The flow and mesh configurations are the same as the previous calculation. As seen from **Figure 5**, the extended length of the serration (from its root to tip) is 5% of the airfoil chord length and the span of the serration is 10% of the chord. The pressure signals obtained from original airfoil and from the serrated airfoil are compared. In **Figure 6**, the two signals are selected at x/c = 0.64 and x/c = 0.9. The curves with legend Org_1 and SE_1 are located at x/c = 0.9. The other two curves are plotted for location x/c = 0.64. As it is seen, the signal closer to TE contains more energy than the one in the upper stream. Also it is observed that near the TE region, the serrated airfoil has less spectral density amplitude which might indicate less noise generation.

3.3. Acoustic Results

The acoustic computation is performed when the turbulent flow is numerically stabilized. The integration of sound source is through Equation (6) where only the loading noise plays the role. The time history data of acoustic pressure is recorded during the simulation. After that, the Sound Pressure Level (SPL) of the original airfoil and the serrated airfoil is compared at a same receiver point which is located at 3-chord distance normal to the suction side of TE. It is interesting to see that the difference of the SPL is quite similar to the one we see in **Figure 7**. The strength of the incompressible pressure fluctuation at TE has the same behavior as the noise spectra. This indicates that the strength of pressure fluctuation at TE is responsible for noise generation. A small change of TE geometry might influence a lot of noise generation.

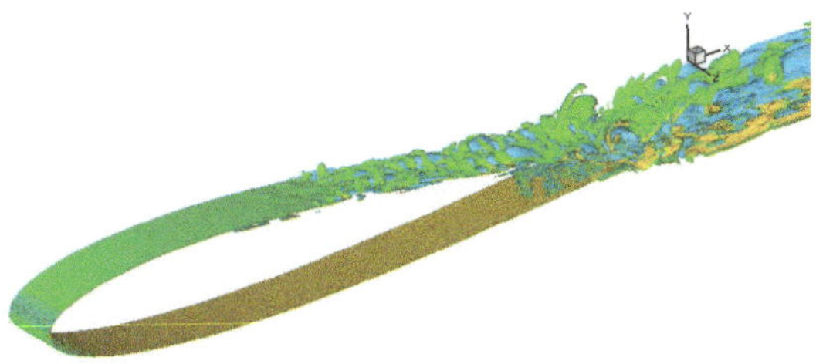

Figure 3. Iso-vorticity plot.

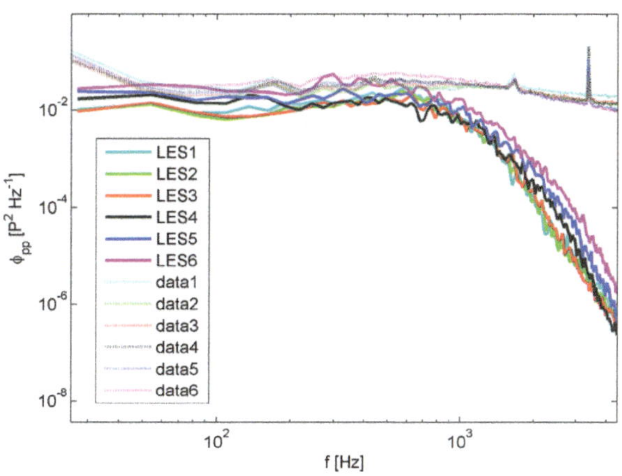

Figure 4. Spectral density of the incompressible pressure at trailing edge suction side.

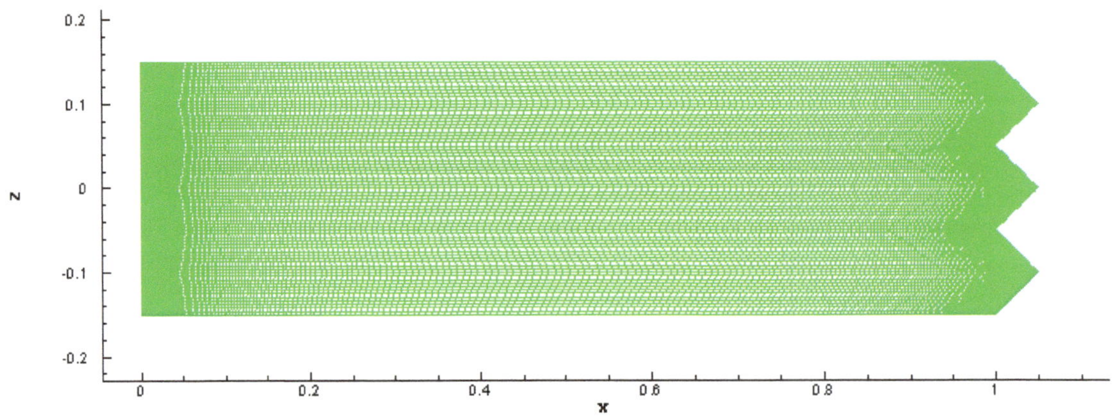

Figure 5. Top view of the wall surface mesh with serrated TE.

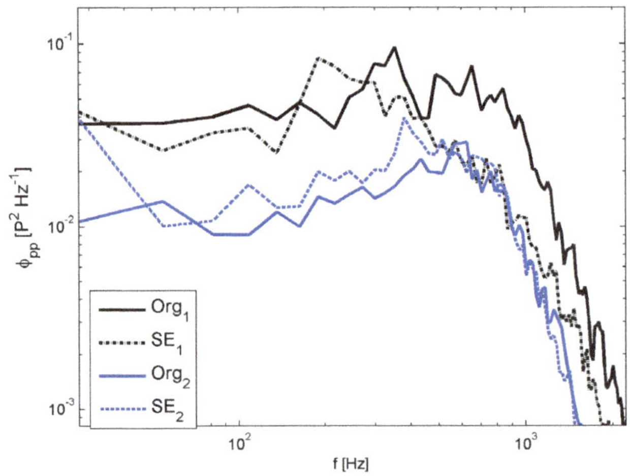

Figure 6. Spectral density of the incompressible pressure at TE suction side. Comparisons with original airfoil and airfoil with TE serration.

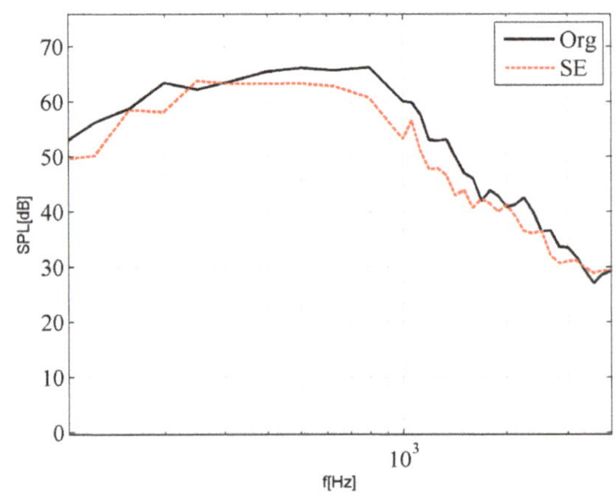

Figure 7. Sound pressure level in 1/3 octave band: comparison with original airfoil and airfoil with TE serration.

4. Conclusion

Numerical simulations of flow and noise generation of a NACA 0015 airfoil at Re = 3×10^6 are carried out. The incompressible pressure signals at TE are compared with wind tunnel experiments where some general agreements are found at low frequencies. The geometry of the original airfoil is then modified with serrated TE. At a relatively low angle of attack of 4°, the computations show that it is the small part of the TE that dominates the noise level.

Acknowledgements

The current numerical study is supported by the European Union's Seventh Programme for research, technological developement and demonstration under the grant agreement No 322449. The auther wish to acknowledge the Energy-Technological Development and Demonstration Program under the project (J.nr. 64012-0146) and the Danish Council for Strategic Research under the project (Sags nr. 0603-00506B).

References

[1] Brooks, T.F., Pope, D.S. and Marcolini, M.A. (1989) Airfoil Self-Noise and Prediction. NASA Reference Publication 1218, National Aeronautics and Space Administration, USA.

[2] Zhu, W.J., Heilskov, N., Shen, W.Z. and Sørensen, J.N. (2005) Modeling of Aerodynamically Generated Noise from Wind Turbines. *Journal of Solar Energy Engineering*, **127**, 517-528. http://dx.doi.org/10.1115/1.2035700

[3] Moriarty, P. (2005) NAFNoise User's Guide. National Renewable Energy Laboratory.

[4] Amiet, R.K. (1975) Acoustic Radiation from an Airfoil in a Turbulent Stream. *Journal of Sound and Vibration*, 407-420.

[5] Lowson, M.V. (1993) Assessment and Prediction Model for Wind Turbine Noise: 1. Basic Aerodynamic and Acoustic Models. Flow Soltion Report 93/06, 1-46.

[6] Parchen, R. (1998) Progress Report DRAW: A Prediction Scheme for Trailing-Edge Noise Based on Detailed Boundary-Layer Characteritics, TNO Rept. HAGRPT 980023, TNO Institute of Applied Physics, The Netherlands.

[7] Bertagnolio, F., Fischer, A. and Zhu, W.J. (2013) Tuning of Turbulent Boundary Layer Anisotropy for Improved Surface Pressure and Trailing-Edge Noise Modeling. *Journal of Sound and Vibration*, **333**, 991-1010.

[8] Wolf, A., Lutz, T., Würz, W., Krämer, E., Stalnov, O., Seifert, A., *et al.* (2014) Trailing Edge Noise Reduction of Wind Turbine Blades by Active Flow Control. *Journal of Wind Energy*. http://dx.doi.org/10.1002/we.1737

[9] Howe, M.S. (1978) A Review of the Theory of Trailing Edge Noise. *Journal of Sound and Vibration*, **61**, 437-465. http://dx.doi.org/10.1016/0022-460X(78)90391-7

[10] Howe, M.S. (1991) Noise Produced by a Sawtooth Trailing Edge. *Journal of the Acoustic Society of America*, **90**, 482-487. http://dx.doi.org/10.1121/1.401273

[11] Lighthill, M.J. (1952) On Sound Generated Aerodynamically. I. General Theory. *Proceedings of the Royal Society of London A*, 211. http://dx.doi.org/10.1098/rspa.1952.0060

[12] Ffowcs Williams, J.E. and Hawkings, D.L. (1969) Sound Generated by Turbulence and Surfaces in Arbitrary Motion. *Philosophical Transactions of the Royal Society*, **A264**, 321-342.

[13] Hardin, J.C. and Pope, D.S. (1994) An Acoustic/Viscous Splitting Technique for Computational Aeroacoustics. *Theoretical and Computational Fluid Dynamics*, **6**, 321-342. http://dx.doi.org/10.1007/BF00311844

[14] Shen, W.Z. and Sørensen, J.N. (1999) Comment on the Aeroacoustic Formulation of Hardin and Pope. *AIAA Journal*, **1**, 141-143. http://dx.doi.org/10.2514/2.682

[15] Zhu, W.J., Shen, W.Z. and Sørensen, J.N. (2011) High-Order Numerical Simulations of Flow Induced Noise. *International Journal for Numerical Methods in Fluids*, **66**, 17-37. http://dx.doi.org/10.1002/fld.2241

[16] Michelsen, J.A. (1992) Basis 3D—A Platform for Development of Multiblock PDE Solvers. *Technical Report*, AFM 92-05, Technical University of Denmark, Denmark.

[17] Sørensen, N.N. (1995) General Purpose Flow Solver Applied to Flow over Hills. *Risø-R-827-(EN)*, Risø National Laboratory, Roskilde, Denmark.

[18] Ta Phuoc, L. (1994) Modèles de Sous Maille Appliqués aux Ecoulements Instationnaires Décollés. *Proceedings of the DRET Conference*: *Aérodynamique Instationnaire Turbulente-Aspects Numériques et Expérimentaux*, Paris, France: DGA/DRET Editors.

[19] Sagaut, P. (2006) Large Eddy Simulation for Incompressible Flows. 3rd Edition, Springer.

[20] Farassat, F. (2007) Derivation of Formulations 1 and 1A of Farassat NASA/TM-2007-214853.

[21] Fischer, A. (2011) Experimental Characterization of Airfoil Boundary Layers for Improvement of Aeroacoustic and Aerodynamic Modeling. Ph.D. Thesis, Technical University of Denmark.

11

Dynamic Model Study and Analysis of DME Auto-Thermal Steam Reforming Reaction

Yan Gao, Cong Li

Shanghai University of Engineering Science, Shanghai, China
Email: 449284019@qq.com

Abstract

Through the study of the kinetics of dimethyl ether steam reforming reaction, a two-dimensional model of the reactor is established. At the same time, rate equations of simplified elementary reactions of DME steam reforming reaction are deduced by the mechanism of Langmuir-Hinshelwood, and the rate constants are obtained by correcting the pre-exponential factor of Arrhenius equation. Finally, the analog data of DME steam reforming reaction under a variety of conditions are obtained, and making a simulation diagram. The two-dimensional model is substantially correct because of the analog data more in line with a variety of knowledge.

Keywords

Dimethyl Ether (DME), Kinetic Model, The Reaction Rate, Reforming Reaction

1. Introduction

Dimethyl ether has the advantage of economy, security, and the sources of diversity, bound from the research stage to the industrial stage [1]. However, due to the engine of DME has a shortcoming of fire earlier, making it smaller work load range [2], while the hydrogen fuel cell and electric vehicles are being tested and toward industrialization in the world, but blocked in hydrogen storage technology [3], so automotive fuel reforming is an important way to solve the problem of the hydrogen energy [4] [5].

DME is an ideal vehicle fuel, but also has capability of chemical hydrogen storage. Therefore, many scholars have been studied to the DME reforming processes [6]-[8], confirming its value of commerce and scientific and finding some technical issues, such as the hydrogen yield. Generally speaking, the main method of DME reforming is the traditional methods and the plasma reforming method. The traditional methods include Autothermal reforming, steam reforming, partial oxidation reforming and so on. However, the study of kinetics model about DME reforming reaction is still relatively lacking in the current. Weiqi Qian *et al.* [9] used the quasi-steady method to establish a simplified chemical kinetic model for combustion of hydrocarbon fuels, effectively reducing the reaction components. Solution variables and differential equations are reduced when it is calculated in the flow field, improving the efficiency of calculation. Yumin Chen *et al.* [10] conducted kinetic analysis to the process of methane auto-thermal reforming reaction. Wang Feng *et al.* [11] conducted kinetic analysis to the process of methanol steam reforming in the micro-reactor. Dongmei Feng *et al.* [12] made the Kinetic Study of the process about dimethyl ether steam reforming reaction using of a catalyst CuO − ZnO − Al_2O_3 + ZSM − 5. This article will use the catalyst $Cu-Ni/\gamma-Al_2O_3$ to make kinetic studies of Dimethyl ether steam reforming reaction, establishing a two-dimensional model of the reactor, deducing the rate equation of simplified elementary reaction, learning the influence of the reaction rate and the mole fraction with temperature and time.

2. Experimental Methods

During the whole experiment, the space velocity of DME steam reforming reactor is 0.3 m/s. The molar ratio of water and DME is 1 to 3. The feed mole fraction of dimethyl ether is between 5 and 15. The flow rate of DME is 10^{-5} mol/s in the feed mouth. The pressure inside the reactor is 0.1 Mpa and the range of temperature is 200°C to 400°C.

3. The Kinetic Models

3.1. Reforming Reactor Model

The kinetic model of DME steam reforming reactor applies to a two-dimensional model. Although the kinetic parameters of the reactor are difficult to be determined in the process of reaction, it can improve the accuracy of the change of amount of substance and energy conversion. The two-dimensional model is non-continuous, axial symmetry, similar to the two-dimensional reactor of Groppi, G. *et al.* [13]. The reaction of DME and H_2O on the catalyst coating layer of the reactor produced H_2 and oxides of carbon. In the course of the reaction, the mass of reactants and product are equal, the mass conservation equation as follows:

$$\frac{dw_i}{dz} m_t RT = S_c S_v k_c P_t M_i \left(Y_{is} - Y_{ig}\right) \tag{1}$$

Because the reactants are gas and in the high-speed motion in the reactor, and so axial diffusion of the gas mass is negligible. Because the heat exchanger conduit wall can isolate gas, so the radial diffusion of the gas mass in the reactor is also negligible. Then the above equation can be simplified to as follows:

$$\rho_a RT \sum v_i r_{js} = S_v k_c P_t \left(Y_{is} - Y_{ig}\right) \tag{2}$$

Further, on the catalyst coating layer of the reactor, the mass conservation equation of each element in the course of the reaction and diffusion is as follows:

$$\frac{d^2 Y_{ic}}{dx^2} = -\left(\sum v_i r_j\right) \frac{\rho_c R T_s}{P_t D_e} \tag{3}$$

In this reactor, if the catalyst coating layer is a uniformly thickness, the boundary conditions are shown in **Table 1**. The average thickness of the catalyst coating layer is about 35 microns, while the effective diffusion coefficient D_e $\left(m^2 \cdot s^{-1}\right)$ is estimated by random pore model [14]. On the catalyst coating layer, according to the mass conservation of the elements, the average reaction rate r_{js} can be obtained:

$$r_{js} = \frac{\int_0^{t_c} r_j dx}{t_c} \tag{4}$$

Similar to the gas phase mass balance of (2), axial and radial dispersions were neglected in the gas-phase heat balance:

Table 1. Boundary conditions for the 2D monolithic reactor wash-coat mass balance and solid heat balance.

Position	Boundary conditions
$x = 0$ (Catalyst coating surface)	$Y_{ic} = Y_{is}$
$x = t_c$ (Catalyst coating boundary)	$\frac{dY_{ic}}{dx} = 0$
$z = 0$ (Inlet)	$dT_s/dz = \varepsilon_m \sigma (T_s^4 - T_g^4)/(\lambda_s - \varepsilon \lambda_s)$
$z = L$ (outlet)	$dT_s/dz = 0$
$r = 0$ (monolith center)	$dT_s/dr = 0$
$r = R$ (monolith outer edge)	$dT_s/dr = (T_w - T_s)/x$

$$\frac{dT_g}{dz} m_t \sum_i \frac{w_i}{M_i} c_{pi} = S_c S_v h(T_s - T_g) \tag{5}$$

External energy transfers to DME and water using of the heat conduction manner of the axial and radial through the reactor wall and the catalyst coating layer, using the equation:

$$\frac{d^2T_s}{dz^2} + \frac{d^2T_s}{dr^2} + \frac{1}{r}\frac{dT_s}{dr} + \frac{\rho_a\left(\sum r_{js}\left(-\Delta H_{Rj}\right)\right)}{(1-\varepsilon)\lambda_s} = \frac{S_v h(T_s - T_g)}{(1-\varepsilon)\lambda_s} \tag{6}$$

Wherein, m_t—The total mass flow rate $(kg \cdot s^{-1})$, M—Molar mass $(kg \cdot mol^{-1})$, P_t—The total pressure (Pa), S_c—Reactor cross-sectional area (m^2), R—$(8.314\ Pa \cdot m^3 \cdot mol^{-1} \cdot K^{-1})$, t_c—Catalyst coating thickness (m), S_v—Catalyst coating surface of the unit volume $(m^2 \cdot m^{-3})$, z—The axial distance (m), R—Radial heat transfer coefficient, ε—Reactor porosity, ε_m—Emissivity, σ—$(5.67 \times 10^{-8}\ W \cdot m^{-2} \cdot K^{-4})$, λ_s—effective thermal conductivity $(W \cdot m^{-1} \cdot K^{-1})$, V—Stoichiometric coefficients, c—Inside catalyst coating, g—gas, x—Axial distance, Y—Mole fraction, i—Species index, j—Response index, h—Heat transfer coefficient $(W \cdot m^{-2} \cdot K^{-1})$, ΔH_R—Reaction enthalpy $(J \cdot mol^{-1})$, k_c—Mass transfer coefficient $(m \cdot s^{-1})$, w—Reactor wall, s—Catalyst coating surface, w—Mass fraction. **Table 1** shows the boundary conditions of conservation of energy and the quality of the reactor model catalyst coating.

3.2. Kinetics Model of DME Steam Reforming Reaction

There are three major reactions in the process of DME steam reforming [15], which can be shown in **Table 2**, and DME oxidation reaction is the heat source. In the ideal case, the inlet of reactor is water, DME and O_2, and the outlet is CO_2 and H_2. When each elementary reaction is in a relatively stable state of the reaction rate, adsorption equilibrium constant of H_2 and CO_2 on the catalyst is relatively small, and it can be ignored as the case in the calculation process. The reaction rate equation takes Langmuir-Hinshelwood model of water molecules and other molecules competitive adsorption, and it is ideal for surface reaction.

3.3. Parameters of the Kinetic Model

Arrhenius equation is used to solve the rate constants, and the expression is:

$$k = A e^{-\frac{E}{RT}} \tag{7}$$

Due to the range of experimental temperature is 200°C to 400°C, so temperature variation range is small, and the pre-exponential factor A is corrected and the activation energy not, then Arrhenius equation:

$$k_j = A_{s_j} e^{-\frac{E_{Aj}}{RT}} \tag{8}$$

where, k_j $(mol \cdot kg^{-1} \cdot s^{-1})$ is the rate constant at temperature T.

Then, the activation energy E_A $\left(\mathrm{J \cdot mol^{-1}}\right)$ can be considered as it of the initial temperature T_{s_j} (200˚C) and unchanged in the whole calculation process. The pre-exponential factor A_{s_j} is calculated as follows:

$$A_{s_j} = \frac{k_{s_j}}{\mathrm{e}^{-\frac{E_{Aj}}{RT_{s_j}}}} \tag{9}$$

The fitting parameter of Adsorption equilibrium constant $\left(K_j\right)$ comprises of K_{mj} of average temperature (300˚C) and adsorption enthalpy $\left(\Delta H_{adsj}\right)$, and its expression is as follows:

$$K_j = K_{mj} \exp\left(\frac{\Delta H_{adsj}}{R}\left(\frac{1}{T_m} - \frac{1}{T}\right)\right) \tag{10}$$

The main elementary reaction rate constant can be obtained by Equation (8) and (9) using ofrate constant k_{s_j} and activation energy E_{A_j} in the **Table 3** [16]. While adsorption equilibrium constant $\left(K_j\right)$ of any temperature can be known by formula (10) using of K_{mj} and the adsorption enthalpy $\left(\Delta H_{adsj}\right)$ in **Table 3**. Finally, the rate formula of major elementary reactions in **Table 2** can be computed for getting a variety of data.

4. Results and Discussion

The legend of all graphics is the same with the legend of **Figure 1(a)**.

Table 2. Kinetic model overall reactions and rate expression.

Reaction Name	Chemical equation	The reaction rate
DME oxidation reaction	$CH_3OCH_3 + 3O_2 \rightarrow 2CO_2 + 3H_2O$	$r_1 = \frac{k_1 y_{DME} y_{O_2}}{\left(1 + K_{H_2O} y_{H_2O}\right)^2}$
DME hydrolysis	$CH_3OCH_3 + H_2O(g) \rightarrow 2CH_3OH$	$r_2 = \frac{k_2 y_{DME} y_{H_2O}}{\left(1 + K_{H_2O} y_{H_2O}\right)^2}$
CH_3OH decomposition	$CH_3OH \rightarrow CO + 2H_2$	$r_3 = \frac{k_3 y_{CH_3OH}}{1 + K_{H_2O} y_{H_2O}}$
Water-gas shift	$H_2O + CO \leftrightarrow CO_2 + H_2$	$r_4 = \frac{k_4 y_{CO} y_{H_2O}}{\left(1 + K_{H_2O} y_{H_2O}\right)^2}\left(1 - \frac{y_{CO_2} y_{H_2O}}{y_{CO} y_{H_2O} K_{eq}}\right)$

Table 3. Fitted kinetic parameter values and 95% confidence intervals.

Reaction type	Parameters	The value	Confidence intervals	Unit
DME oxidation reaction	k_{s_1}	53.5	±6	$\mathrm{mol \cdot kg^{-1} \cdot s^{-1}}$
	E_{A_1}	104.1	±5	$\mathrm{kJ \cdot mol^{-1}}$
DME hydrolysis	k_{s_2}	60.00	±14.03	$\mathrm{mol \cdot kg^{-1} \cdot s^{-1}}$
	E_{A_2}	173.53	±24.01	$\mathrm{kJ \cdot mol^{-1}}$
CH_3OH decomposition	k_{s_3}	135.42	±29.75	$\mathrm{mol \cdot kg^{-1} \cdot s^{-1}}$
	E_{A_3}	107.36	±44.5	$\mathrm{kJ \cdot mol^{-1}}$
Water-gas shift	k_{s_4}	356.19	±170.4	$\mathrm{mol \cdot kg^{-1} \cdot s^{-1}}$
	E_{A_4}	180.84	±25.3	$\mathrm{kJ \cdot mol^{-1}}$
Adsorption reaction water	K_m	3.88	1.15	$\mathrm{mol \cdot L^{-1}}$
	ΔH_{ads}	−31.71	0.72	$\mathrm{kJ \cdot mol^{-1}}$

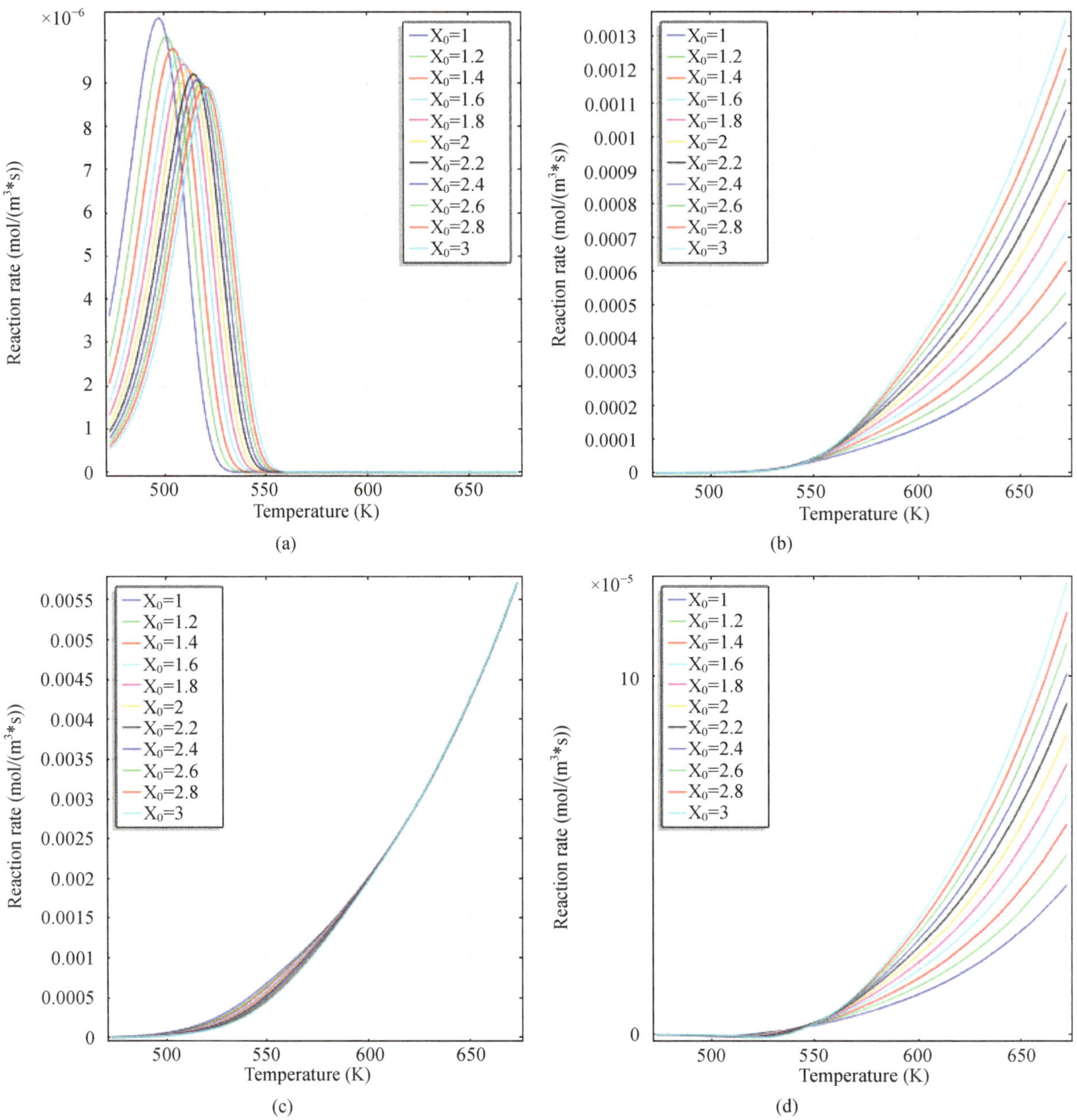

Figure 1. (a) The variation diagram of r_1 and temperature; (b) The variation diagram of r_2 and temperature; (c) The variation diagram of r_3 and temperature; (d) The variation diagram of r_4 and temperature.

Figure 1 is the graph of rate variation of each reaction, and the equation of temperature changing with time is $473 + 200t$. **Figure 1(a)** is the variation diagram of rate and temperature for the oxidation reaction of DME, with increasing temperature, r_1 first adding and then reduced. The r_1 reduces due to insufficient of O_2, so it can guarantee the hydrogen yield. **Figure 1(b)** is rate-temperature diagram of DME hydrolysis reaction, and r_2 increases with increasing temperature, while it increases with ratio of water and DME (X_0), indicating that this reaction is an endothermic reaction. **Figure 1(c)** is rate-temperature diagram of CH_3OH decomposition reaction, and r_3 increases with increasing temperature. **Figure 1(d)** is rate-temperature curve of water-gas reaction, and r_4 increases with increasing temperature and X_0. From **Figure 1**, at 550 K, r_1 reducing to near 0, r_2 and r_4 beginning to rapidly increase, r_2 being moment of greatest impact by X_0, this indicates DME reacts with O_2 preferentially and it will quickly decompose and produce H_2 in case of lack of oxygen and sufficient energy.

When r_1 rises or O_2 is enough, r_2, r_3 and r_4 basically keep zero. When r_1 reaches the maximum, DME begins to decompose and produce CH_3OH and it has decomposed at the same time. In short, the ratio of water-DME and the ratio of oxygen-DME are important step of controlling the hydrogen yield.

Figure 2(a) is the molar flow rate variation graph of H_2 with time, which rapidly increases to stabilize with increasing time, and it increases with increasing X_0 at the same time. **Figure 2(b)** is the molar flow rate variation graph of DME with time, which rapidly declines to stabilize with increasing time and it reduces with increasing X_0 at the same time. **Figure 2(c)** is the mole fraction graph of H_2 with temperature variation, which increases with increasing temperature, and it is reduced at the same temperature with increasing X_0. **Figure 2(d)** is the mole fraction curve of DME with temperature variation, and the mole fraction decreases with increasing temperature. According to **Figure 2(c)** and **Figure 2(d)**, at 550 K, the mole fraction derivative of DME

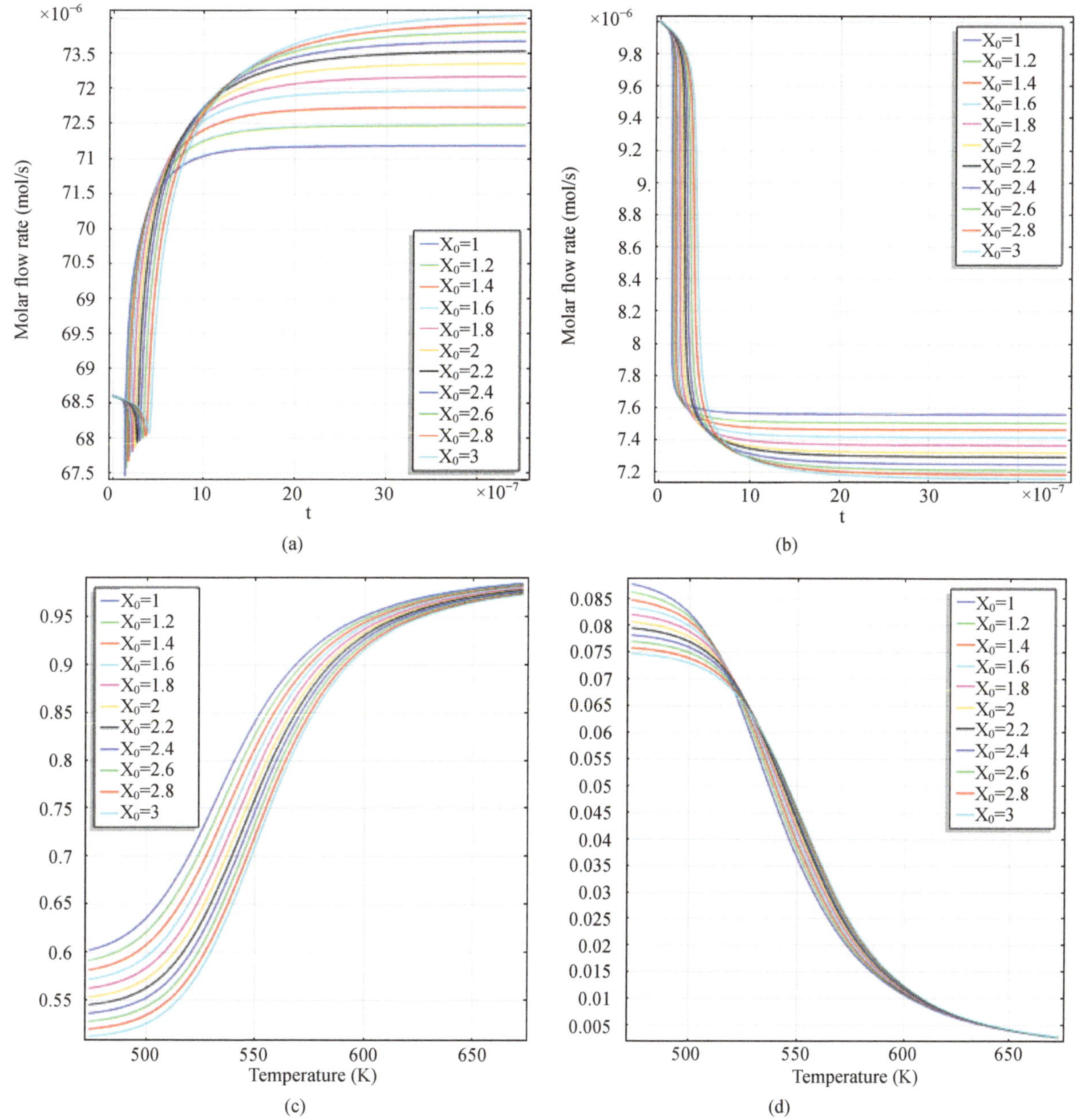

Figure 2. (a) The molar flow rate variation graph of H_2 with time; (b) The molar flow rate variation graph of DME with time; (c) The mole fraction graph of H_2 with temperature variation; (d) The mole fraction graph of DME with temperature variation.

and H_2 to temperature is the maximum, indicating that their speed of response here has a jump, which can be proven from **Figure 1(b)**, **Figure 1(c)** and **Figure 1(d)**. When the curve of **Figure 2(a)** and **Figure 2(b)** trend to steady, the curve of **Figure 2(c)** and **Figure 2(d)** are becoming more stable, indicating that the reactants at the inlet of the reactor achieves to conditions optimally. That is, if this state is changed, the reactants conditions of the reactor at the inlet need to change.

5. Conclusion

The two-dimensional model is substantially correct because of the analog data more in line with a variety of knowledge. For example, the ratio of water-DME and the ratio of oxygen-DME are important step of controlling the hydrogen yield. Finally, this article can provide important theoretical references to the kinetics of dimethyl ether steam reforming reaction.

References

[1] Yu, Y.-F. and Jing, Y.-H. (2010) The Research Analysis of the Application of the Vehicle DME Fuel. *Northwest Coal*, **6**.

[2] Shudo, T. and Yamada, H. (2007) Hydrogen as an Ignition-Controlling Agent for HCCI Combustion Engine by Suppressing the Low-Temperature Oxidation. *International Journal of Hydrogen Energy*, **32**, 3066-3072. http://dx.doi.org/10.1016/j.ijhydene.2006.12.002

[3] Sun, D.-L. (2011) Development and Challenge for On-Board Hydrogen Storage. *Chinese Journal of Nature*, **33**, 13-18.

[4] Zhang, W.G., Lu, W.D. and Yan, Y. (2010) Experiment on the Performance of Steam Reforming by Dimethyl Ether. *Transactions of the Chinese Society for Agricultural Machinery*, **41**, 6-16.

[5] Ji, C.W., Ju, B.J. and Dai, X.X. (2013) Mixed Combustion Performance Experiment of Syngas Produced by Ethanol Steam Reforming for Engine. *Transactions of the Chinese Society for Agricultural Machinery*, **44**, 27-32.

[6] Feng, D.M., Wang, J.F. and Wang, D.Z. (2007) Production of Hydrogen from Dimethyl Ether over Catalysts. Sciencepaper Online.

[7] Li, C., Wang, Y.S. and Fan, P.Q. (2012) Numerical Analysis and Experimental Study of Hydrogen Production from Dimethyl Ether Steam Reforming. *Science China Chemistry*, **9**.

[8] Zheng, Z.L. and Li, S.L. (2013) Thermodynamics Analysis of Hydrogen Production in Vehicle DME Steam Reforming Reaction System. *Transactions of the Chinese Society for Agricultural Machinery*, **9**.

[9] Qian, W.-Q., Yang, S.-H., Xiao, B.-G. and Le, J.-L. (2007) Development of Reduced Chemical Reaction Kinetic Model for Hydro-Carbon Fuel Combustion with Quasi-Steady-State Approximation Method. *Acta Aerodynamica Sinica*, **3**.

[10] Chen, Y.-M., Zhao, Y.-C., *et al.* (2011) Thermodynamic and Kinetic Analyses for Hydrogen Production via Methane Autothermal Reforming. *Journal of Fuel Chemistry and Technology*, **8**.

[11] Wang, F., Zhou, J., Wang, G.J. and Zhou, X.J. (2011) Kinetics Study of Methanol Steam Reforming Hydrogen Production in a Microreactor. *Acta Aerodynamica Sinica*, **4**.

[12] Feng, D.M., Wang, Y.Y., Wang, D.Z. and Wang, J.F. (2009) Steam Reforming of Dimethyl Ether over $CuO-ZnO-Al_2O_3+ZSM-5$: A Kinetic Study. *Chemical Engineering Journal*, **2**.

[13] Groppi, G. and Tronconi, E. (2001) Simulation of Structured Catalytic Reactors with Enhanced Thermal Conductivity for Selective Oxidation Reactions. *Catalysis Today*, **69**, 63-73. http://dx.doi.org/10.1016/S0920-5861(01)00356-X

[14] Froment, G.F. and Bischoff, K.B. (1990) Chemical Reactor Analysis and Design. 2nd Edition, John Wiley & Sons, New York.

[15] Daubert, T.E. and Danner, R.P. (1989) Physical and Thermodynamic Properties of Pure Chemicals (Data Compilation). Hemisphere Publishing Corporation, New York.

[16] Creaser, D. and Creaser, D. (2010) Kinetic Modeling of Autothermal Reforming of Dimethyl Ether. *Industrial Engineering Chemistry Research*, **49**, 9712-9719.

12

Effects of Hydrogen Addition on Power and Emissions Outputs from Diesel Engines

Momen Sughayyer

Mechanical Engineering Department, Palestine Polytechnic University, Hebron, Palestine
Email: sughayyer@ppu.edu

Abstract

Energy efficiency and environmental impact have become dominant topics in internal combustion engines development. Among many strategies to improve power and emissions outputs from diesel engines is the partial mix of hydrogen and air as fresh charge components to form extremely lean and homogenous mixture, which resist the spontaneous combustion, while diesel fuel is injected directly inside combustion chamber using the conventional fuel injection systems. This contribution presents an analytical and experimental investigation for the effects of adding hydrogen on diesel engines power output and the reduction of emissions. Parametric analysis is used based on lamped parameters modeling of intake manifold to estimate in cylinder trapped charge. The fuel energy flow to engine cylinders is compared for a range of loads and concentrations to simulate relevant case studies. Diesel fuel reduction for significant range of part-load operation can be achieved by introducing hydrogen, along with power improvement emission reductions are affected positively as well. This is achievable without compromising the engine maximum efficiency, given that most engines are operated at small and part-load during normal driving conditions, which allow for introducing more hydrogen instead of large quantities of excess air during such operation conditions that also can be further improved by charge boosting.

Keywords

Energy Efficiency, Power and Emissions, Diesel Engines, Hydrogen, Air-Fuel Ratio

1. Introduction

Currently, thermal systems are required to become energy efficient and produce less emission. Diesel engines are one of these systems that need to be considered for improvement in this respect. Hydrogen also is expected to become one of the most important fuels for reducing greenhouse gas emission problem in the near future. Based on the characteristics of both fuels, this contribution provides an insight into the possibility of improving diesel engines power and emissions outputs by hydrogen addition. The effectiveness of the proposed technique will be demonstrated analytically and experimentally.

Diesel engines are characterized by their carbon oxides, nitrogen oxides, soot, and particulate emissions. Related literature shows that introducing hydrogen to partially substitute diesel fuel charge will enhance engine performance [1]. It mainly helps to reduce carbon oxides emissions as a consequence to the optimized mixture characteristics due to hydrogen gaseous state. Other researchers have used Computational Fluid Dynamics to explore the changes in nitrogen oxides emissions in hydrogen-enriched diesel combustion over a range of operating conditions. Their results confirm that there is a notable reduction in exhaust emissions with increasing of hydrogen percentage [2]. Other authors show an example of measured in-cylinder pressure for hydrogen-diesel system. The obtained data indicates normal combustion of pilot fuel without pre-ignition and premixed combustion. But the pressure has the tendency to sudden increase since hydrogen with high auto-ignition temperature ignites after pilot ignition of diesel. This allows for using hydrogen without serious modification of already on-board installed diesel engines [3]. In different study [4], the lean combustion using hydrogen fuel has been considered. The result of that contribution indicates that hydrogen helps to optimize combustion process, which allows for extending the lean limit significantly due to the improvement of lean misfire limit. In addition, it shows that cycle-to-cycle variations are reduced due to better mixing and faster combustion. Another investigation presented in [5] concerns both pure hydrogen combustion under homogeneous charge compression ignition conditions and hydrogen-diesel combustion in a compression ignition engine. Various hydrogen doses are used to provide insight related to the impact on combustion knock intensity, mass fraction burned and heat release rate, in addition to engine durability. The presented results show that up to seventeen percentage of hydrogen can give positive effect on performance. The experimental investigations [6]-[9] have targeted the effect of using hydrogen on engine parameters such as injection timing, maximum brake thermal efficiency, and oxides of nitrogen. It concludes that port-injected hydrogen helps to increase in brake thermal efficiency of the engine with a greater reduction in emissions.

In the present contribution, the performance of diesel engine is considered using addition of hydrogen and based on mean value engine modeling oriented for power and emissions outputs improvement. Hydrogen gas is introduced into the intake manifold using the available gas fuelling techniques. The concept is based on continuous injection inside intake manifold where a central metering valve and distributor deliver vaporized hydrogen into the intake runners. For this purpose, engine coolant heat is used for hydrogen evaporation and to control its temperature. Allowing relatively low-temperature hydrogen gas into the intake manifold helps to cool the fresh charge, which increases the charge density and helps to reduce engine emissions. Lean mixture prevents backfiring into the intake ducts. Diesel fuel is injected directly inside the combustion chamber using the current fuelling technologies with some minor modifications to reduce diesel fuel flow, which will be replaced by hydrogen fuel flow [10]. Mixture combustion is realized by diesel combustion, given that, the hydrogen mixture is relatively very lean and exceeds misfire limit in Otto cycle based engines. Thus, it is possible to assume that combustion of hydrogen synchronized to that of the diesel. The trapped fuel energy is compared for a range of loads and concentrations. The analysis and experimental results bring about a clear insight into the proposed possibility of partially and economically improving diesel engines outputs by partial addition of hydrogen at part-loads.

2. System Modeling

For the purpose of this contribution, a lamped parameter based model for lean air-hydrogen mixture formation inside the intake manifold is developed. The presented modeling allows also for considering the possibility of exhaust gas recirculation. As for starting, it is enough accurate to consider that intake manifold as single capacity connected to the boosting system whenever there is one on the input side along with the EGR system. It is connected from the other side as output to the engine cylinders. This capacity exchanges heat to the surroundings to allow for isothermal conditions, which allows for the assumption that single pressure and single temperature values can be used to characterize the whole capacity state changes. Closed volume around the intake manifold to account for mass flow to the engine cylinders is used. Therefore, the rate of mass change inside this capacity can be given by the continuity equation as following:

$$\frac{\mathrm{d}m_m}{\mathrm{d}t} = \dot{m}_i - \dot{m}_o \quad (1)$$

where m_m is the intake manifold mass content, $\dot{m}_i$ is the input mass flow to the intake manifold, and $\dot{m}_o$ is the outflow to engine cylinders.

Assuming that all intake manifold air, hydrogen, re-circulated exhaust gas are perfect gases, the mass in this case can be expressed Equation (2).

$$m_m = \frac{P_m V_m}{R_m T_m} \tag{2}$$

where P_m is the pressure, T_m is the temperature, and R_m is the gas constant of the manifold mixture, while V_m is the manifold volume. Accordingly, the pressure rate of change inside the intake manifold is given by

$$\frac{\mathrm{d}P_m}{\mathrm{d}t} = \frac{R_m T_m}{V_m}\left(\dot{m}_i - \dot{m}_o\right) \tag{3}$$

R_m and T_m of the mixture at the engine inlet valve can be found by Equation (4) and Equation (5), respectively, using the mixture corresponding variables of the air, hydrogen, and exhaust gas. The two equations are based on the implicit assumption that perfect and adiabatic mixing takes place in the intake manifold.

$$R_m = \frac{\dot{m}_a R_a + \dot{m}_h R_h + \dot{m}_e R_e}{\dot{m}_a + \dot{m}_h + \dot{m}_e} \tag{4}$$

$$T_m = \frac{\dot{m}_a c_{p,a} T_a + \dot{m}_h c_{p,h} T_h + \dot{m}_e c_{p,e} T_e}{\dot{m}_a c_{p,a} + \dot{m}_h c_{p,h} + \dot{m}_e c_{p,e}} \tag{5}$$

where $\dot{m}_a$ is the air mass flow rate, $\dot{m}_h$ is the hydrogen mass flow, and $\dot{m}_e$ is the re-circulated exhaust gas mass flow. T_a, T_h, and T_e are the air, hydrogen, and exhaust gas sources' temperatures, respectively. $c_{p,a}$, $c_{p,h}$, and $c_{p,e}$ are the air, hydrogen, and exhaust gas constant pressure specific heats, respectively. R_a, R_h, and R_e are the air, hydrogen, and exhaust gas constants, respectively. The total mass flow into the intake manifold capacity depends mainly of the pressure differences between the source and the intake manifold, in addition, the flow restriction areas, which is given by

$$\dot{m}_i = \dot{m}_{i,a} + \dot{m}_{i,h} + \dot{m}_{i,e} \tag{6}$$

where $\dot{m}_{i,a}$ is the air mass flow rate, $\dot{m}_{i,h}$ is the hydrogen mass flow, and $\dot{m}_{i,e}$ is the re-circulated exhaust gas mass flow. For the purpose of this case modeling, it is sufficiently accurate to adopt subsonic flow through restriction modeling to account for the flow into the intake manifold. For subsonic flow, the airflow, hydrogen flow, and exhaust flow into the intake capacitance can be given by the following well known equations, respectively, which are given in [11] [12].

$$\dot{m}_{i,a} = \frac{C_{d,a} A_{v,a} P_a}{\sqrt{R_a T_a}} \left(\frac{P_m}{P_a}\right)^{\frac{1}{\gamma_a}} \left(\frac{2\gamma_a}{\gamma_a - 1}\left[1 - \left(\frac{P_m}{P_a}\right)^{\frac{\gamma_a - 1}{\gamma_a}}\right]\right)^{\frac{1}{2}} \tag{7}$$

$$\dot{m}_{i,h} = \frac{C_{d,h} A_{v,h} P_h}{\sqrt{R_h T_h}} \left(\frac{P_m}{P_h}\right)^{\frac{1}{\gamma_h}} \left(\frac{2\gamma_h}{\gamma_h - 1}\left[1 - \left(\frac{P_m}{P_h}\right)^{\frac{\gamma_h - 1}{\gamma_h}}\right]\right)^{\frac{1}{2}} \tag{8}$$

$$\dot{m}_{i,e} = \frac{C_{d,e} A_{v,e} P_a}{\sqrt{R_e T_e}} \left(\frac{P_m}{P_e}\right)^{\frac{1}{\gamma_e}} \left(\frac{2\gamma_e}{\gamma_e - 1}\left[1 - \left(\frac{P_m}{P_e}\right)^{\frac{\gamma_e - 1}{\gamma_e}}\right]\right)^{\frac{1}{2}} \tag{9}$$

where $C_{d,a}$, $C_{d,h}$, and $C_{d,e}$ are coefficient of discharge for air, hydrogen and exhaust gas, respectively. $A_{v,a}$, $A_{v,h}$, and $A_{v,e}$ are restrictions flow areas for air, hydrogen and exhaust gas, respectively. P_a, P_h, and P_e are the air, hydrogen, and exhaust sources' pressures, respectively. γ_a, γ_h, and γ_e are the air, hydrogen, and exhaust specific heat ratios, respectively. For mass flow out of the intake manifold to engine cylinders, we can be based on mean value modeling, which is given by Equation (10) for the four stroke systems.

$$\dot{m}_o = \eta_v \rho_m V_d \frac{N}{120} \tag{10}$$

where ρ_m is the intake manifold density, η_v is the volumetric efficiency, V_d is the displacement volume, and N is the engine speed given in revolution per minute.

$$\rho_m = \frac{P_m}{R_m T_m} \tag{11}$$

For compression ignition engines and gasoline direct injection engines the volumetric efficiency, η_v, can be given by Equation (12) as indicated in [11]. Assuming perfect gases with constant specific heat ratio and isentropic processes, the volumetric efficiency can be considered pressure dependent.

$$\eta_v = \frac{V_c + V_d}{V_d} - \frac{V_c}{V_d}\left(\frac{P_e}{P_m}\right)^{\frac{1}{\gamma}} \tag{12}$$

where V_c is the combustion chamber volume. Now, by substituting Equation (4), Equation (5), and Equation (11) in Equation (10) the following relation can be obtained:

$$\dot{m}_o = \frac{\eta_v V_d N}{120}\left(\frac{P_m}{\left(\frac{\dot{m}_a R_a + \dot{m}_h R_h + \dot{m}_e R_e}{\dot{m}_a + \dot{m}_h + \dot{m}_e}\right)\left(\frac{\dot{m}_a c_{p,a} T_a + \dot{m}_h c_{p,h} T_h + \dot{m}_e c_{p,e} T_e}{\dot{m}_a c_{p,a} + \dot{m}_h c_{p,h} + \dot{m}_e c_{p,e}}\right)}\right) \tag{13}$$

Rearranging Equation (13), the following quadratic relationship can be obtained:

$$\begin{aligned} &\dot{m}_a^2\left[c_{p,a} R_a T_a\right] + \dot{m}_a\left[\dot{m}_h\left(c_{p,h} R_a T_h + c_{p,a} R_h T_a\right) + \dot{m}_e\left(c_{p,e} R_a T_e + c_{p,a} R_e T_a\right) - \frac{\eta_v V_d P_m N}{120} c_{p,a}\right] \\ &+ \left[\dot{m}_h^2 c_{p,h} R_h T_h + \dot{m}_h \dot{m}_e\left(c_{p,e} R_h T_e + c_{p,h} R_e T_h\right) + \dot{m}_e^2 c_{p,e} R_e T_e - \dot{m}_h c_{p,h}\frac{\eta_v V_d P_m N}{120}\right] = 0 \end{aligned} \tag{14}$$

For the purpose of this study, it is assumed that there is no exhaust re-circulation, $\dot{m}_e = 0$. Thus, it is possible to relate hydrogen flow to airflow inside the intake manifold by the following relation:

$$\dot{m}_a = AFR_h \dot{m}_h = \lambda_h AFR_{s,h} \dot{m}_h \tag{15}$$

where AFR_h is the air/hydrogen fuel mass flow ratio, $AFR_{s,h}$ is the air/fuel stoichiometric ratio, and λ_h is the excess air factor inside the intake manifold.

For physical parameters, the expression of the air mass flow can be given by Equation (16), if the fuel's temperature, specific heat, and gas constant are known.

$$\dot{m}_a = \frac{\frac{\eta_v V_d P_m N}{120}\left(c_{p,a} + \frac{c_{p,h}}{AFR_h}\right)}{c_{p,a} R_a T_a + \frac{c_{p,h} R_a T_h + c_{p,a} R_h T_a}{AFR_h} + \frac{c_{p,h} R_h T_h}{AFR_h^2}} \tag{16}$$

To account for air mass trapped inside engine cylinder, which is the available portion for the combustion of the injected diesel fuel. Therefore, it is possible to assume that all introduced hydrogen will be burned according to stoichiometric conditions. Based on that, all the rest air trapped could be used for diesel combustion as follows:

$$\dot{m}_a = AFR_d \dot{m}_d + AFR_{s,h} \dot{m}_h \tag{17}$$

To account for diesel fuel flow, which is injected inside cylinders, the following relation can be obtained by rearranging Equation (17).

$$\dot{m}_d = \dot{m}_a \left(\frac{\lambda_h - 1}{\lambda_h \lambda_d AFR_{s,d}} \right) \quad (18)$$

where λ_d is the excess air factor inside the engine cylinder excluding the portion required for stoichiometric hydrogen combustion. While to account for the total trapped fuel energy content, Equation (19) can be used

$$\dot{Q}_c = \dot{m}_d LHV_d + \dot{m}_h LHV_h \quad (19)$$

Equation (19) helps to give insight into the available energy for producing useful power from introducing hydrogen and reducing diesel, though increasing this amount will increase the efficiency. Of course, maintaining or increasing this amount by replacing diesel by hydrogen will satisfy the sought target.

3. Simulation

In this section, the developed modeling will be used to analysis the effects of partially adding hydrogen in the induction system on the charge concentrations and the total fuel energy flow to the diesel engine. At first, the effect on charge trapped quantity is considered at various hydrogen concentrations to give insight on the effect of its gaseous state on the fresh charge composition. Then the effect of varying relative air-fuel ratio for both fuels on diesel mass flow is considered to allow for relating the three working fluids at various engine loads. After that, fuel energy flow to engine cylinders at various loads in the form of excess air factor is considered, followed by considering the effect of manifold pressure variation to give insight into possibility of increasing hydrogen flow.

For the purpose of this study, engine speed is kept constant as its main effect is on volumetric efficiency and for not complicating results analysis. **Table 1** gives the main parameters used for simulation.

Figure 1 show the variation of hydrogen mixing increases for constant speed and diesel air-fuel ratio of in the form of excess air factor, $\lambda_d = 4$, $N = 2000$ rpm, $P_m = 1$ bar, $T_a = 300$ K, $T_h = 300$ K. The addition of hydrogen gas reduces the fresh air mass flow to engine cylinders, which reduces the possibility of burning diesel fuel, given the fact that diesel engines always operate with excess air due to their smoke limit it will not affect fresh air needs at part-loads. In fact, experimental results available in literature proved that the presence of hydrogen mixing process with air, the mixture properties improve and helps to reduce the duration of combustion process. The volumetric efficiency decreases with load as the percentage of hydrogen increases for constant speed due to the effect of hydrogen gaseous state, which reduces airflow through the inlet manifold.

Therefore, specific diesel fuel consumption decreases with load as the percentage of hydrogen partial mixing increases at constant engine speed. Due to the mixing of hydrogen with air, more efficient burning process of fuel mixture is achievable, which could increases engine power, given the brake mean effective pressure increases with load when the percentage of hydrogen mixing increases.

Figure 2 shows the variation of hydrogen relative air-fuel ratio with diesel mass flow at various excess air factors. According to the presented literature review, introducing hydrogen could help in enhancing fuel and air

Table 1. Diesel engine and working fluids parameters [10] [12].

Parameter	Symbol	Value
Engine	V_d (cm^3)	2151
	N (rpm)	2000
	r_c	19:1
Diesel Fuel	AFR_s (kg/kg)	14.5
	$C_{10.8}H_{18.7}$	
	LHV (MJ/kg)	42.5
Hydrogen Fuel	AFR_s (kg/kg)	34.3
	C_p (kJ/kg·K)	14.21
	LHV (MJ/kg)	120

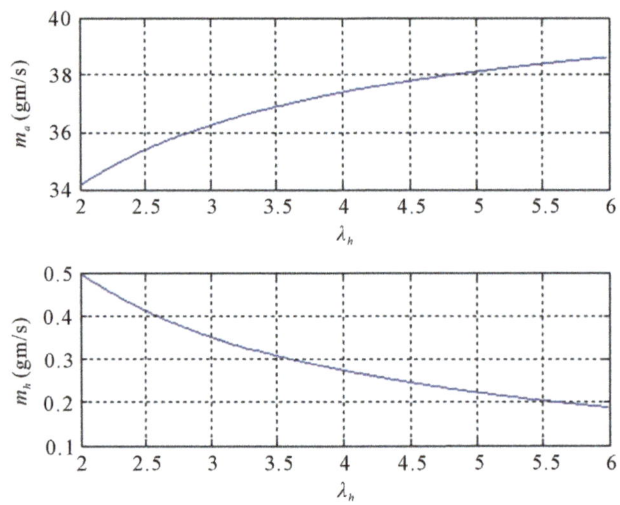

Figure 1. Variation of air and hydrogen mass flow with hydrogen relative air fuel ratio.

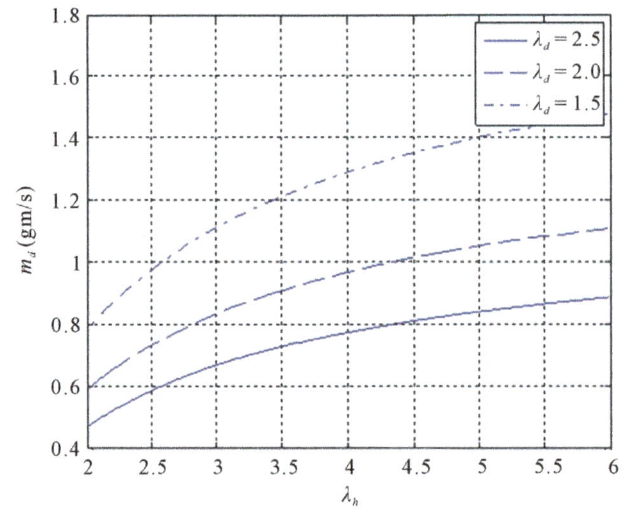

Figure 2. Variation of diesel mass flow with hydrogen and diesel relative air fuel ratio.

mixing, which gives better combustion characteristics; therefore, it becomes possible to use lower excess air factor for diesel. This allows for higher hydrogen induction rate for various loading of diesel engine running at constant speed. For duel fuel system, the equivalence ratio decreased as hydrogen induction rate increased for a given load and speed.

Figure 3 shows the variation of trapped fuel energy with hydrogen and diesel flow to engine cylinders at various engines loads in relation to diesel excess air factors. The results are for $\lambda_h = 4$, $N = 2000$ rpm, $P_m = 1$ bar, $T_a = 300$ K, $T_h = 300$ K. The induction rate for various loading of diesel engine at constant speed shows that it is possible at part-load to maintain or reduce diesel flow at lower excess air factor given that it will be higher than that for smoke limit. This allows for more introduction of hydrogen reducing the diesel consumption for the same load as the energy balance can be shifted to hydrogen flow.

The general trend is consistent with that observed in **Figure 4**. For diesel-hydrogen duel fuel system, the energy flow variation with the intake manifold pressure. The presented results are for $\lambda_h = 4$, $\lambda_d = 4$, $N = 2000$ rpm, $T_a = 300$ K, $T_h = 300$ K, for certain diesel mass flow increasing manifold pressure allow for higher hydrogen mass flow rate. This could be very useful technique to improve the whole operating range of the diesel engine in question.

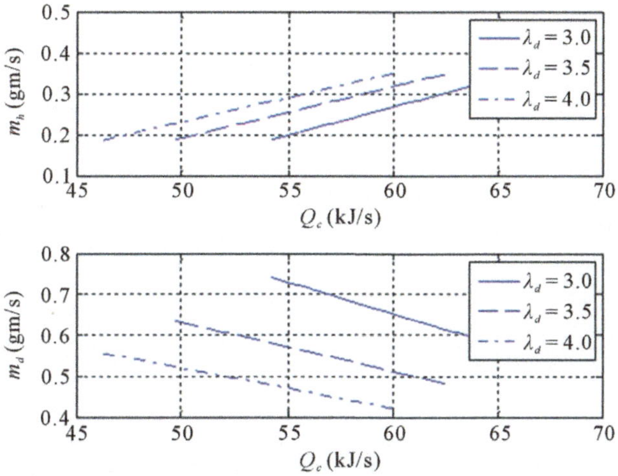

Figure 3. Variation of in-cylinder fuel energy with fuels mass flow and relative air-fuel ratio.

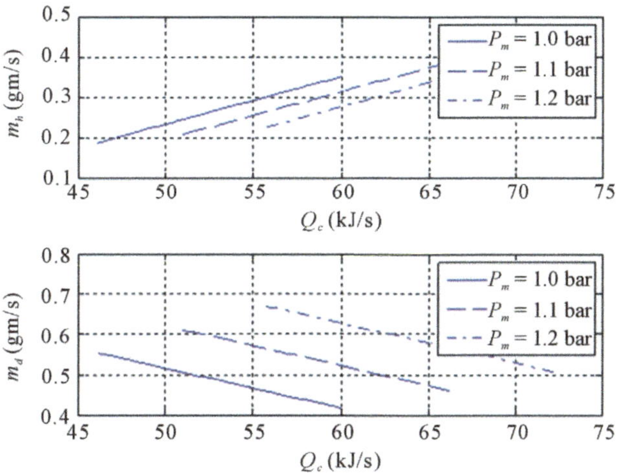

Figure 4. Variation of in-cylinder fuel energy with fuels mass flow with intake manifold pressure.

4. Experimentation

Figure 5 shows the testing setup that is used to validate the proposed method. It is composed of a multi cylinder diesel engine with an added on hydrogen gas supply system. It allows for manual regulating the hydrogen flow to the engine intake manifold. The setup also includes a hydraulic dynamometer for loading the engine in addition to engine speed measurement device.

The experimental procedure is designed to validate that using hydrogen gas a part-load could improve diesel engine power with less emissions. To achieve hydrogen this goal, hydrogen was introduced at constant engine speed and constant diesel flow. By increasing the load using the dynamometer, the difference of extra power generated from hydrogen addition is then measured. The experimental results for different tests, which are presented in **Table 2**, are obtained by the author and his students as part of this study [13].

5. Results Discussion

To increase cylinder energy content from fuel without increasing diesel fuel flow, it is possible to achieve that in two ways. The first one is to lower the excess air factor at part-load operating conditions allowing for more hydrogen mass flow. **Figure 3** shows that $m_d = 0.5$ gm/s, $Q_c = 52$ kJ/s at $m_h = 0.21$ gm/s, $\lambda_h = 4$, $\lambda_d = 4$, $N = 2000$ rpm, $T_a = 300$ K, $T_h = 300$ K. But, when $\lambda_d = 3.5$, $Q_c = 61$ kJ/s and $m_h = 0.32$ gm/s, with other parameters

Figure 5. Test setup with the loading unit on right and hydrogen supply system on left.

Table 2. Experimental results for diesel engine with and without hydrogen.

Speed (rpm)	Diesel Only		Diesel and Hydrogen Mixture		
	Torque (N·m)	Power (kW)	Hydrogen (cm^3/s)	Torque (N·m)	Power (kW)
950	20	1.98	2.1	24	2.38
			5.25	28	2.78
			7.36	33	3.28
1200	34	4.27	2.1	39	4.9
			5.25	41	45
			7.36	58	9.71
1600	45	7.53	2.1	49	8.21
			5.25	56	9.38
			7.36	58	9.71

are kept constant. This indicates that more power could be achieved at part-load condition by increasing hydrogen flow. Of course, this is very positive issue as most of engines installed onboard are operated within its part-load range. To have more insight into the effect of increasing hydrogen flow at part-load conditions, adiabatic combustion is considered. It is used to evaluate the adiabatic flame temperature T_c, given it is a direct indicator of the engine indicated power; a higher obtained temperature means a higher indicated power output. Based on this fact, constant pressure system is considered as given by

$$\sum_P n_i \tilde{h}_i(T_c) = \sum_R n_i \tilde{h}_i(T_1) \tag{20}$$

Equation (20) can be rearranged to obtain Equation (21), which is more suitable for using JANAF thermodynamic properties tables.

$$\sum_P n_i\left(\tilde{h}_i(T_c) - \tilde{h}_i(298\text{ K})\right) = \sum_R n_i\left(\tilde{h}_i(T_1) - \tilde{h}_i(298\text{ K})\right) - \left[\sum_P n_i \tilde{h}^o_{f,i} - \sum_R n_i \tilde{h}^o_{f,i}\right] \tag{21}$$

The chemical reaction is given by the flowing equation, which is based on complete combustion of the introduced fuel (Diesel and Hydrogen) in the presence of excess air.

$$n_d C_{10.8}H_{18.7} + n_h H_2 + \left\{\left(10.8 + \frac{18.7}{4}\right) n_d \lambda_d + n_h\right\}\left(O_2 + 3.773N_2\right)$$

$$\rightarrow 10.8 n_d CO_2 + \left(\frac{18.7}{4} n_d + n_h\right) H_2O + 3.773\left\{\left(10.8 + \frac{18.7}{4}\right) n_d \lambda_d + n_h\right\} N_2 + \left(\lambda_d - 1\right)\left(10.8 + \frac{18.7}{4}\right) n_d O_2$$

From the equation above, it is very clear that excess air adds many moles of diatomic molecules (O_2 and N_2) into the products that does not contribute to heat release just soaks it up. Therefore; the above discussed case is also considered for evaluating adiabatic flame temperature. Look up enthalpy values from JANAF tables, and iteration gives for $\lambda_d = 4$, the obtained $T_c = 1920$ K, but when excess air is reduced to $\lambda_d = 3.5$, $T_c = 1980$ K. It is obvious that allowing for more hydrogen flow at part-load helps to increase power output and reduce carbon emissions for certain engine load and speed. For evaluating T_1, isentropic compression is assumed for calculations simplicity.

The second way is to increase manifold pressure while keeping diesel flow constant, this allows to increase hydrogen flow as can be seen in **Figure 4**. If $m_d = 0.5$ gm/s at $P_m = 1$ bar, $Q_c = 52$ kJ/s at part load with $\lambda_h = 4$, $\lambda_d = 4$, $N = 2000$ rpm, $T_a = 300$ K, $T_h = 300$ K, for this case it needs $m_h = 0.25$ gm/s, but if manifold pressure increased to become $P_m = 1.1$ bar, more hydrogen could be introduced for the presence of more fresh air, so $Q_c = 62$ kJ/s using same diesel quantity while we are able to use $m_h = 0.42$ gm/s of hydrogen fuel, which increase significantly the use of clean energy in diesel powered vehicles.

Experimental results shows obvious increase in power output with introduction of hydrogen, which means that diesel consumption rate could be decreased with increase in hydrogen flow. This leads to decrease in carbon oxides emission by a very significant percentage. Further, to reduce the risk of hydrogen mixture auto-ignition, it is possible to benefit from the introduction of hydrogen at low temperature in the intake manifold; this could easily lower the mixture temperature before the compression process

6. Conclusion

The presented analytical and experimental investigation for the effects of adding hydrogen on diesel engines power output and the reduction of emissions shows obvious positive signs to follow-up the obtained results. Engine power to weight ratio can be improved at optimized hydrogen-diesel operating conditions. Carbon oxides emissions can be reduced along with the increasing of hydrogen percentage. Increasing the intake manifold pressure can improve the engine filling, however, knock limit remains an important concern. This is achievable without compromising the engine maximum efficiency, given that most engines are operated at small and part-load during normal driving conditions, which allow for introducing more hydrogen instead of the large quantity of excess air during such operation conditions that also further are improved by charge boosting. In view of the above observations, experimental investigations are further needed to be carried out to revile all technical issues and demonstrate practical solutions viability.

Acknowledgements

Acknowledgement is given to the technicians and automotive engineering students from Mechanical Engineering Department in Palestine Polytechnic University, for technical support and help in conducting the experimentation, respectively.

References

[1] Varde, K.S. and Frame, G.A. (1983) Hydrogen Aspiration in a Direct Injection Type Diesel Engine—Its Effects on Smoke and Other Engine Performance Parameters. *International Journal of Hydrogen Energy*, **8**, 549-555. http://dx.doi.org/10.1016/0360-3199(83)90007-1

[2] Zhang, H., Lilik, G.K., Boehman, A.L. and Haworth, D.C. (2009) Effects of Hydrogen Addition on NOx Emissions in Hydrogen-Assisted Diesel Combustion. *International Multidimensional Engine Modeling Users Group Meeting Detroit*, MI, 19 April 2009.

[3] Adnana, R., Masjukib, H.H. and Mahlia, T.M.I. (2010) Experimental Investigation on In-Cylinder Pressure and Emissions of Diesel Engine with Port Injection Hydrogen System. *International Journal of Mechanical and Materials Engineering* (*IJMME*), **5**, 136-141.

[4] Goldwitz, J.A. (2004) Combustion Optimization in Hydrogen-Enhanced Lean Burn SI Engine. Thesis, Massachusetts Institute of Technology, USA.

[5] Szwaja, S. and Rogalinski, K.G. (2009) Hydrogen Combustion in a Compression Ignition Diesel Engine. *International Journal of Hydrogen Energy*, **34**, 441-4421.

[6] Saravanana, N., Nagarajana, G., Dhanasekaranb, C. and Kalaiselvanb, K.M. (2007) Experimental Investigation of Hydrogen Port Fuel Injection in DI Diesel Engine. *International Journal of Hydrogen Energy*, **32**, 4071-4080.

http://dx.doi.org/10.1016/j.ijhydene.2007.03.036

[7] Saravanan, N., Nagarajan, G. and Kalaisselvan, K.M. (2007) Experimental Investigation of Hydrogen Fuel Injection in DI Dual Fuel Diesel Engine. SAE, 2007-01-1465.

[8] Maki, D.F. and Prabhakaran, P. (2011) An Experimental Investigation on Performance and Emissions of a Multi Cylinder Diesel Engine Fueled with Hydrogen-Diesel Blends. *World Renewable Energy Congress*, Vol. 13, 8-13 May 2011. http://dx.doi.org/10.3384/ecp110573557

[9] Shahad, H.A.K. and Abdul-Hadi, N. (2011) Experimental Investigation of the Effect of Hydrogen Manifold Injection on the Performance of Compression Ignition Engines. World Academy of Science, Engineering and Technology, 277-282.

[10] Robert Bosch GmbH (2000) Automotive Handbook. 5th Edition, SAE Inc.

[11] Heywood, J. (1988) Fundamentals of Internal Combustion Engines. MacGrow Hill Int., 304-307.

[12] Guzzella, L. and Onder, C.H. (2010) Introduction to Modeling and Control of Internal Combustion Engine Systems. Springer-Verlag, Berlin Heidelberg.

[13] Jalal, R. and Adnan, S. (2012) Hydrogen Aspiration in Diesel Engine. Graduation Report, Palestine Polytechnic University, Hebron.

13

On the Maximum of Wind Power Efficiency

Gerhard Kramm[1*], Gary Sellhorst[2], Hannah K. Ross[3], John Cooney[4], Ralph Dlugi[5], Nicole Mölders[6]

[1]Engineering Meteorology Consulting, Fairbanks, USA
[2]Geophysical Institute, University of Alaska Fairbanks, Fairbanks, USA
[3]Department of Mechanical Engineering, University of Washington, Seattle, USA
[4]Department of Atmospheric Sciences, Texas A & M University, College Station, USA
[5]Arbeitsgruppe Atmosphärische Prozesse (AGAP), Munich, Germany
[6]College of Natural Science and Mathematics and Geophysical Institute, University of Alaska Fairbanks, Fairbanks, USA
Email: [*]gerhardkramm46@gmail.com

Abstract

In our paper we demonstrate that the filtration equation used by Gorban' *et al.* for determining the maximum efficiency of plane propellers of about 30 percent for free fluids plays no role in describing the flows in the atmospheric boundary layer (ABL) because the ABL is mainly governed by turbulent motions. We also demonstrate that the stream tube model customarily applied to derive the Rankine-Froude theorem must be corrected in the sense of Glauert to provide an appropriate value for the axial velocity at the rotor area. Including this correction leads to the Betz-Joukowsky limit, the maximum efficiency of 59.3 percent. Thus, Gorban' *et al.*'s 30% value may be valid in water, but it has to be discarded for the atmosphere. We also show that Joukowsky's constant circulation model leads to values of the maximum efficiency which are higher than the Betz-Joukowsky limit if the tip speed ratio is very low. Some of these values, however, have to be rejected for physical reasons. Based on Glauert's optimum actuator disk, and the results of the blade-element analysis by Okulov and Sørensen we also illustrate that the maximum efficiency of propeller-type wind turbines depends on tip-speed ratio and the number of blades.

Keywords

Wind Power, Power Efficiency, General Momentum Theory, Axial Momentum Theory, Blade Element Analysis, Betz-Joukowsky Limit, Joukowsky's Constant Circulation Model, Glauert's Optimum Actuator Disk, Balance Equation for Momentum, Equation of Continuity, Balance Equation for Kinetic Energy, Reynolds' Average, Hesselberg's Average, Favre's Average, Bernoulli's Equation, Integral Equations

[*]Corresponding author.

1. Introduction

In 2001, Gorban' *et al.* [1] challenged the Betz limit (in our study called the Betz-Joukowsky limit [2]-[6] even though it has also been denoted as Lanchester-Betz-Joukowsky limit [2] [6]), *i.e.*, the maximum power efficiency of 59.3 percent for a propeller-type turbine, where the power efficiency is generally defined by

$$C_P = \frac{P}{P_\infty}. \tag{1.1}$$

Here, P is the extracted (or consumed) power, and P_∞ is the power carried by the flow through the projection of the turbine section region onto the plane perpendicular to it. Gorban' *et al.* [1] argued that the maximum efficiency of the plane propeller is about 30 percent for free fluids. Meanwhile, their paper has been cited numerous times in conference papers (e.g., [7] [8]), but recently van Kuik *et al.* [9] rejected their method and pointed out that the main problem of Gorban' *et al.* "*is their lack of comprehension of the working principles how the turbine operates*". Since this argument is very harsh, it is indispensable to show that the result of Gorban' *et al.* is based on an equation that may be acceptable for water fluids of low flow velocity, but that is indeed not very suitable for flows as they are typical in the atmospheric boundary layer (ABL, the lowest layer of the troposphere, with a thickness of the order of $h = 10^3$ m), especially at heights between 30 m to 150 m above the Earth's surface.

According to Gorban' *et al.* [1], the filtration equation,

$$-\nabla p = r\boldsymbol{v}, \tag{1.2}$$

holds in an open domain V_c (denoted by [1] as Ω) with a smooth or piecewise smooth boundary together with the equation of continuity, $\nabla\cdot\boldsymbol{v} = 0$ (only valid for an incompressible stationary flow), where p and $\boldsymbol{v}$ denote the pressure and the velocity of the flow, respectively. The shape of V_c is considered as a semi-penetrable obstacle for the stream with a resistance density r inside. Let A_c the cross section of V_c perpendicular to the flow axis. The power carried by the flow through A_c is then given by

$$P_\infty = \frac{1}{2}\rho A_c v_\infty^3. \tag{1.3}$$

Following these authors, v_∞ is the velocity of a uniform laminar current. Gorban' *et al.* [1] argued that the power, P, consumed by the turbine is given by

$$P = \int_{V_c} \boldsymbol{v}\cdot\nabla p \,\mathrm{d}V = \int_{V_c} \frac{1}{r}(\nabla p)^2\,\mathrm{d}V = \int_{V_c} r\boldsymbol{v}^2\mathrm{d}V, \tag{1.4}$$

where the filtration Equation (1.2) has been inserted. In accord with Equation (1.1), they obtained

$$C_P = \frac{P}{P_\infty} = \frac{\int_{V_c} \boldsymbol{v}\cdot\nabla p\,\mathrm{d}V}{\frac{1}{2}\rho A_c v_\infty^3}. \tag{1.5}$$

Gorban' *et al.* [1] claimed:

"*The efficiency coefficient can be maximized by optimizing the resistance density. The optimal ratio between the streamlining current and the current passing through the turbines can also be obtained from this model. This parameter can be measured experimentally to determine how close a real turbine is to the theoretically optimal one.*"

Obviously, the maximum of the efficiency coefficient deduced by Gorban' *et al.* [1] depends on the filtration equation. This equation, however, plays no role in the description of ABL flows. In addition, the ABL is mainly governed by turbulent motions. If we assume, for instance, a wind speed of $|\boldsymbol{v}| = 7.5\ \mathrm{m\cdot s^{-1}}$ at a hub height of $z = 80$ m and a kinematic viscosity of $\nu = 1.5\times10^{-5}\ \mathrm{m^2\cdot s^{-1}}$ we will obtain a Reynolds number of about

$$Re = \frac{|\boldsymbol{v}|z}{\nu} = \frac{7.5\ \mathrm{m\cdot s^{-1}}\times 80\ \mathrm{m}}{1.5\times10^{-5}\ \mathrm{m^2\cdot s^{-1}}} = 4\times10^7. \tag{1.6}$$

This Re value is far beyond the critical Reynolds number at which the transition from a laminar to a turbulent flow occurs.

In the following section, we will present the governing equations for macroscopic and turbulent systems relevant for wind power studies: (a) the local balance equations for momentum (also called the Navier-Stokes equation), (b) total mass (also called the equation of continuity), and (c) kinetic energy. It is shown that the Bernoulli equation for an incompressible flow can simply be derived from the local balance equation of kinetic energy. Furthermore, we will derive the simplified integral balance equations recently used by Sørensen [6] in his review on the aerodynamic aspects of wind energy conversion to incorporate his results in our discussion. Additionally, we will demonstrate that Equation (1.4) derived by Gorban' *et al.* [1] is incomplete for the ABL so that their maximum efficiency calculation for plane propellers of about 30 percent for free fluids has to be discarded, as suggested by van Kuik *et al.* [9]. This means that the filtration Equation (1.2) is meritless if the maximum efficiency of wind power has to be determined. In Section 3, we will discuss the main characteristics of propeller-type wind turbines. Our discussion will include the basics of the axial momentum theory, Joukowsky's constant circulation model, Glauert's infinite-bladed actuator disk model, and finite-bladed rotor models. We will show that the Betz-Joukowsky limit is, indeed, the maximum of the wind power efficiency, even though some results of Joukowsky's constant circulation model might exceed it because of physically inadequate conditions. Glauert's [10] optimum actuator disk and finite-bladed rotors [6] [11] [12] tend to this maximum, if the tip-speed ratio, λ, increases.

2. Theoretical Background

2.1. The Governing Equations for the Macroscopic System

In order to outline the generation of electricity by extracting kinetic energy from the wind field we consider the local balance equations for momentum (*i.e.*, Newton's 2nd axiom), Equation (2.1), and total mass, Equation (2.2), for a macroscopic system given by (e.g., [13]-[16]):

$$\frac{\partial(\rho \boldsymbol{v})}{\partial t} + \nabla \cdot \left(\rho \boldsymbol{v}\boldsymbol{v} + p\boldsymbol{E} + \boldsymbol{J}\right) = -\rho \nabla \varphi - 2\rho\left(\boldsymbol{\Omega} \times \boldsymbol{v}\right) \tag{2.1}$$

and

$$\frac{\partial \rho}{\partial t} + \nabla \cdot \left(\rho \boldsymbol{v}\right) = 0\,. \tag{2.2}$$

Here, ρ is the air density, t is time, $\boldsymbol{v}$ is the velocity of the flow, $\boldsymbol{J}$ is the Stokes stress tensor given by

$$\boldsymbol{J} = \rho\nu\left(\nabla \boldsymbol{v} + \left(\nabla \boldsymbol{v}\right)^{\mathrm{T}}\right) + \left(\mu_d - \frac{2}{3}\rho\nu\right)\left(\nabla \cdot \boldsymbol{v}\right)\boldsymbol{E}\,. \tag{2.3}$$

where μ_d is the bulk viscosity (near zero for most gases), $\boldsymbol{E}$ is the identity tensor, φ is the gravity potential, and $\boldsymbol{\Omega}$ is the angular velocity of the Earth. Both $\boldsymbol{J}$ and $\boldsymbol{E}$ are symmetric second-rank tensors. Furthermore, the 1st term of the left-hand side of Equation (2.1) describes the local temporal change of momentum, and the 2nd term represents the exchange of momentum between the system under study and its surroundings, where $p\,\boldsymbol{E} + \boldsymbol{J}$ exerts on the boundary of this system. The 1st term on the right-hand side of this equation represents the gravity force, and the 2nd one the Coriolis force. Equation (2.2) is the equation of continuity. In addition, local balance equations for various energy forms (*i.e.*, internal energy, kinetic energy, potential energy, and total energy), various water phases (*i.e.*, water vapor, liquid water, and ice), and gaseous and particulate atmospheric trace constituents exist. All these local balance equations can be derived from integral balance equations (e.g., [13] [14]). Since $D(\rho \boldsymbol{v})/Dt = \partial(\rho \boldsymbol{v})/\partial t + \nabla \cdot (\rho \boldsymbol{v}\boldsymbol{v}) = \rho\, \mathrm{d}\boldsymbol{v}/\mathrm{d}t$ and $\nabla \cdot (p\boldsymbol{E}) = \nabla p$, Equation (2.1) is often written as (e.g., [17]-[19])

$$\frac{\partial \boldsymbol{v}}{\partial t} + \boldsymbol{v} \cdot \nabla \boldsymbol{v} + \frac{1}{\rho}\nabla \cdot \boldsymbol{J} = -\frac{\nabla p}{\rho} - \nabla \varphi - 2\left(\boldsymbol{\Omega} \times \boldsymbol{v}\right), \tag{2.4}$$

where $\mathrm{d}\boldsymbol{v}/\mathrm{d}t = \partial \boldsymbol{v}/\partial t + \boldsymbol{v} \cdot \nabla \boldsymbol{v}$ is the substantial derivative with respect to time. This equation form not only disguises its origin, namely the corresponding integral balance equation, but also is unfavorable if it has to be averaged, for instance, in the sense of Reynolds [20] to find a tractable equation for turbulent atmospheric layers. Nevertheless, in accord with Lamb's transformation (e.g., [21])

$$\boldsymbol{v}\cdot\nabla\boldsymbol{v}=\nabla\left(\frac{v^2}{2}\right)-\boldsymbol{v}\times\left(\nabla\times\boldsymbol{v}\right), \tag{2.5}$$

Equation (2.4) may be written as

$$\frac{\partial\boldsymbol{v}}{\partial t}+\nabla\left(\frac{v^2}{2}\right)-\boldsymbol{v}\times\left(\nabla\times\boldsymbol{v}+2\boldsymbol{\Omega}\right)=-\frac{1}{\rho}\nabla\cdot\boldsymbol{J}-\frac{1}{\rho}\nabla p-\nabla\varphi. \tag{2.6}$$

The curl of Equation (2.6) leads to the prognostic equation for the vorticity

$$\frac{\partial}{\partial t}(\nabla\times\boldsymbol{v})+\nabla\times\nabla\left(\frac{v^2}{2}\right)-\nabla\times\left(\boldsymbol{v}\times\left(\nabla\times\boldsymbol{v}+2\boldsymbol{\Omega}\right)\right)=-\nabla\times\left(\frac{1}{\rho}\nabla\cdot\boldsymbol{J}\right)-\nabla\times\left(\frac{1}{\rho}\nabla p\right)-\nabla\times\nabla\varphi, \tag{2.7}$$

As the curl of the gradient of a scalar field is equal to zero, Equation (2.7) can be written as

$$\frac{\partial}{\partial t}(\nabla\times\boldsymbol{v})+\nabla\times\left(\left(\nabla\times\boldsymbol{v}+2\boldsymbol{\Omega}\right)\times\boldsymbol{v}\right)=-\nabla\times\left(\frac{1}{\rho}\nabla\cdot\boldsymbol{J}\right)-\nabla\times\left(\frac{1}{\rho}\nabla p\right). \tag{2.8}$$

This equation plays an important role in the description of rotational flows as occurred in the wake of the wind turbine. If the friction effect is negligible and the density is considered as spatially constant like in case of an incompressible fluid we will obtain

$$\frac{\partial}{\partial t}(\nabla\times\boldsymbol{v})+\nabla\times\left(\left(\nabla\times\boldsymbol{v}+2\boldsymbol{\Omega}\right)\times\boldsymbol{v}\right)=0. \tag{2.9}$$

To deduce the local balance equation for the kinetic energy of the flow, Equation (2.1) has to be scalarly multiplied by the velocity vector $\boldsymbol{v}$. Using the identities

$$\boldsymbol{v}\cdot\nabla\cdot\left(p\boldsymbol{E}+\boldsymbol{J}\right)=\nabla\cdot\left\{p\boldsymbol{v}+\boldsymbol{v}\cdot\boldsymbol{J}\right\}-\nabla\boldsymbol{v}:\left(p\boldsymbol{E}+\boldsymbol{J}\right)=\nabla\cdot\left\{p\boldsymbol{v}+\boldsymbol{v}\cdot\boldsymbol{J}\right\}-\left(p\boldsymbol{E}+\boldsymbol{J}\right):\nabla\boldsymbol{v} \tag{2.10}$$

and

$$p\boldsymbol{E}:\nabla\boldsymbol{v}=p\nabla\cdot\boldsymbol{v} \tag{2.11}$$

leads to

$$\frac{\partial}{\partial t}\left(\rho\frac{v^2}{2}\right)+\nabla\cdot\left\{\left(\rho\frac{v^2}{2}+p\right)\boldsymbol{v}+\boldsymbol{v}\cdot\boldsymbol{J}\right\}=-\rho\boldsymbol{v}\cdot\nabla\varphi+p\nabla\cdot\boldsymbol{v}+\boldsymbol{J}:\nabla\boldsymbol{v}. \tag{2.12}$$

The colon expresses the double-scalar product (also called the double dot product) of the tensor algebra. Furthermore, $\boldsymbol{v}\cdot\left(\boldsymbol{\Omega}\times\boldsymbol{v}\right)=\boldsymbol{\Omega}\cdot\left(\boldsymbol{v}\times\boldsymbol{v}\right)=0$. The 1st term of the left-hand-side of Equation (2.12) describes the local temporal change of kinetic energy, and the 2nd term is the energy exchange of the system with its surroundings which is performed by the surrounding air on the boundary of the system. The 1st term of the right-hand-side represents the conversion of potential energy into kinetic energy and vice versa, the 2nd term describes the reversible work rate of expansion, $\nabla\cdot\boldsymbol{v}>0$, or contraction, $\nabla\cdot\boldsymbol{v}<0$, and the 3rd term represents the irreversible work rate owing to viscous friction. This term represents the dissipation of kinetic energy into the reservoir of heat. The term of our primary interest reads

$$\boldsymbol{S}_{kin}=\rho\frac{v^2}{2}\boldsymbol{v}. \tag{2.13}$$

It describes the transport of kinetic energy by the flow, and it may be called the kinetic energy stream density, but it is also denoted as wind power density. Inserting the definition of the total pressure,

$$H=p+\rho\frac{v^2}{2}, \tag{2.14}$$

into Equation (2.12) yields

$$\frac{\partial H}{\partial t}+\nabla\cdot\left(H\boldsymbol{v}+\boldsymbol{v}\cdot\boldsymbol{J}\right)=\frac{\partial p}{\partial t}-\rho\boldsymbol{v}\cdot\nabla\varphi+p\nabla\cdot\boldsymbol{v}+\boldsymbol{J}:\nabla\boldsymbol{v}. \tag{2.15}$$

2.2. The Governing Equations for the Turbulent System

Since the ABL is mainly governed by turbulent motion, the use of the macroscopic balance Equations (2.1), (2.2), and (2.12) is rather impracticable. Therefore, these balance equations are customarily averaged in the sense of Reynolds [20]. However, conventional Reynolds averaging will lead to various short-comings in the set of governing equations for turbulent atmospheric flow, even if these averaging techniques can be performed accurately [22]. If we ignore, for instance, density fluctuation terms, the possibility to describe physical processes as a whole will clearly be restricted (see [23] [24]). The key questions that still remain are (a) how to average the governing macroscopic equations in the case of turbulent atmospheric flows and (b) what are the consequences of such an averaging, not only for momentum and total mass, but also for various energy forms like kinetic energy, potential energy, internal energy, and total energy, consisting of the sum of these three energy forms. In the terrestrial atmosphere, the total energy is conserved. As sketched in **Figure 1** for a turbulent system (Hesselberg fluid), there are various ways of energy conversion.

As argued by various authors [22]-[34], the density-weighted averaging procedure suggested by Hesselberg [35] is very appropriate to formulate the balance equation for turbulent systems. It is given by

$$\hat{\chi} = \frac{\overline{\rho\chi}}{\overline{\rho}}, \tag{2.16}$$

where χ is a field quantity like the wind vector, $\boldsymbol{v}$, the specific internal energy, *e*, and the specific enthalpy, *h*. Furthermore, the overbar $(\overline{\ldots})$ characterizes the conventional Reynolds mean. Whereas the hat $(\widehat{\ldots})$ denotes the density-weighted average according to Hesselberg, and the double prime (") marks the departure from that. It is obvious that $\overline{\rho}\,\widehat{\chi''} = \overline{\rho\chi''} = 0$. The Hesselberg mean of the wind vector, for instance, is given by $\hat{\boldsymbol{v}} = \overline{\rho\boldsymbol{v}}/\overline{\rho}$. Note that intensive quantities like the pressure, *p*, and the density, ρ, of air are averaged in the sense of Reynolds. Arithmetic rules can be found, for instance, in [25]-[27] [29] [31]. As pointed out by Kramm and Meixner [22] and Lumley and Yaglom [36], Hesselberg's average is sometimes misnamed the Favre average.

In comparison with that of Reynolds, Hesselberg's averaging calculus leads to several prominent advantages [22] [25]-[27] [29]-[31] [35]: (a) The equation of continuity,

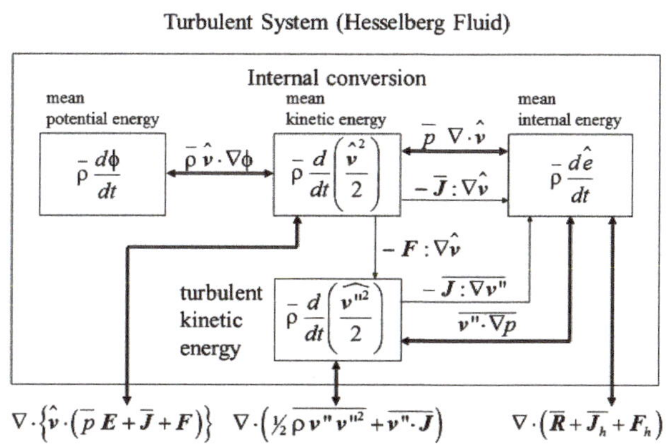

Figure 1. Schematic representation of the energy conversion within a turbulent system (Hesselberg fluid) and the exchange of energy with its surroundings which is performed by the surrounding air on the boundary of the system. As illustrated in this sketch, there is no direct conversion of mean internal energy into mean potential energy and vice versa. Note that $\boldsymbol{R} = \boldsymbol{R}_S + \boldsymbol{R}_{IR}$ is the total irradiance, where $\boldsymbol{R}_S$ is solar irradiance and $\boldsymbol{R}_{IR}$ is the infrared irradiance. Furthermore, $\overline{\boldsymbol{J}_h}$ and $\boldsymbol{F}_h = \overline{\rho\boldsymbol{v}''h}$ are the mean molecular and the turbulent enthalpy flux densities, respectively. All other symbols are explained in the text.

$$\frac{\partial \overline{\rho}}{\partial t} + \nabla \cdot \left(\overline{\rho} \hat{\boldsymbol{v}} \right) = 0 , \tag{2.17}$$

keeps its form, and (b) the mean value of kinetic energy can exactly be split into the kinetic energy of the mean motion and mean value of the kinetic energy of the eddying motion, according to

$$\frac{1}{2}\overline{\rho \boldsymbol{v}^2} = \frac{1}{2}\overline{\rho}\hat{\boldsymbol{v}}^2 + \frac{1}{2}\overline{\rho \boldsymbol{v}''^2} = \frac{\overline{\rho}}{2}\left(\hat{\boldsymbol{v}}^2 + \widehat{\boldsymbol{v}''^2} \right) . \tag{2.18}$$

This advantage is especially important in the theoretical description of the extraction of the kinetic energy from the wind field for generating electricity. The use of density-weighted averages is the common way to define averages in studies of highly compressible turbulent flows (see also [29] [32]), probably the most natural way to define averages. The kinetic energy of the mean motion is usually abbreviated by MKE, and the kinetic energy of the eddying motion is usually called the turbulent kinetic energy abbreviated by TKE.

Hesselberg's average procedure will be applied within the framework of this contribution. It can be related to that of Reynolds by (e.g., [22] [26] [30] [31] [37])

$$\hat{\chi} = \overline{\chi} + \frac{\overline{\rho' \chi'}}{\overline{\rho}} = \overline{\chi}\left\{ 1 + \frac{\overline{\rho' \chi'}}{\overline{\rho}\,\overline{\chi}} \right\} . \tag{2.19}$$

Here, the prime (') denotes the deviation from the Reynolds mean. Obviously, the different means, $\hat{\chi}$ and $\overline{\chi}$, are nearly equal if $\left| \overline{\rho' \chi'} / \left\{ \overline{\rho}\, \overline{\chi} \right\} \right| \ll 1$ as used, for instance, in case of the Boussinesq approximation. In case of a nearly incompressible fluid, the distinction between $\hat{\boldsymbol{v}}$ and $\overline{\boldsymbol{v}}$ is not necessary because the condition $\left| \overline{\rho' \boldsymbol{v}'} / \left\{ \overline{\rho}\, \overline{\boldsymbol{v}} \right\} \right| \ll 1$ is clearly fulfilled. However, to avoid any kind of confusion, we keep our notation.

In the averaged form, the local balance equation for momentum of the turbulent atmosphere reads (e.g., [22] [25]-[27] [29] [35] [38])

$$\frac{\partial \left(\overline{\rho} \hat{\boldsymbol{v}} \right)}{\partial t} + \nabla \cdot \left(\overline{\rho} \hat{\boldsymbol{v}} \hat{\boldsymbol{v}} + \overline{p} \boldsymbol{E} + \overline{\boldsymbol{J}} + \boldsymbol{F} \right) = - \overline{\rho} \nabla \overline{\varphi} - 2 \overline{\rho} \left(\boldsymbol{\Omega} \times \hat{\boldsymbol{v}} \right) , \tag{2.20}$$

where $\boldsymbol{F} = \overline{\rho \boldsymbol{v}'' \boldsymbol{v}''}$ is the Reynolds stress tensor. It results from averaging the term $\rho \boldsymbol{v}\boldsymbol{v}$ in Equation (2.1) leading to $\overline{\rho \boldsymbol{v}\boldsymbol{v}} = \overline{\rho}\hat{\boldsymbol{v}}\hat{\boldsymbol{v}} + \overline{\rho \boldsymbol{v}'' \boldsymbol{v}''}$. Similar local balance equations can be derived for various energy forms (*i.e.*, internal energy, kinetic energy, potential energy, and total energy), various water phases (*i.e.*, water vapor, liquid water, and ice), and gaseous and particulate atmospheric trace constituents [22] [25]-[31] [38] [39].

Averaging Equation (2.12) provides the corresponding local balance equation for the kinetic energy

$$\begin{aligned} &\frac{\partial}{\partial t}\left(\frac{\overline{\rho}}{2}\left(\hat{\boldsymbol{v}}^2 + \widehat{\boldsymbol{v}''^2} \right) \right) + \nabla \cdot \left\{ \frac{\overline{\rho}}{2}\left(\hat{\boldsymbol{v}}^2 + \widehat{\boldsymbol{v}''^2} + \overline{p} \right) \hat{\boldsymbol{v}} + \hat{\boldsymbol{v}} \cdot \left(\overline{\boldsymbol{J}} + \boldsymbol{F} \right) + \frac{1}{2}\overline{\rho \boldsymbol{v}'' \boldsymbol{v}''^2} + \overline{\boldsymbol{v}'' \cdot \boldsymbol{J}} \right\} \\ &= - \overline{\boldsymbol{v}'' \cdot \nabla p} - \overline{\rho} \hat{\boldsymbol{v}} \cdot \nabla \overline{\varphi} + \overline{p} \nabla \cdot \hat{\boldsymbol{v}} + \overline{\boldsymbol{J}} : \nabla \hat{\boldsymbol{v}} + \overline{\boldsymbol{J} : \nabla \boldsymbol{v}''} \end{aligned} \tag{2.21}$$

Obviously, the local derivative with respect to time not only contains the MKE, but also the TKE as outlined by Equation (2.18). Assuming, for instance, steady-state condition leads to

$$\frac{\partial}{\partial t}\left(\frac{\overline{\rho}}{2}\left(\hat{\boldsymbol{v}}^2 + \widehat{\boldsymbol{v}''^2} \right) \right) = 0 \Rightarrow \frac{\overline{\rho}}{2}\left(\hat{\boldsymbol{v}}^2 + \widehat{\boldsymbol{v}''^2} \right) = \text{const.} \tag{2.22}$$

This means that the total kinetic energy is time-invariant, but MKE can be converted into TKE. In the inertial range, for instance, the TKE is transferred from lower to higher wave numbers until the far-dissipation range is reached, where kinetic energy is converted into heat energy by direct dissipation, $\overline{\boldsymbol{J}} : \nabla \hat{\boldsymbol{v}}$, and turbulent dissipation, $\overline{\boldsymbol{J} : \nabla \boldsymbol{v}''}$. Even though the fluctuations of the wind vector are usually small as compared to the mean wind vector, $\left| \boldsymbol{v}'' \right| \ll \left| \hat{\boldsymbol{v}} \right|$, the opposite is true for their gradients, $\left| \nabla \boldsymbol{v}'' \right| \gg \left| \nabla \hat{\boldsymbol{v}} \right|$. This phenomenon is connected with a great intensity of rotation and is characteristic for all turbulent flows. Except for the immediate vicinity of rigid walls, turbulent dissipation exceeds direct dissipation by several orders of magnitude depending on the Reynolds number (e.g., [18] [22]). Furthermore, the mean kinetic energy stream density reads

$$\overline{\mathbf{S}_{kin}} = \frac{\bar{\rho}}{2}\left(\hat{\mathbf{v}}^2 + \widehat{\mathbf{v}''^2}\right)\hat{\mathbf{v}} + \frac{1}{2}\overline{\rho\mathbf{v}''\mathbf{v}''^2}\,. \tag{2.23}$$

This equation describes the transfer of MKE and TKE by the mean wind field and the transfer of TKE by the eddying wind field. Ignoring the turbulent effects yields

$$\overline{\mathbf{S}_{kin}} = \frac{1}{2}\bar{\rho}\hat{\mathbf{v}}^2\hat{\mathbf{v}}\,, \tag{2.24}$$

i.e., $\overline{\mathbf{S}_{kin}}$ is approximated by the MKE stream density. The magnitude of $\overline{\mathbf{S}_{kin}}$ is given by

$$\left|\overline{\mathbf{S}_{kin}}\right| = \frac{1}{2}\bar{\rho}\hat{v}^2\left|\hat{\mathbf{v}}\right| = \frac{1}{2}\bar{\rho}\hat{v}^3\,, \tag{2.25}$$

where $\left|\hat{\mathbf{v}}\right| = \hat{v}$. Apparently, this quantity expresses that the wind power density is proportional to the cube of the wind speed. The rotor of a wind turbine causes a divergence effect expressed by $\nabla\cdot\overline{\mathbf{S}_{kin}} \neq 0$.

Unfortunately, there is a notable inconsistency regarding the role of the turbulence intensity. According to de Vries [40], for instance, this quantity is $Ti_u = \sigma_u/\hat{u}$, where σ_u is the standard deviation of the horizontal wind speed and $\sigma_u^2 = \widehat{u''^2}$ is the corresponding variance. If we assume that only a horizontal component of the mean wind field exists, for the purpose of convenience, in the direction of the x-axis of a Cartesian coordinate frame, Equation (2.23) would provide

$$\left.\overline{\mathbf{S}_{kin}}\right|_x = \frac{\bar{\rho}}{2}\hat{u}^3\left(1 + \frac{\widehat{\mathbf{v}''^2}}{\hat{u}^2} + \frac{\overline{\rho u''\mathbf{v}''^2}}{\bar{\rho}\hat{u}^3}\right) \tag{2.26}$$

or

$$\widehat{u^3} = \hat{u}^3\left(1 + \frac{\widehat{\mathbf{v}''^2}}{\hat{u}^2} + \frac{\overline{\rho u''\mathbf{v}''^2}}{\bar{\rho}\hat{u}^3}\right), \tag{2.27}$$

i.e. we have still to consider the fluctuations of all components in this coordinate frame. On the other hand, de Vries [40] argued that the instantaneous value is given by $u = \hat{u} + u''$, and, hence, $\widehat{u^3} = \widehat{\left(\hat{u} + u''\right)^3}$, *i.e.*,

$$\widehat{u^3} = \widehat{\left(\hat{u} + u''\right)^3} = \hat{u}^3 + 3\hat{u}^2\widehat{u''} + 3\hat{u}\widehat{u''^2} + \widehat{u''^3}\,. \tag{2.28}$$

Since $\widehat{u''} = 0$, we have

$$\widehat{u^3} = \hat{u}^3\left(1 + 3\frac{\sigma_u^2}{\hat{u}^2} + \frac{\widehat{u''^3}}{\hat{u}^3}\right). \tag{2.29}$$

The term $\widehat{u''^3}/\hat{u}^3$ is only equal to zero when the probability distribution of u is symmetrical. Nevertheless, for estimating the effect owing to turbulence this term is ignored which leads to

$$\widehat{u^3} = \hat{u}^3\left(1 + 3Ti_u^2\right). \tag{2.30}$$

Ignoring the similar term in Equation (2.27) yields

$$\widehat{u^3} = \hat{u}^3\left(1 + \frac{\sigma_u^2}{\hat{u}^2} + \frac{\sigma_v^2}{\hat{u}^2} + \frac{\sigma_w^2}{\hat{u}^2}\right), \tag{2.31}$$

where $\sigma_v^2 = \widehat{v''^2}$ and $\sigma_w^2 = \widehat{w''^2}$ are the variances with respect to the *y*- and *z*-axis of a Cartesian *coordinate* frame. Thus, only in case of $\sigma_u^2 = \sigma_v^2 = \sigma_w^2$ Equations (2.30) and (2.31) become identical, but such an equality does not generally exist. **Figure 2** shows that the mean and the median of the turbulence intensity depending at the height of 90 m [41]. The observations were performed at the offshore measurement platform FINO1 which is located 45 km north of the island of Borkum in the German Bight. For wind speeds ranging from $3\ \mathrm{m \cdot s^{-1}}$ to $25\ \mathrm{m \cdot s^{-1}}$ the mean and the median of the turbulence intensity are smaller than 0.1. This means that according to Equation (2.30), the effects of the turbulence intensity are smaller than 3 percent. As reported by Türk and Emeis [41], the same is true for this wind speed range at the 30 m height. The effect by turbulence may become more influential in case of aerodynamically rougher landscapes covered, for instance, with vegetation. In case of wind farms the effect by turbulence may considerably increase inside the array of wind turbines [42] [43].

To obtain the local balance equation of MKE, Equation (2.20) has to be scalarly multiplied by $\hat{\boldsymbol{v}}$ leading to

$$\frac{\partial}{\partial t}\left(\overline{\rho}\frac{\hat{\boldsymbol{v}}^2}{2}\right) + \nabla \cdot \left\{\left(\overline{\rho}\frac{\hat{\boldsymbol{v}}^2}{2} + \overline{p}\right)\hat{\boldsymbol{v}} + \hat{\boldsymbol{v}} \cdot \left(\overline{\boldsymbol{J}} + \boldsymbol{F}\right)\right\} = -\overline{\rho}\hat{\boldsymbol{v}} \cdot \nabla\varphi + \overline{p}\nabla \cdot \hat{\boldsymbol{v}} + \left(\overline{\boldsymbol{J}} + \boldsymbol{F}\right) : \nabla\hat{\boldsymbol{v}} \tag{2.32}$$

or

$$\frac{\partial \overline{H}}{\partial t} + \nabla \cdot \left\{\overline{H}\hat{\boldsymbol{v}} + \hat{\boldsymbol{v}} \cdot \left(\overline{\boldsymbol{J}} + \boldsymbol{F}\right)\right\} = \frac{\partial \overline{p}}{\partial t} - \overline{\rho}\hat{\boldsymbol{v}} \cdot \nabla\varphi + \overline{p}\nabla \cdot \hat{\boldsymbol{v}} + \left(\overline{\boldsymbol{J}} + \boldsymbol{F}\right) : \nabla\hat{\boldsymbol{v}} \tag{2.33}$$

with

$$\overline{H} = \overline{\rho}\frac{\hat{\boldsymbol{v}}^2}{2} + \overline{p}. \tag{2.34}$$

The quantity *H* may be considered as the mean total pressure. Subtracting Equation (2.32) from Equation (2.21) yields

$$\frac{\partial}{\partial t}\left(\frac{\overline{\rho}}{2}\widehat{\boldsymbol{v}''^2}\right) + \nabla \cdot \left(\frac{\overline{\rho}}{2}\widehat{\boldsymbol{v}''^2}\,\hat{\boldsymbol{v}} + \frac{1}{2}\overline{\rho\boldsymbol{v}''\boldsymbol{v}''^2} + \overline{\boldsymbol{v}'' \cdot \boldsymbol{J}}\right) = -\overline{\boldsymbol{v}'' \cdot \nabla p} + \overline{\boldsymbol{J} : \nabla\boldsymbol{v}''} - \boldsymbol{F} : \nabla\hat{\boldsymbol{v}} \tag{2.35}$$

or

$$\frac{\partial}{\partial t}\left(\frac{\overline{\rho}}{2}\widehat{\boldsymbol{v}''^2}\right) + \nabla \cdot \left(\frac{\overline{\rho}}{2}\hat{\boldsymbol{v}}\widehat{\boldsymbol{v}''^2} + \frac{1}{2}\overline{\rho\boldsymbol{v}''\boldsymbol{v}''^2} + \overline{\boldsymbol{v}'' \cdot \boldsymbol{J}}\right) = \overline{\boldsymbol{J} : \nabla\boldsymbol{v}''} - \boldsymbol{F} : \nabla\hat{\boldsymbol{v}}\left(1 - S_f\right), \tag{2.36}$$

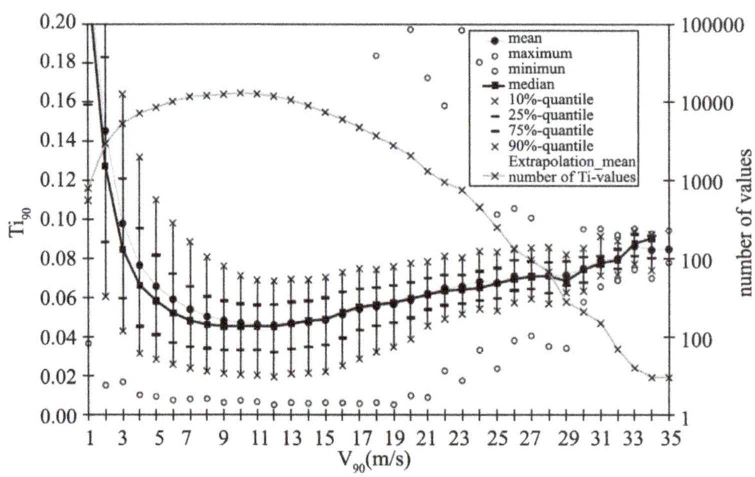

Figure 2. Turbulence intensity depending on wind speed at 90 m height for the period September 2003-August 2007 (taken from Türk and Emeis, [41]). The observations were performed at the offshore measurement platform FINO1 which is located 45 km north of the island of Borkum in the German Bight.

where

$$S_f = \frac{\overline{\boldsymbol{v}'' \cdot \nabla p}}{-\boldsymbol{F} : \nabla \hat{\boldsymbol{v}}} \tag{2.37}$$

is a non-dimensional parameter characterizing the thermal stability of a turbulent flow. This stability parameter expresses the relative importance of the two TKE-terms. It may be interpreted as a generalized Richardson number. The difference between the well-known flux-Richardson number and the generalized Richardson number results from the parameterization of $\overline{\boldsymbol{v}'' \cdot \nabla p}$ [22] [27]. Besides the vertical effects also horizontal effects have to be regarded under certain circumstances. In case of $S_f > 0 \Leftrightarrow \overline{\boldsymbol{v}'' \cdot \nabla p} > 0$, mechanically produced TKE is mainly consumed by Archimedean effects. Consequently, there exists a critical S_f-value given by $S_{f,cr} = 1$. It characterizes that the mechanical gain of TKE is equal to the thermal loss of TKE, *i.e.*, the term $\boldsymbol{F} : \nabla \hat{\boldsymbol{v}} \left(1 - S_f\right)$ becomes equal to zero, and the net production rate of TKE vanishes. As the turbulent dissipation still acts as a sink of TKE, the turbulent flow will become more and more viscous (laminar). In case of $S_f < 0 \Leftrightarrow \overline{\boldsymbol{v}'' \cdot \nabla p} < 0$, TKE is generated mechanically and thermally. If the mechanically generated TKE is much smaller than the thermal gain of TKE, and, hence, negligible, free convective conditions, $S_f \le S_{f,fc}$, will occur. In the remaining range, forced convective conditions may prevail, $S_{f,fc} < S_f < 0$. Thermally neutral stratification is characterized by $S_f = 0$. The 2nd-order balance equation (2.35) is the only balance equation that additionally arises from averaging a macroscopic balance equation (e.g., [18] [23]). In meteorological models, the balance equation of TKE (2.35) serves to derive the eddy diffusivities for momentum and—via the turbulent Prandtl number and the species-dependent turbulent Schmidt numbers—the eddy diffusivities for sensible heat, and water vapor. This method of parameterization is known as one-and-a-half-order closure (e.g. [22] [44] [45]). In the mesoscale model of the National Centers for Environmental Prediction (NCEP) and the Weather Research and Forecasting (WRF) model, it is realized with respect to the level 2.5 of Mellor and Yamada [46]-[48].

The local balance equation for the mean total energy $\widehat{e_{tot}} = \hat{e} + \phi + \frac{1}{2}\left(\hat{\boldsymbol{v}}^2 + \widehat{\boldsymbol{v}''^2}\right)$ can be deduced from **Figure 1** leading to

$$\frac{\partial\left(\bar{\rho}\widehat{e_{tot}}\right)}{\partial t} + \nabla \cdot \left\{\overline{\boldsymbol{R}} + \overline{\boldsymbol{J}_h} + \boldsymbol{F}_h + \hat{\boldsymbol{v}} \cdot \left\{\left(\bar{\rho}\widehat{e_{tot}} + \bar{p}\right)\boldsymbol{E} + \overline{\boldsymbol{J}} + \boldsymbol{F}\right\} + \frac{1}{2}\overline{\rho \boldsymbol{v}'' \boldsymbol{v}''^2} + \overline{\boldsymbol{v}'' \cdot \boldsymbol{J}}\right\} = 0. \tag{2.38}$$

This equation demonstrates that no production or destruction of mean total energy within any given fixed volume exists (e.g., [22] [26] [27] [29]). Obviously, contributions of energy of different orders of magnitude are summed, where only a very small fraction of the total potential energy (or total internal energy), $\hat{e} + \phi$, is available for conversion into kinetic energy (e.g., [22] [49]-[52]).

From the perspective of the generation of electricity by extracting kinetic energy from the wind field, Equations (2.17), (2.20), and (2.32) play the dominant role. To obtain a tractable set of equations, effects caused by molecular and turbulent friction, $\overline{\boldsymbol{J}} + \boldsymbol{F}$, $\hat{\boldsymbol{v}} \cdot \left(\overline{\boldsymbol{J}} + \boldsymbol{F}\right)$, and $\left(\overline{\boldsymbol{J}} + \boldsymbol{F}\right) : \nabla \hat{\boldsymbol{v}}$, are usually ignored. In addition, incompressibility ($\nabla \cdot \hat{\boldsymbol{v}} = 0$) and steady state ($\partial(\ldots)/\partial t = 0$) conditions are presupposed. In doing so, the set of approximated equations reads

$$\nabla \cdot \left(\bar{\rho}\hat{\boldsymbol{v}}\right) = 0, \tag{2.39}$$

$$\nabla \cdot \left(\bar{\rho}\hat{\boldsymbol{v}}\hat{\boldsymbol{v}} + \bar{p}\boldsymbol{E}\right) = -\bar{\rho}\nabla\bar{\varphi} - 2\bar{\rho}\left(\boldsymbol{\Omega} \times \hat{\boldsymbol{v}}\right), \tag{2.40}$$

and

$$\nabla \cdot \left\{\left(\bar{\rho}\frac{\hat{\boldsymbol{v}}^2}{2} + \bar{p}\right)\hat{\boldsymbol{v}}\right\} = -\bar{\rho}\hat{\boldsymbol{v}} \cdot \nabla\varphi. \tag{2.41}$$

2.3. The Bernoulli Equation

Because of the condition of incompressibility, $\nabla \cdot \hat{\boldsymbol{v}} = 0$, Equation (2.41) may also be written as

$$0 = \hat{v} \cdot \nabla\left(\overline{p} + \overline{\rho}\frac{\hat{v}^2}{2}\right) + \overline{\rho}\,\hat{v} \cdot \nabla\varphi = \hat{v} \cdot \nabla\left(\overline{p} + \overline{\rho}\frac{\hat{v}^2}{2} + \overline{\rho\varphi}\right). \quad (2.42)$$

Based on this condition, Bernoulli's equation, which plays an important role in describing the conversion of wind energy, can simply be derived by considering this condition along a streamline. In accord with the natural coordinate frame for streamlines, the Nabla operator reads

$$\nabla(\ldots) = \boldsymbol{t}_s \frac{\partial(\ldots)}{\partial s} + \boldsymbol{n}_s \frac{\partial(\ldots)}{\partial n_s} + \boldsymbol{b}_s \frac{\partial(\ldots)}{\partial b_s}. \quad (2.43)$$

Here, we consider a natural coordinate frame with the unit vectors $\boldsymbol{t}_s$, $\boldsymbol{n}_s$, and $\boldsymbol{b}_s$ that form a right-handed rectangular coordinate system at any given point of a curve in space (moving trihedron) like a trajectory or a streamline (see **Figure 3**), where the subscript s characterizes the streamline-related quantities. The velocity vector at a given point along the streamline is given by $\boldsymbol{v} = V\boldsymbol{t}_s = V\boldsymbol{t}_t$, where V is its magnitude, $\boldsymbol{t}_s$ is the unit tangent of the streamline, and $\boldsymbol{t}_t$ is the unit tangent of the corresponding trajectory. The unit vectors $\boldsymbol{n}_s$ and $\boldsymbol{b}_s$ are the principal normal and the binormal, respectively (e.g., [53] [54]). The different meaning of trajectories and streamlines is explained in the **Appendix A**.

With respect to Equation (2.43), the condition (2.42) results in

$$0 = \hat{V}\boldsymbol{t}_s \cdot \boldsymbol{t}_s \frac{\partial}{\partial s}\left(\overline{p} + \overline{\rho}\frac{\hat{v}^2}{2} + \overline{\rho\varphi}\right) = \hat{V}\frac{\partial}{\partial s}\left(\overline{p} + \overline{\rho}\frac{\hat{v}^2}{2} + \overline{\rho\varphi}\right). \quad (2.44)$$

This means that for any value of $V \neq 0$, the condition

$$\overline{p} + \overline{\rho}\frac{\hat{v}^2}{2} + \overline{\rho\varphi} = \text{const.} \quad (2.45)$$

is fulfilled along a streamline. Equation (2.45) is Bernoulli's equation (e.g., [14] [15] [17] [55] [56]). Even though air density is considered as spatially constant, Bernoulli's equation can often be applied to atmospheric flows. If the streamlines are mainly horizontally oriented and the variation of the gravity potential with height is small like in case of the swept area of a wind turbine, the variation of the gravity effect may be considered as negligible

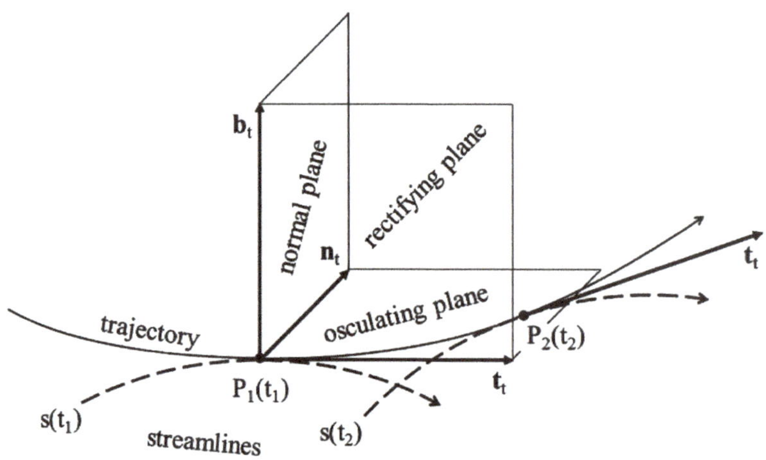

Figure 3. Chronologically ordered streamlines (dashed lines) enveloped by a trajectory (solid line). The trihedron at any point of the trajectory is given by the unit tangent, $\boldsymbol{t}_t$, the principal normal, $\boldsymbol{n}_t$, and the binormal, $\boldsymbol{b}_t$. The $\boldsymbol{t}_t - \boldsymbol{n}_t$ plane is called the osculating plane, the $\boldsymbol{t}_t - \boldsymbol{b}_t$ plane is the rectifying plane, and the $\boldsymbol{n}_t - \boldsymbol{b}_t$ plane is the normal plane (e.g., [53] [54]). The corresponding unit vectors of a streamline are $\boldsymbol{t}_s$, $\boldsymbol{n}_s$, and $\boldsymbol{b}_s$, where at a given point $\boldsymbol{t}_s$ and $\boldsymbol{t}_t$ are identical.

so that Equation (2.45) results in (e.g., [21] [56])

$$\overline{H} = \overline{p} + \overline{\rho}\frac{\hat{\boldsymbol{v}}^2}{2} = \text{const.} \tag{2.46}$$

This approximation of Bernoulli's equation customarily serves as the foundation of, and is used to derive the Rankine-Froude theorem.

2.4. The Integral Equations

The integration of Equations (2.39) to (2.41) over a time-independent control volume, encompassing the rotor of the wind turbine, yields [6] [56]

$$\int_{V_C} \nabla \cdot \left(\overline{\rho}\hat{\boldsymbol{v}}\right) \mathrm{d}V = 0, \tag{2.47}$$

$$\int_{V_C} \nabla \cdot \left(\overline{\rho}\hat{\boldsymbol{v}}\hat{\boldsymbol{v}} + \overline{p}\boldsymbol{E}\right) \mathrm{d}V = -\int_{V_C} \left(\overline{\rho}\nabla\overline{\varphi} + 2\overline{\rho}\left(\boldsymbol{\Omega}\times\hat{\boldsymbol{v}}\right)\right) \mathrm{d}V, \tag{2.48}$$

and

$$\int_{V_C} \nabla \cdot \left\{\left(\overline{\rho}\frac{\hat{\boldsymbol{v}}^2}{2} + \overline{p}\right)\hat{\boldsymbol{v}}\right\} \mathrm{d}V = -\int_{V_C} \left(\overline{\rho}\,\hat{\boldsymbol{v}}\cdot\nabla\varphi\right) \mathrm{d}V. \tag{2.49}$$

In accord with Gauss' integral theorem, Equation (2.47) and the left-hand side of Equation (2.48) can be written as

$$\int_{V_C} \nabla \cdot \left(\overline{\rho}\hat{\boldsymbol{v}}\right) \mathrm{d}V = \int_{A(V_C)} \overline{\rho}\hat{\boldsymbol{v}}\cdot\boldsymbol{n}\mathrm{d}A = 0 \tag{2.50}$$

and

$$\int_{V_C} \nabla \cdot \left(\overline{\rho}\hat{\boldsymbol{v}}\hat{\boldsymbol{v}} + \overline{p}\boldsymbol{E}\right) \mathrm{d}V = \int_{A(V_C)} \left(\overline{\rho}\hat{\boldsymbol{v}}\hat{\boldsymbol{v}} + \overline{p}\boldsymbol{E}\right)\cdot\boldsymbol{n}\mathrm{d}A = \boldsymbol{T}, \tag{2.51}$$

where $\boldsymbol{T}$ is the thrust. Since $\overline{\rho}\hat{\boldsymbol{v}}\hat{\boldsymbol{v}}$ is a second-rank tensor, it is advantageous to scalarly multiply Equation (2.48) by the unit vector $\boldsymbol{e}_x$ from the left to get the more tractable equation

$$\int_{A(V_C)} \overline{\rho}\widehat{u_x}\hat{\boldsymbol{v}}\cdot\boldsymbol{n}\mathrm{d}A + \int_{A(V_C)} \overline{p}\boldsymbol{e}_x\cdot\boldsymbol{n}\mathrm{d}A = T_x = -\int_{V_C} \left(\overline{\rho}\boldsymbol{e}_x\cdot\nabla\overline{\varphi} + 2\overline{\rho}\boldsymbol{e}_x\cdot\left(\boldsymbol{\Omega}\times\hat{\boldsymbol{v}}\right)\right) \mathrm{d}V. \tag{2.52}$$

Here, $T_x = \boldsymbol{e}_x\cdot\boldsymbol{T}$ is the axial force acting on the rotor (e.g., [6]). If we assume that the axial direction coincides with any horizontal direction, the term $\boldsymbol{e}_x\cdot\nabla\overline{\varphi}$ will be nearly equal to zero. Since the Coriolis acceleration is given by $2\boldsymbol{\Omega}\times\hat{\boldsymbol{v}} = 2\Omega\left(\hat{w}\cos\phi - \hat{v}\sin\phi\right)\boldsymbol{i} + 2\hat{u}\Omega\sin\phi\boldsymbol{j} - 2\hat{u}\Omega\cos\phi\boldsymbol{k}$, where ϕ is the latitude, and $\hat{u}$, $\hat{v}$, and $\hat{w}$ are the components of the mean wind vector in west-east direction (characterized by the unit vector $\boldsymbol{i}$), south-north direction (characterized by the unit vector $\boldsymbol{j}$), and the vertical direction (characterized by the unit vector $\boldsymbol{k}$), respectively; the term $2\left|\boldsymbol{e}_x\cdot\left(\boldsymbol{\Omega}\times\hat{\boldsymbol{v}}\right)\right| = 2\Omega\left|\left(\hat{w}\cos\phi - \hat{v}\sin\phi\right)\boldsymbol{e}_x\cdot\boldsymbol{i} + \hat{u}\sin\phi\boldsymbol{e}_x\cdot\boldsymbol{j}\right|$ is very small for any wind speed smaller than the cut-out wind speed because $\Omega = 7.27\times10^{-5}\ \mathrm{rad}\cdot\mathrm{s}^{-1}$. Thus, this term is negligible, and Equation (2.52) may be approximated by

$$\int_{A(V_C)} \overline{\rho}\widehat{u_x}\hat{\boldsymbol{v}}\cdot\boldsymbol{n}\mathrm{d}A + \int_{A(V_C)} \overline{p}\boldsymbol{e}_x\cdot\boldsymbol{n}\mathrm{d}A = T_x. \tag{2.53}$$

The second term of the left-hand side of this equation is usually ignored in the blade element momentum (BEM) theory. However, this term is not zero [6] [10] [57].

The velocity vector may be expressed by $\hat{\boldsymbol{v}} = \widehat{v_x}\boldsymbol{e}_x + \widehat{v_r}\boldsymbol{e}_r + \widehat{v_\theta}\boldsymbol{e}_\theta$, where $\widehat{v_x}$, $\widehat{v_r}$, and $\widehat{v_\theta} = \hat{\omega}r$ are the cylindrical polar coordinates, respectively; and $\boldsymbol{e}_x$, $\boldsymbol{e}_r$, and $\boldsymbol{e}_\theta$ are the corresponding unit vectors pointing in axial, radial, and azimuthal direction. The azimuthal velocity component acting on the rotor at a certain radius r causes a torque given by

$$\int_{A(V_C)} \overline{\rho} r \widehat{v_\theta} \hat{\boldsymbol{v}} \cdot \boldsymbol{n} \mathrm{d}A = Q . \tag{2.54}$$

In accord with Gauss' integral theorem, the left-hand side of Equation (2.49)reads

$$\int_{V_C} \nabla \cdot \left\{ \left(\overline{\rho} \frac{\hat{\boldsymbol{v}}^2}{2} + \overline{p} \right) \hat{\boldsymbol{v}} \right\} \mathrm{d}V = \int_{A(V_C)} \left(\overline{\rho} \frac{\hat{\boldsymbol{v}}^2}{2} + \overline{p} \right) \hat{\boldsymbol{v}} \cdot \boldsymbol{n} \mathrm{d}A = P . \tag{2.55}$$

This term represents the power extracted by the rotor of the wind turbine. In case of a quasi-horizontal flow, the right-hand side of Equation (2.49) can be neglected because $\nabla \varphi$ is quasi-perpendicular to $\hat{\boldsymbol{v}}$. The effect of the gravity potential was already considered as negligible in Bernoulli's Equation (2.45). The integral relation (2.55) underlines the importance of Bernoulli's equation in wind power studies.

Rearranging the left-hand side of Equation (2.49) yields

$$P = \int_{V_C} \nabla \cdot \left\{ \left(\overline{\rho} \frac{\hat{\boldsymbol{v}}^2}{2} + \overline{p} \right) \hat{\boldsymbol{v}} \right\} \mathrm{d}V = \int_{V_C} \nabla \cdot \left(\overline{\rho} \frac{\hat{\boldsymbol{v}}^2}{2} \hat{\boldsymbol{v}} \right) \mathrm{d}V + \int_{V_C} \nabla \cdot \left(\overline{p} \hat{\boldsymbol{v}} \right) \mathrm{d}V . \tag{2.56}$$

Because of $\nabla \cdot \hat{\boldsymbol{v}} = 0$, the divergence term $\nabla \cdot \left(\overline{p} \hat{\boldsymbol{v}} \right)$ can be expressed by $\nabla \cdot \left(\overline{p} \hat{\boldsymbol{v}} \right) = \hat{\boldsymbol{v}} \cdot \nabla \overline{p}$ leading to

$$P = \int_{V_C} \nabla \cdot \left\{ \left(\overline{\rho} \frac{\hat{\boldsymbol{v}}^2}{2} + \overline{p} \right) \hat{\boldsymbol{v}} \right\} \mathrm{d}V = \int_{V_C} \nabla \cdot \left(\overline{\rho} \frac{\hat{\boldsymbol{v}}^2}{2} \hat{\boldsymbol{v}} \right) \mathrm{d}V + \int_{V_C} \hat{\boldsymbol{v}} \cdot \nabla \overline{p} \mathrm{d}V . \tag{2.57}$$

Obviously, the first term on the right-hand side of this equation is missing in that of Gorban' *et al.* [1], repeated here by Equation (1.4). This means that the filtration Equation (1.2) that leads to Equation (1.4) is meritless in determining the maximum efficiency of propeller-type wind turbines. Thus, the argument of van Kuik *et al.* [9] seems to be justified by Equation (2.57).

3. Wind Turbine Characteristics

3.1. The Axial Momentum Theory

3.1.1. The Rankine-Froude Theorem

In the following, we assume a pure axial flow (one-dimensional problem), *i.e.*, the undisturbed wind speed far upstream of the wind turbine, $\widehat{\boldsymbol{v}_\infty}$, the wind speed at the rotor area, $\widehat{\boldsymbol{v}_R}$, and the undisturbed wind speed far downstream of the wind turbine, $\widehat{\boldsymbol{v}_w}$, have the same direction so that we may consider only the magnitude of these wind vectors expressed by $\widehat{v_\infty}$, $\widehat{v_R}$, and $\widehat{v_w}$, respectively. Doing so agrees with the so-called stream-tube model sketched in **Figure 4**, in which an "actuator disk" is representing the axial load on a rotor (e.g., [58]). This axial momentum theory was developed by Rankine [59], W. Froude [60], and R.E. Froude [61].

To derive the Rankine-Froude theorem we consider the variation of wind speed and pressure by approaching and leaving the rotor area as sketched in **Figure 4**, part A. In accord with Bernoulli's equation in its approximated form (see Equation (2.46)), the former can be expressed by

$$\overline{H_\infty} = \overline{p_\infty} + \overline{\rho} \frac{\widehat{v_\infty}^2}{2} = \overline{p_R} + \overline{\rho} \frac{\widehat{v_R}^2}{2} . \tag{3.1}$$

Whereas the latter is given by

$$\overline{H_w} = \overline{p_R} - \delta p + \overline{\rho} \frac{\widehat{v_R}^2}{2} = \overline{p_w} + \overline{\rho} \frac{\widehat{v_w}^2}{2} . \tag{3.2}$$

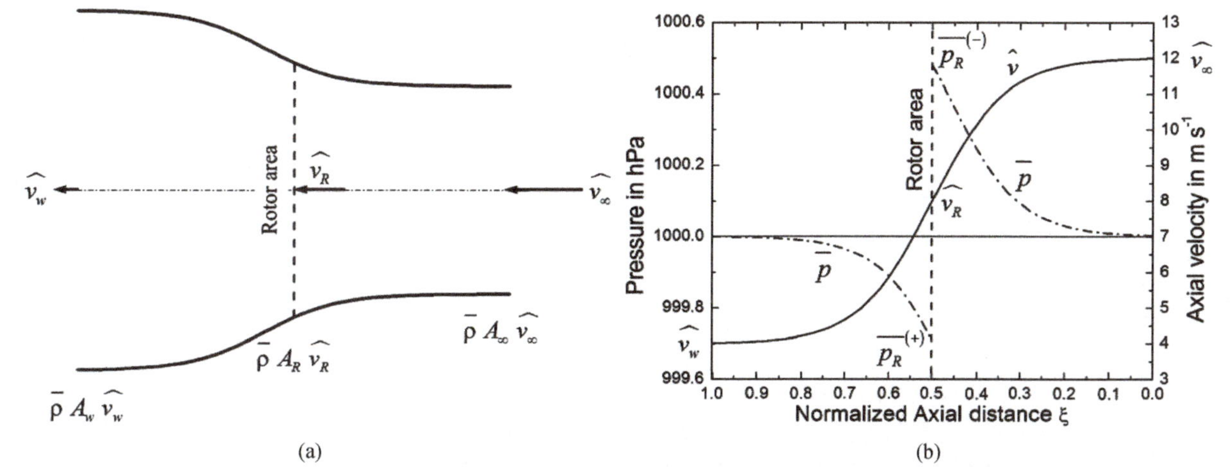

Figure 4. (a) Sketch of the stream-tube model; (b) Wind speed and pressure variations by approaching and leaving the rotor area (with respect to Betz [58]). The stream-tube model is based on the equation of continuity expressed by Equation (3.7), where the mean axial velocity is approximated by a sigmoidal function $\hat{v} = \widehat{v_\infty} - \left(\widehat{v_\infty} - \widehat{v_w}\right) / \left(1 + \exp\left((\xi_0 - \xi)/\delta\xi\right)\right)$ to guarantee that $\widehat{v_R} = 0.5\left(\widehat{v_\infty} + \widehat{v_w}\right)$. Note that $\overline{p_w} = \overline{p_\infty} = 1000\ \text{hPa}$, $\xi_0 = 0.5$, and $\delta\xi = 0.0833$ have been chosen.

Here, $\overline{p_\infty}$ is the static air pressure far upstream of the wind turbine, $\overline{p_w}$ the static air pressure far downstream of the wind turbine, and $\overline{p_R}$ and $\overline{p_R} - \delta p$ are the static air pressures directly in front and directly behind the rotor area, respectively. Thus, the jump in the Bernoulli constant, $\Delta H = H_\infty - H_w$, caused by the wind turbine is given by

$$\Delta H = \delta p = \overline{p_\infty} - \overline{p_w} + \frac{1}{2}\overline{\rho}\left(\widehat{v_\infty}^2 - \widehat{v_w}^2\right). \tag{3.3}$$

Assuming that $\overline{p_\infty} = \overline{p_w}$ yields

$$\Delta H = \delta p = \frac{1}{2}\overline{\rho}\left(\widehat{v_\infty}^2 - \widehat{v_w}^2\right) = \frac{1}{2}\overline{\rho}\left(\widehat{v_\infty} + \widehat{v_w}\right)\left(\widehat{v_\infty} - \widehat{v_w}\right). \tag{3.4}$$

The thrust force acting on the rotor is then given by (the subscript x that occurs in Equations (2.52) and (2.53) is ignored in this section because a pure axial flow is presupposed so that $\boldsymbol{n} = \boldsymbol{e}_x$)

$$T = A_R \Delta \overline{p_R} = A_R \frac{1}{2}\overline{\rho}\left(\widehat{v_\infty} + \widehat{v_w}\right)\left(\widehat{v_\infty} - \widehat{v_w}\right). \tag{3.5}$$

On the other hand, the thrust force experienced by the rotor can also be expressed by

$$T = \overline{\rho} A_\infty \widehat{v_\infty}^2 - \overline{\rho} A_w \widehat{v_w}^2. \tag{3.6}$$

According to **Figure 4**, the equation of continuity (as outlined by Equation (2.47)) can be expressed by

$$\overline{\rho} A_\infty \widehat{v_\infty} = \overline{\rho} A_R \widehat{v_R} = \overline{\rho} A_w \widehat{v_w} = \text{const.}, \tag{3.7}$$

i.e., the mass flow rate through the wind turbine is $\overline{\rho} A_R \widehat{v_R} = \text{const.}$ With Equation (3.7), the thrust force (see Equation (3.6)) may be written as

$$T = \overline{\rho} A_\infty \widehat{v_\infty}^2 - \overline{\rho} A_w \widehat{v_w}^2 = \underbrace{\overline{\rho} A_\infty \widehat{v_\infty}}_{\overline{\rho} A_R \widehat{v_R}} \widehat{v_\infty} - \underbrace{\overline{\rho} A_w \widehat{v_w}}_{\overline{\rho} A_R \widehat{v_R}} \widehat{v_w} = \overline{\rho} A_R \widehat{v_R}\left(\widehat{v_\infty} - \widehat{v_w}\right). \tag{3.8}$$

Thus, combining Equations (3.5) and (3.8) provides

$$T = \frac{1}{2} A_R \overline{\rho}\left(\widehat{v_\infty} + \widehat{v_w}\right)\left(\widehat{v_\infty} - \widehat{v_w}\right) = \overline{\rho} A_R \widehat{v_R}\left(\widehat{v_\infty} - \widehat{v_w}\right). \tag{3.9}$$

Rearranging yields

$$\overline{\rho} A_R \widehat{v_R}\left(\widehat{v_\infty}-\widehat{v_w}\right)=\frac{1}{2} A_R \overline{\rho}\left(\widehat{v_\infty}+\widehat{v_w}\right)\left(\widehat{v_\infty}-\widehat{v_w}\right) \tag{3.10}$$

or

$$\widehat{v_R}=\frac{1}{2}\left(\widehat{v_\infty}+\widehat{v_w}\right), \tag{3.11}$$

i.e., the axial velocity at the rotor disk corresponds to the arithmetic mean of the axial velocities far upstream and far downstream of the wind turbine. Equation (3.11) is the Rankine-Froude theorem (e.g., [10] [40] [58] [62]-[64]).

3.1.2. The Betz-Joukowsky Limit

According to Equation (2.55), the total wind power of the undisturbed wind field far upstream of the wind turbine is given by

$$P_\infty=\frac{1}{2} \overline{\rho} A_\infty \widehat{v_\infty}^3+A_\infty \overline{p_\infty} \widehat{v_\infty} \tag{3.12}$$

and that of the undisturbed wind field far downstream of the wind turbine is given by

$$P_w=\frac{1}{2} \overline{\rho} A_w \widehat{v_w}^3+A_w \overline{p_w} \widehat{v_w} . \tag{3.13}$$

Again, we assume that $\overline{p_\infty}=\overline{p_w}$. Thus, the power extracted by the wind turbine is given by

$$\begin{aligned} P=P_\infty-P_w &=\frac{1}{2} \overline{\rho} A_\infty \widehat{v_\infty}^3+A_\infty \overline{p_\infty} \widehat{v_\infty}-\frac{1}{2} \overline{\rho} A_w \widehat{v_w}^3-A_w \overline{p_w} \widehat{v_w} \\ &=\frac{1}{2}\left(\overline{\rho} A_\infty \widehat{v_\infty}^3-\overline{\rho} A_w \widehat{v_w}^3\right)+\overline{p_\infty}\left(A_\infty \widehat{v_\infty}-A_w \widehat{v_w}\right) \\ &=\frac{1}{2}\left(\overline{\rho} A_R \widehat{v_R} \widehat{v_\infty}^2-\overline{\rho} A_R \widehat{v_R} \widehat{v_w}^2\right)+\frac{\overline{p_\infty}}{\overline{\rho}}\left(\overline{\rho} A_R \widehat{v_R}-\overline{\rho} A_R \widehat{v_R}\right) \\ &=\frac{1}{2} \overline{\rho} A_R \widehat{v_R}\left(\widehat{v_\infty}^2-\widehat{v_w}^2\right)=\frac{1}{2} \overline{\rho} A_R \widehat{v_R}\left(\widehat{v_\infty}+\widehat{v_w}\right)\left(\widehat{v_\infty}-\widehat{v_w}\right) \end{aligned} \tag{3.14}$$

Inserting Equation (3.11) into Equation (3.14) yields

$$P=\frac{1}{4} \overline{\rho} A_R\left(\widehat{v_\infty}+\widehat{v_w}\right)\left(\widehat{v_\infty}^2-\widehat{v_w}^2\right) \tag{3.15}$$

or

$$P=\frac{1}{4} \overline{\rho} A_R \widehat{v_\infty}^3\left(1+\frac{\widehat{v_w}}{\widehat{v_\infty}}\right)\left(1-\left(\frac{\widehat{v_w}}{\widehat{v_\infty}}\right)^2\right). \tag{3.16}$$

Defining the power efficiency by $C_P=P / P_\infty$ leads to

$$C_P=\frac{P}{P_\infty}=\frac{1}{2}(1+X)\left(1-X^2\right), \tag{3.17}$$

where $X=\widehat{v_w} / \widehat{v_\infty}$ and $P_\infty=1/2\, A_R \overline{\rho} \widehat{v_\infty}^3$. Trivially, $X=0$ leads to $C_p=1/2$, and for $X=1$ we obtain $C_p=0$. To determine the maximum of C_P, we have to consider the first derivative test, $\mathrm{d}C_P / \mathrm{d}X=0$, and the second derivative test, $\mathrm{d}^2 C_P / \mathrm{d}X^2<0$. The first derivative test leads to $X=\widehat{v_w} / \widehat{v_\infty}=1/3$, for which the second derivative becomes negative, *i.e.*, for $X=1/3$, the wind power efficiency reaches its maximum (see **Figure 5**). Inserting $X=1/3$ into Equation (3.17) yields

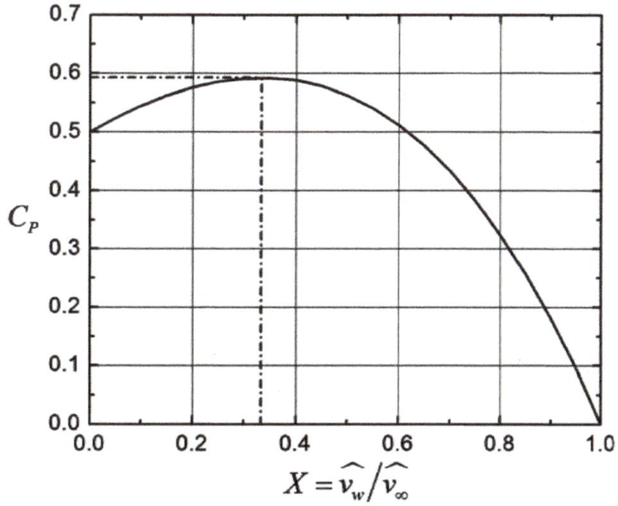

Figure 5. The Betz-Joukowsky limit. The solid line represents Equation (3.17) and the dash-dotted lines characterize the maximum of the power efficiency (with respect to Betz [58]).

$$C_P = \frac{16}{27} \cong 0.593 \,. \tag{3.18}$$

According to Betz [3], and Joukowsky [4], this value is the maximum wind power efficiency (see also [6] [11] [40] [58] [62]-[64]).

Sometimes, the axial interference factor, a, defined by (e.g., [6] [11] [12] [62] [63])

$$a = \frac{\widehat{v_\infty} - \widehat{v_R}}{\widehat{v_\infty}} = 1 - \frac{\widehat{v_R}}{\widehat{v_\infty}}, \tag{3.19}$$

is inserted. Using this factor leads to

$$\widehat{v_R} = \widehat{v_\infty}(1 - a) \tag{3.20}$$

and

$$\widehat{v_w} = \widehat{v_\infty}(1 - 2\,a) = \widehat{v_\infty}(1 - b) \tag{3.21}$$

with $b = 2a$. The axial interference factor measures the impact of the wind turbine on the air flow. In accord with the definition of this factor, the wind power efficiency and the thrust force can be expressed by

$$C_P = 4a(1 - a)^2 \tag{3.22}$$

and

$$T = \overline{\rho} A_R \widehat{v_R}\left(\widehat{v_\infty} - \widehat{v_w}\right) = 4a(1 - a)\frac{1}{2}\overline{\rho} A_R \widehat{v_\infty}^2 \,. \tag{3.23}$$

The latter may be used to define the thrust coefficient, C_T, by (e.g., [6] [11])

$$C_T = \frac{T}{\frac{1}{2}\overline{\rho} A_R \widehat{v_\infty}^2} = 4a(1 - a) \,. \tag{3.24}$$

Thus, we have $C_P = (1 - a)C_T$.

3.2. General Momentum Theory

The result of the Betz-Jowkowsky limit is based on simplified description of the flow field. Even though the flow field exhibits a pure axial behavior in front of the rotor, the exertion of a torque on the rotor disk by the air passing through it causes an equal, but opposite torque to be imposed on the air. Because of this reaction torque, the air starts to rotate in a direction opposite to that of the rotor; the air gains angular momentum and so in the wake of the rotor disk the air particles have a velocity component in a direction which is tangential to the rotation as well as having an axial velocity component [65]. Since the stream tube is opening behind the propeller, there is also a velocity component in the radial direction. Thus, by interacting with the rotor also velocity components in radial and azimuthal directions occur. The velocity vector at the rotor may be expressed by cylindrical polar co-ordinates; the velocity vector in the wake behind the rotor may be expressed in a similar manner.

To consider these rotational effects, Glauert [10] developed a simple model for the optimum rotor. In his approach, the rotor is a rotating axisymmetric actuator disk, corresponding to a rotor with an infinite number of blades [6] [11] [12].

As outlined in **Appendix B**, the general equations of the General Momentum Theory lead to (see Equation (B.24))

$$\frac{1}{2}\left(\widehat{v_w} - \widehat{v_\infty}\right)^2 = \left(\frac{\Omega_R + \frac{1}{2}\omega_w}{\widehat{v_w}} - \frac{\Omega_R + \frac{1}{2}\omega}{\widehat{v_x}}\right)\widehat{v_w}\omega_w r_w^2 , \tag{3.25}$$

where $\widehat{v_\infty}$ is, again, the undisturbed wind speed far upstream of the wind turbine, Ω_R is the angular velocity of the rotor, $\widehat{v_x}$ is the axial velocity through the propeller disk, $\widehat{\omega}$ is the angular velocity imparted to the slipstream, $\widehat{v_w}$ the axial velocity in the final wake, and ω_w the corresponding angular velocity at radial distance r_w from the axis of the slipstream. Equation (3.25) already derived by Glauert [10] for an engine-driven propeller and by Wilson and Lissaman [62] for propeller-type wind turbines suffice to determine the relationship between the thrust and torque of the propeller and the flow in the slipstream. Owing to the complexity of the equations, however, it is customary to adopt certain approximations based on the fact that the rotational velocity in the slipstream is generally very small.

3.2.1. Joukowsky's Constant Circulation Model

An exact solution of the general equations of the General Momentum Theory can be obtained when the flow in the slipstream is irrotational except along the axis [10]. This condition implies that the rotational momentum ωr^2 has the same value *k* for all radial elements, *i.e.*,

$$\omega_w r_w^2 = \omega r^2 = k . \tag{3.26}$$

Here, *r* is the radial distance of any annular element of the propeller disk. Equation (3.26) is the basis for Joukowsky's constant circulation model [10] [65].

On the basis of Equation (B.19) of **Appendix B**,

$$\frac{1}{2}\frac{\mathrm{d}}{\mathrm{d}r_w}\left(\widehat{v_\infty}^2 - \widehat{v_w}^2\right) = \left(\Omega_R + \omega_w\right)\frac{\mathrm{d}}{\mathrm{d}r_w}\left(\omega_w r_w^2\right), \tag{3.27}$$

we can deduce that the axial velocity $\widehat{v_w}$ is constant across the wake because $\mathrm{d}\left(\omega_w r_w^2\right)/\mathrm{d}r_w = 0$ and, hence, $\mathrm{d}\widehat{v_w}/\mathrm{d}r_w = 0$. Furthermore, Equation (3.25) is satisfied by a constant value of the axial velocity $\widehat{v_x}$ across the propeller disk. If $\widehat{v_w}$ and $\widehat{v_x}$ are constant, we will obtain from the equation of continuity (see Equation (B.1) of **Appendix B**),

$$\widehat{v_w} r_w \mathrm{d}r_w = \widehat{v_x} r \mathrm{d}r = \text{const.} , \tag{3.28}$$

and the conservation of angular momentum (see Equation (B.6) of **Appendix B**),

$$\omega_w r_w^2 = \omega r^2 , \tag{3.29}$$

the following relationship [10]

$$\frac{\widehat{v_w}}{\widehat{v_x}} = \frac{\omega_w}{\omega} = \frac{r^2}{r_w^2} = \frac{R^2}{R_w^2}. \quad (3.30)$$

In accord with Equation (3.30), Equation (3.25) becomes

$$\frac{1}{2}\left(\widehat{v_w} - \widehat{v_\infty}\right)^2 = \frac{\widehat{v_x} - \widehat{v_w}}{\widehat{v_x}} \Omega_R k. \quad (3.31)$$

If we assume again that $\Delta p = \overline{p_\infty} - \overline{p_w} = 0$, we will obtain (see Equation (B.14) of **Appendix B**)

$$\frac{1}{2}\left(\widehat{v_w}^2 - \widehat{v_\infty}^2\right) = -\left(\Omega_R + \frac{1}{2}\omega_w\right)k. \quad (3.32)$$

Using the definitions $\widehat{v_x} = \widehat{v_\infty}(1-a)$ and $\widehat{v_w} = \widehat{v_\infty}(1-b)$ yields (see Equations (C.10) and (C.20) of **Appendix C**)

$$\frac{1}{2}b^2\widehat{v_\infty}^2 = \frac{b-a}{1-a}\Omega_R k \quad (3.33)$$

and

$$a = \frac{1}{2}b\left(1 - \frac{(1-a)b^2}{4\lambda^2(b-a)}\right), \quad (3.34)$$

where

$$\lambda = \frac{\Omega_R R}{\widehat{v_\infty}} \quad (3.35)$$

is the tip speed ratio. Thus, $\widehat{v_x} \geq \left(\widehat{v_\infty} + \widehat{v_w}\right)/2$, *i.e.*, Equation (3.11) is not generally valid. Formula (3.34) was already derived by Wilson and Lissaman [62] for a propeller-type wind turbine; a similar formula was given by Glauert [10] for an engine-driven propeller. Obviously, Equation (3.34) can only be solved iteratively. Results of such a solution are shown in **Figure 6**. As illustrated, for tip speed ratios in the range of $0 < \lambda < 0.25$, the axial interference factor, *a*, becomes negative. These results have to be discarded because they disagree with observations. For tip speed ratios $\lambda \geq 2$, the condition $b = 2a$, derived in the matter of the axial momentum theory, is nearly fulfilled [10] [62]. From Equation (3.34) we can infer that the condition $b = 2a$ is exactly fulfilled if λ becomes infinite.

In accord with Equation (2.54), the torque, $\mathrm{d}Q$, experienced by this annular stream tube element between r and $r + \mathrm{d}r$ reads

$$\mathrm{d}Q = \rho\widehat{v_x}\widehat{\omega} r^2 \mathrm{d}A = 2\pi\rho\widehat{v_\infty}(1-a)kr\mathrm{d}r, \quad (3.36)$$

where the area of the stream tube element is $\mathrm{d}A = 2\pi r\mathrm{d}r$, and k is given by Equation (3.26). Since the power caused by the rotor is the product of the angular velocity and this annulus torque, *i.e.*, $\mathrm{d}P = \Omega_R \mathrm{d}Q$, the integration over the total blade span provides

$$P = \pi\rho\widehat{v_\infty}(1-a)\Omega_R k R^2. \quad (3.37)$$

Replacing $\Omega_R k$ with the aid of Equation (3.33) yields

$$P = \frac{1}{2}\pi R^2 \rho \widehat{v_\infty}^3 \frac{b^2(1-a)^2}{b-a}. \quad (3.38)$$

Thus, in contrast to the axial momentum theory, the wind power efficiency is given by [60] [62]

$$C_P = \frac{P}{\frac{1}{2}\pi R^2 \rho \widehat{v_\infty}^3} = \frac{b^2(1-a)^2}{b-a}. \quad (3.39)$$

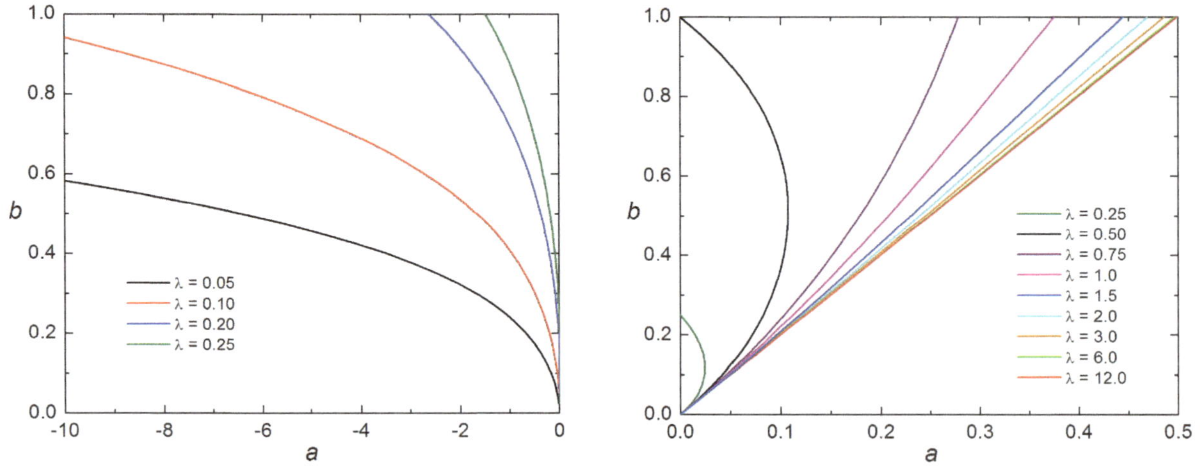

Figure 6. Effect of the tip speed ratio λ, defined by Equation (3.35), on the induced velocities for flow with an irrotational wake. The diagram on the right side is based on Figure 3.3 of Wilson and Lissaman [62].

Inserting $b = 2a$ into this formula provides Equation (3.22). This means that the power efficiency for the irrotational wake tends to that for the axial momentum theory (see Equation (3.22)) if the tip speed ratio exceeds $\lambda = 2$ [62] [65]. Since the ratio a/b varies with λ (**Figure 6**), different tip speed ratios provide different curves for the power efficiency. The maximum power efficiency that is close to the Betz-Joukowsky limit of 0.593 occurs around $b = 2/3$ if the tip speed ratio exceeds $\lambda = 2$ (cf. **Figure 7**). Thus, for $X = \widehat{v_w}/\widehat{v_\infty} = 1 - b$ we obtain $X = 1/3$. This is the value for which the Betz-Joukowsky limit was determined (see **Figure 5** and **Figure 7**). Consequently, in case of an irrotational wake, the axial momentum theory provides reasonable results for tip speed ratios larger than $\lambda = 2$.

Glauert [10] already argued:

The condition of constant circulation k along the blade, which has been the basis of the preceding calculations, cannot be fully realized in practice since it implies that near the roots of the blades the angular velocity imparted to the air is greater than the angular velocity of the propeller itself. In any practical application of the analysis it is therefore necessary to assume that the effective part of the propeller blades commences at a radial distance not less than $\sqrt{k/\Omega_R}$ *at which* $\widehat{\omega}$ *is equal to* Ω_R.

It implies that, near the roots of the blades, the angular velocity imparted to the air is greater than the angular velocity of the propeller itself [6]. Wilson & Lissaman [62] and de Vries [40] shared Glauert's viewpoint that the solution is unphysical as it results in infinite values of power and circulation when the tip-speed ratio tends to zero.

From Equations (3.30) and (3.33) we can derive

$$\frac{k}{\Omega_R} = \frac{1}{2}\left(\frac{bR_w}{\lambda}\right)^2 \frac{1}{\left(\frac{R_w}{R}\right)^2 - 1} \tag{3.40}$$

or

$$\left(\frac{R_w}{R}\right)^2 = \frac{v_x}{v_w} = \frac{1-a}{1-b} = 1 + \frac{1}{2}\frac{\Omega_R}{k}\left(\frac{bR_w}{\lambda}\right)^2. \tag{3.41}$$

The maximum values of the power efficiency for various tip speed ratios are also illustrated in **Figure 8**. This diagram shows that for $\lambda < 2$ the maximum power efficiency notably exceeds the Betz-Joukowsky limit. However, these results must be assessed with care. For $\lambda = 0.5$, for instance, we obtain $C_{P,\max} = 1$ that occurs at $b = 1$. According to **Figure 6**, for $b = 1$ the axial interference factor amounts to $a = 0$. Consequently, $\widehat{v_x} = \widehat{v_\infty}(1-a) = \widehat{v_\infty}$ and $\widehat{v_w} = \widehat{v_\infty}(1-b) = 0$. These results seem to be unlikely because $\widehat{v_x} = \widehat{v_\infty}$ would only

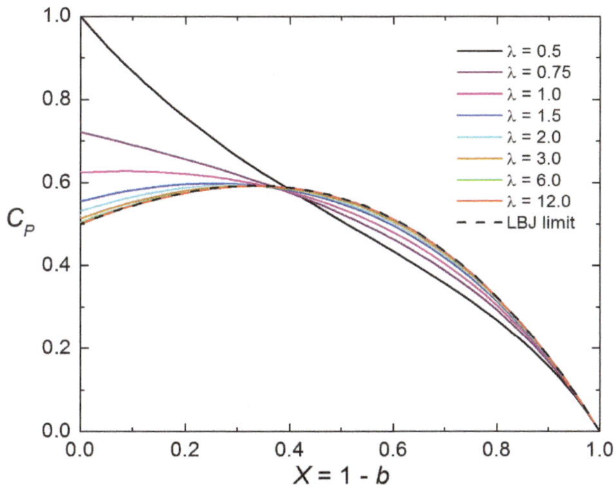

Figure 7. Effect of the tip speed ratio λ, defined by Equation (3.35), on the power efficiency for a flow with an irrotational wake.

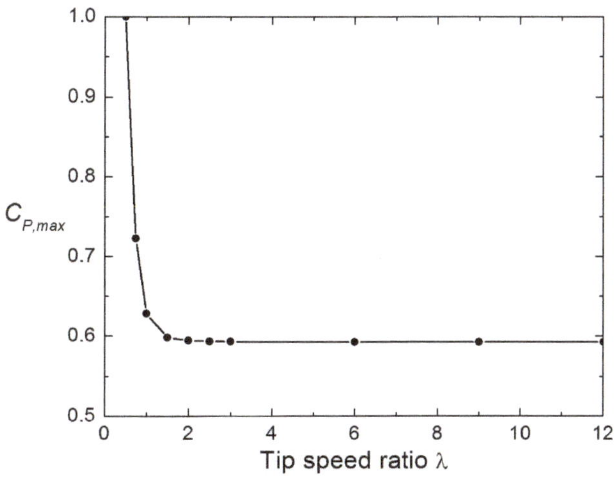

Figure 8. Maximum power efficiency, $C_{P,max}$, taken from **Figure 7** versus tip speed ratio λ defined by Equation (3.35).

be adequate in case of no wind turbine and $\widehat{v_w} = 0$ would require, in accord with Equation (3.30), that the radius of the wake, R_w, must tend to infinity.

3.2.2. Glauert's Optimum Rotor

For wind turbines, Glauert [10] derived an approximate solution on the basis of Equation (3.25). The angular velocity ω imparted to the slipstream is, in general, very small compared with the angular velocity Ω_R of the rotor. Therefore, it is possible to simplify the general equations by neglecting certain terms involving ω^2. Because of this simplification the pressure p_w in the wake becomes equal to the initial pressure p_∞ of the fluid, and the decrease of static pressure across the propeller disk is equal to the decrease of total pressure head, *i.e.*, $\delta p = H_\infty - H_w = \Delta H$. The relationships connecting the thrust and axial velocity are then the same as in the simple axial momentum theory, the axial velocity $\widehat{v_x}$ at the propeller disk is the arithmetic mean of the axial velocity $\widehat{v_w}$ and the slipstream velocity $\widehat{v_w}$. Thus, in accord with Equation (B.20) of **Appendix B**, the element of thrust becomes

$$\mathrm{d}T = \rho\widehat{v_w}\left(\widehat{v_\infty} - \widehat{v_w}\right)\mathrm{d}A_w = 2\rho\widehat{v_x}\left(\widehat{v_\infty} - \widehat{v_x}\right)\mathrm{d}A = 4\pi\rho\widehat{v_x}\left(\widehat{v_\infty} - \widehat{v_x}\right)r\mathrm{d}r\,. \tag{3.42}$$

Now, the torque experienced by this annular stream tube element is given by

$$\mathrm{d}Q = \rho\widehat{v_x}\widehat{\omega}r^2\mathrm{d}A = 2\pi\rho\widehat{v_x}\widehat{\omega}r^3\mathrm{d}r\,. \tag{3.43}$$

Inserting $\widehat{v_x} = \widehat{v_\infty}(1-a)$, λ and $a' = \widehat{\omega}/(2\Omega_R)$ into these equation yields [10]

$$\mathrm{d}T = 4\pi a(1-a)\overline{\rho}\widehat{v_\infty}^2 r\mathrm{d}r \tag{3.44}$$

and

$$\mathrm{d}Q = 4\pi\Omega_R a'(1-a)\overline{\rho}\widehat{v_\infty}r^3\mathrm{d}r\,. \tag{3.45}$$

Since the related power is given by $\mathrm{d}P = \Omega_R\mathrm{d}Q$, the integration over the total blade span provides [10] [62]

$$P = \int_0^R \Omega_R\mathrm{d}Q = \int_0^R 4\pi\Omega_R^2 a'(1-a)\overline{\rho}\widehat{v_\infty}r^3\mathrm{d}r\,. \tag{3.46}$$

Defining $\chi = \Omega r/\widehat{v_\infty}$ leads to

$$P = \frac{1}{2}A_R\overline{\rho}\widehat{v_\infty}^3\,\frac{8}{\lambda^2}\int_0^\lambda a'(1-a)\chi^3\mathrm{d}\chi\,. \tag{3.47}$$

Thus, the power efficiency is given by [10] [62]

$$C_P = \frac{P}{\frac{1}{2}A_R\overline{\rho}\widehat{v_\infty}^3} = \frac{8}{\lambda^2}\int_0^\lambda a'(1-a)\chi^3\mathrm{d}\chi\,. \tag{3.48}$$

Alternatively, defining $x = r/R$ provides [6] [56]

$$C_P = 8\lambda^2\int_0^1 a'(1-a)x^3\mathrm{d}x\,. \tag{3.49}$$

This formula is equivalent to Equation (3.48). Obviously, the power efficiency strongly depends the tip-speed ratio, but weighted by the integral expression. Unfortunately, Equations (3.48) and (3.49) contain the two unknowns a and a'. Thus, we need additional information for determining C_P.

The pressure increment at the propeller disk is given by [62]

$$\mathrm{d}T = \delta p\mathrm{d}A = \rho\left(\Omega_R + \frac{1}{2}\widehat{\omega}\right)\widehat{\omega}r^2\mathrm{d}A = 4\pi\rho\Omega_R^2(1+a')a'r^3\mathrm{d}r\,. \tag{3.50}$$

From Equations (3.44) and (3.50) we obtain

$$\chi^2 a'(1+a') = a(1-a)\,. \tag{3.51}$$

To obtain the maximum power for a given tip-speed ratio λ, the factors a and a' must be related by [10]

$$(1-a)\frac{\mathrm{d}a'}{\mathrm{d}a} = a' \tag{3.52}$$

and

$$\chi^2(1+2a')\frac{\mathrm{d}a'}{\mathrm{d}a} = 1-2a\,. \tag{3.53}$$

Thus, combining Equations (3.51) to (3.53) provides

$$a' = \frac{1-3a}{4a-1} \tag{3.54}$$

and

$$a'\chi^2 = (1-a)(4a-1). \tag{3.55}$$

The quantities a', $a'\chi^2$, and χ as a function of a are illustrated in **Figure 9**. In case of $a = 0.333$, we have a large value of $\chi = 8.57$, while the azimuthal interference factor, $a' = 0.003$, is very small. The opposite is true in the case of $a = 0.253$. The quantities χ and a' amount to $\chi = 0.021$ and $a' = 20.1$, respectively. Inserting these values of the interference factors into Equation (3.48) provides the power efficiency of the wind turbine. The relationship between C_p and the tip-speed ratio λ is illustrated in **Figure 10**. Obviously, the maximum power efficiency depends on the tip-speed ratio. It approaches the Betz-Joukowsky limit at large tip-speed ratio only [6] [11] [62].

3.3. Finite-Bladed Rotor Models

In case of finite-bladed rotor Equations (3.42) and (3.43) are imprecise. Based on the vortex theory, each of the rotor blades has to be replaced by a lifting line on which the radial distribution of bound vorticity is represented

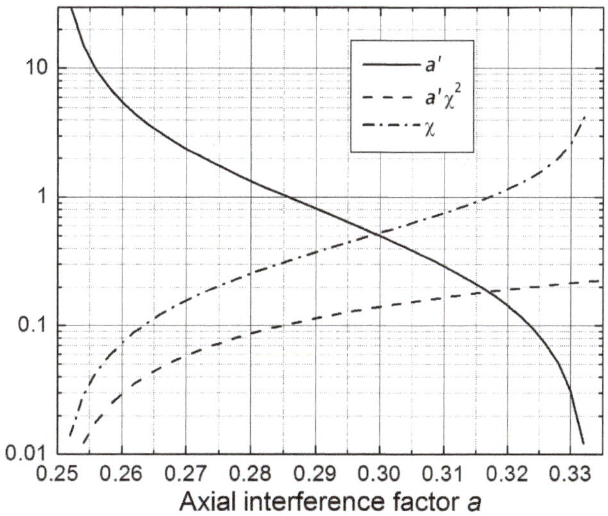

Figure 9. The quantities a', $a'\chi^2$, and χ versus the axial interference factor a.

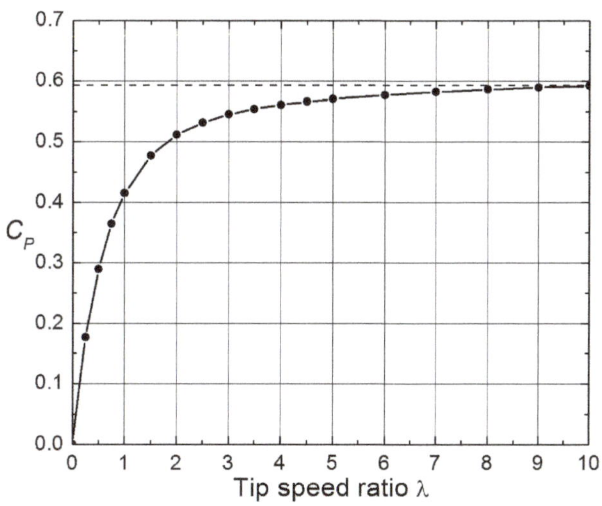

Figure 10. Power coefficient, C_P, vs. tip-speed ratio, λ for Glauert's [10] optimum actuator disk.

by the circulation $\Gamma = \Gamma(r)$ depending on the radial distance along the blade [11] [12]. This results in a free vortex system consisting of helical trailing vortices, as sketched in **Figure 11**. With respect to the vortex theory, the bound vorticity serves to produce the local lift on the blades while the trailing vortices induce the velocity field in the rotor plane and the wake [11] [12]. The velocity vector in the rotor plane is made up by the rotor angular velocity, Ω_R, the undisturbed wind speed, $\widehat{v_\infty}$, the axial and circumferential velocity components, u_{z_R}, and u_{θ_R}, respectively. These velocity components induced at a blade element in the rotor plane by the tip vortices (see **Figure 12**). Another circumferential velocity, v_{θ_R} is induced by the hub vortex (see **Figure 12**). To determine the velocity field given by v_{θ_R}, u_{z_R}, and u_{θ_R} induced at a blade element in the rotor plane, the free half-infinite helical vortex system behind the rotor is replaced by 'an associated vortex system' that extends to infinity in both directions [12]. Neglecting deformations or changes in the wake, the vortex system is uniquely described by the far wake properties in the Trefftz plane [66]. It is defined as the plane normal to the relative wind far downstream of the rotor. In accordance with Helmholtz' vortex theorem, the bound circulation Γ around a blade element is uniquely related to the circulation of a corresponding vortex in the Trefftz plane [12]. By symmetry, the induced velocities at a point in the rotor plane equals half the induced velocity at a corresponding point in the Trefftz plane [67]-[70], *i.e.*, $v_{\theta_R} = v_\theta/2$, $u_{z_R} = u_z/2$, and $u_{\theta_R} = u_\theta/2$ (see also [12]).

Okulov and Sørensen [12] distinguished between two different concepts that dominated the conceptual interpretation of the optimum rotor: (a) Joukowsky [67] defined the optimum rotor as one having constant circulation along the blades, such that the vortex system for an N_b-bladed rotor consists of N_b helical tip vortices of strength Γ and an axial hub vortex of strength $-N_b\Gamma$. A simplified model of this vortex system can be obtained by representing it as a rotating horseshoe vortex (**Figure 11(a)**). Betz and Prandtl [71] argued that optimum

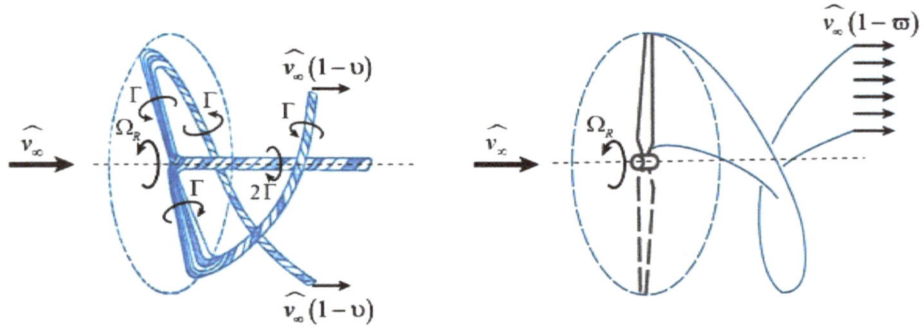

(a) (b)

Figure 11. Sketch of the vortex system corresponding to lifting line theory of the ideal propeller of (a) Joukowsky and (b) Betz (from Okulov and Sørensen [12], but with respect to Sørensen [6]).

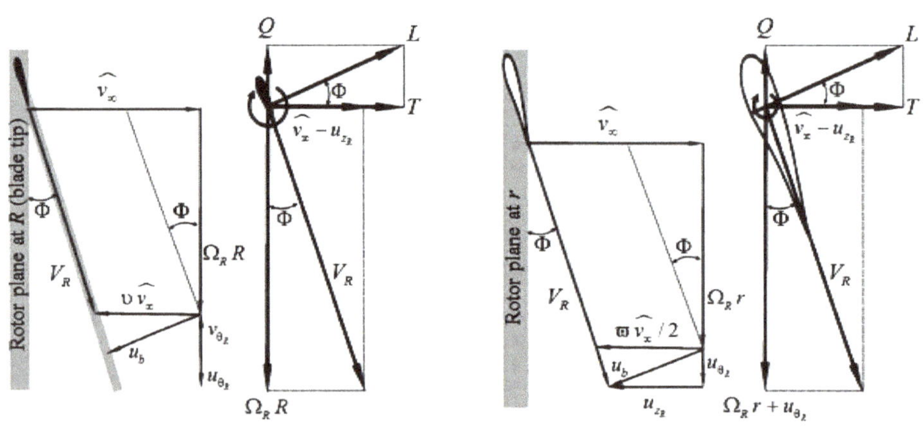

(a) (b)

Figure 12. Velocity and power triangles in the rotor plane of (a) Joukowsky rotor and (b) Betz rotor (adopted from Okulov and Sørensen [12], but some symbols have changed to fit the text).

efficiency is obtained when the distribution of circulation along the blades generates a rigidly helicoidal wake that moves in the direction of its axis with a constant velocity. Betz used a vortex model of the rotating blades based on the lifting-line technique of Prandtl in which the vortex strength varies along the wing-span (**Figure 11(b)**). This distribution, usually referred to as the Goldstein circulation function, is rather complex and difficult to determine accurately [12] [72].

Using the Kutta-Joukowsky-theorem

$$\mathrm{d}\boldsymbol{L} = \rho \boldsymbol{V}_R \times \boldsymbol{\Gamma} \mathrm{d}r , \tag{3.56}$$

where $\mathrm{d}\boldsymbol{L}$ is the lift force on a blade element of radial dimension $\mathrm{d}r$, V_R is the resultant relative velocity and $\boldsymbol{\Gamma}$ is the bound circulation, Okulov and Sørensen [11] deduced the local thrust and the local torque of a rotor blade given by

$$\mathrm{d}T = \bar{\rho}\Gamma\left(\Omega_R r + u_{\theta_R}\right) r\mathrm{d}r \tag{3.57}$$

and

$$\mathrm{d}Q = \bar{\rho}\Gamma\left(\widehat{v_\infty} - u_{z_R}\right) r\mathrm{d}r . \tag{3.58}$$

Here, we only discuss the torque. Since the related power is given by $\mathrm{d}P = \Omega_R \mathrm{d}Q$, the integration over the total blade span provides for N_b blades yields

$$P = \bar{\rho} N_B \Omega_R \int_0^R \Gamma\left(\widehat{v_\infty} - u_{z_R}\right) r\mathrm{d}r . \tag{3.59}$$

Using the analytical solution to the induction of helical vortex filaments developed by Okulov [73], Okulov & Sørensen [11] extended Goldstein's [72] original formulation by a simple modification to handle heavily loaded rotors in accord with the general momentum theory. Assuming that the induction in the rotor plane equals half the induction in the Trefftz plane in the far wake, as described before, they found for the power efficiency

$$C_P = 2\varpi\left(1 - \frac{\varpi}{2}\right)\left(I_1 - \frac{\varpi}{2} I_3\right). \tag{3.60}$$

Here, ϖ is the dimensionless translational velocity of the vortex sheet (see **Figure 11(b)**)

$$I_1 = 2\int_0^1 G(x, l) x\mathrm{d}x \tag{3.61}$$

and

$$I_3 = 2\int_0^1 G(x,l) \frac{x^3}{x^2 + l^2} \mathrm{d}x , \tag{3.62}$$

where $G(x,l) = N_b \Gamma(x,l)/(h\varpi)$ is the Goldstein [72] circulation function, $x = r/R$, $l = h/(2\pi R)$, $h = 2\pi r \tan\Phi$ is the pitch of the vortex sheet, and Φ is the angle between the vortex sheet and the rotor plane. Thus, l may be expressed by $l = x\tan\Phi$. The first derivative test yields

$$\varpi = \frac{2}{3I_3}\left(I_1 + I_3 - \sqrt{I_1^2 - I_1 I_3 + I_3^2}\right). \tag{3.63}$$

The result of the second derivative test shows that Equation (3.63) characterizes the maximum of C_P. As pointed out by Okulov and Sørensen [11], for a rotor with infinitely many blades, both functions, I_1 and I_3, tend to unity when the pitch tends to zero. In this case, Equation (3.60) degenerates to the expression $C_P = 2\varpi(1-\varpi/2)^2$. This result is completely consistent with the axial momentum theory because $\varpi = 2a$ leads to Equation (3.22). Furthermore, Equation (3.63) provides $\varpi = 2/3$ and, hence, $a = 1/3$. This result is in agreement with $X = \widehat{v_w}/\widehat{v_\infty} / = 1 - 2a = 1/3$ that designates the maximum of C_P in **Figure 5**.

In the vortex theory of the Joukowsky rotor [67]-[70], each of the blades is replaced by a lifting line about which the circulation associated with the bound vorticity is constant, resulting in a free vortex system consisting of helical vortices trailing from the tips of the blades and a rectilinear hub vortex [12]. As sketched in **Figure 11(a)**, the vortex system may be interpreted as consisting of rotating horseshoe vortices with cores of finite size, where the radius of the core is ε. The "associated vortex system" consists of a multiplet of helical tip vortices of finite vortex cores ($\varepsilon \ll R$) with constant pitch h and circulation Γ. The multiplet moves downwind (in case of a propeller) or upwind (in case of a wind turbine) with a constant velocity $\widehat{v_\infty}(1 \pm \upsilon)$ in the axial direction, where υ is the difference between the wind speed and axial translational velocity of the vortices [12]. Using the analytical solution to the induction of helical vortex filaments developed by Okulov [73] again, Okulov & Sørensen [12] derived for the power efficiency

$$C_P = 2a\left(1 - \frac{1}{2}aJ_1\right)\left(1 - \frac{1}{2}aJ_3\right) \tag{3.64}$$

where $J_1 = 1 + \sigma$ and

$$J_3 = 2\int_0^1 \widetilde{u_z}(x, 0)\,x\mathrm{d}x\,. \tag{3.65}$$

Here, $\sigma = \varepsilon/R$ is the non-dimensional radius of the vortex core, and $\widetilde{u_z}(x, 0)$ is a non-dimensional axial velocity. For a given helicoidal wake structure, the power coefficient is seen to be uniquely determined, except for the parameter a. The first derivative test yields

$$a = \frac{2}{3J_1J_3}\left(J_1 + J_3 - \sqrt{J_1^2 - J_1J_3 + J_3^2}\right). \tag{3.66}$$

The result of the second derivative test shows that for this value of a characterizes the maximum of the power efficiency.

Figure 13 illustrates that the power coefficient computed by Okulov and Sørensen [11] [12] for various number of blades depends on the tip-speed ratio λ given by Equation (3.35). Also shown is the result provided by Glauert's [10] optimum actuator disk. Obviously, the optimum power coefficient of the Joukowsky rotor for all number of blades is larger than that for the Betz rotor, but the efficiency of the Betz rotor is larger if we compare it for the same deceleration of the wind speed [12]. The difference, however, vanishes for $\lambda \to \infty$ or for $N_b \to \infty$, where in both models tend towards the Betz-Joukowsky limit [12].

Since neither the axial interference factor a nor λ explicitly occurs in Equations (3.60) and (3.63), λ and a have to be connected to the helical pitch l and the generic parameter w by [12]

$$\lambda = \frac{\Omega_R R}{\widehat{v_\infty}} = \frac{R}{l}\left(1 - \frac{w}{2}\right) \tag{3.67}$$

and

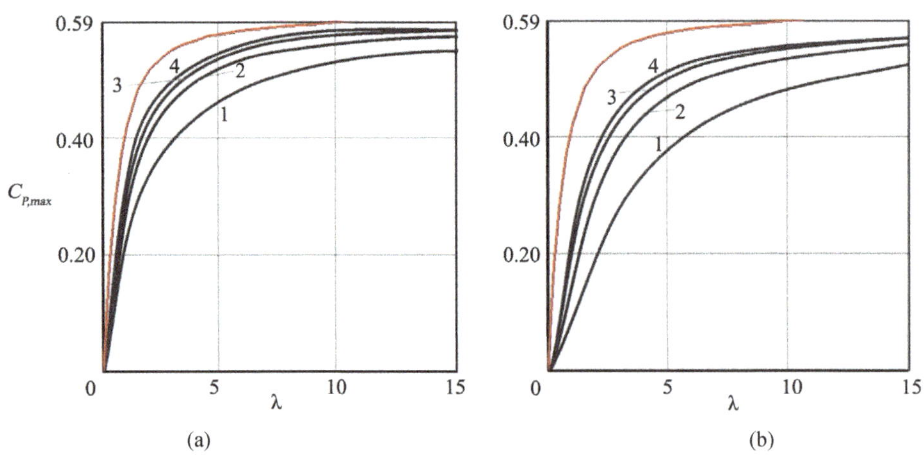

Figure 13. Power coefficients, C_P, of an optimum rotor as a function of tip speed ratio and number of blades. (a) Joukowsky rotor and (b) Betz rotor (adopted from Okulov and Sørensen [12]). The red lines are added. They illustrate the solution of Glauert's optimum actuator disk shown in **Figure 10**.

$$a = w\int_0^1 G(x, l)\,\mathrm{d}x\,, \tag{3.68}$$

In case of the Joukowsky rotor the tip-speed ratio can be expressed by [12]

$$\lambda = \frac{\Omega_R R}{\widehat{v_\infty}} = \frac{R}{l}\left(1 - \frac{a}{2}\left(1 + \frac{\varepsilon}{R}\right)\right). \tag{3.69}$$

3.4. The Efficiency of Real Wind Turbines

Figure 14 shows the power curves of seven wind turbines of different rated power listed in **Table 1**. The power curves were determined by considering the listed values (only the Enercon machines) or by taking discrete values from the power curves illustrated in the actual brochures found at the manufacturers' websites. Based on these discrete values, the parameters *A*, *K*, *Q*, *B*, *M*, and *u* of the generalized logistic function (e.g., [56])

$$P(v) = A + \frac{K - A}{\left(1 + Q\exp\{-B(v - M)\}\right)^{\frac{1}{u}}}. \tag{3.70}$$

were numerically determined for each of the seven wind turbines (see **Table 2**). The function $P(v)$ represents the power generated by the corresponding wind turbine at the wind speed *v*.

Figure 15 illustrates the wind power density, $\left|\overline{\mathbf{S}_{kin}}\right|$, for (a) a flow far upstream to the wind turbine given Equation (2.25) with $\rho = 1.2\ \mathrm{kg\cdot m^{-3}}$, (b) an "ideal" wind turbine generating wind power by obeying the Betz-

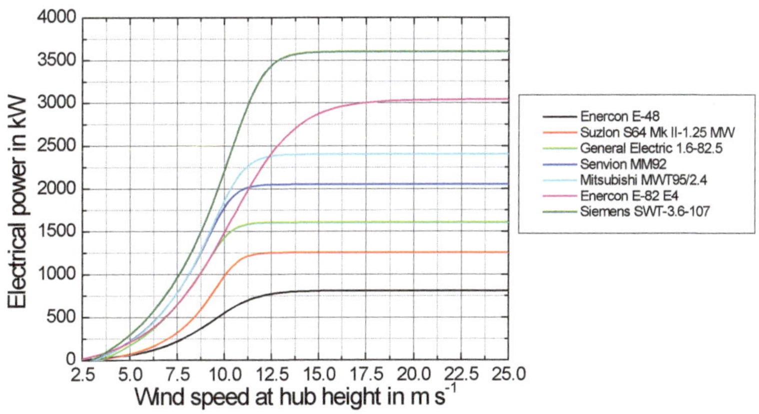

Figure 14. Wind power density of seven wind turbines of different rated power considered in this study. They are based on the parameters of the generalized logistic function (see Equation (3.70)) listed in **Table 2**.

Table 1. Specifications of the wind turbines considered in this study.

Wind turbine	Hub height (m)	Swept area (m^2)	Cut-in wind speed (m·s^{-1})	Rated wind speed (m·s^{-1})	Cut-out wind speed (m·s^{-1})	Rated power (kW)	Wind Class
Enercon E-48	76	1810	2-3	13.5	25	800	IEC IIa
Suzlon S64 Mark II-1.25 MW	74.5	3217	4	12.0	25	1250	IIa
General Electric 1.6 - 82.5	80	5345	3.5	11.5	25	1600	IEC IIIb
Senvion MM92	78 - 80	6720	3	12.5	24	2050	IEC IIa
Mitsubishi MWT95/2.4	80	7088	3	12.5	25	2400	IEC IIa
Enercon E-82 E4	78/84	5281	2-3	16	25	3000	IEC IIa
Siemens SWT-3.6 - 107	80	9000	3	14.0	25	3600	IEC Ia

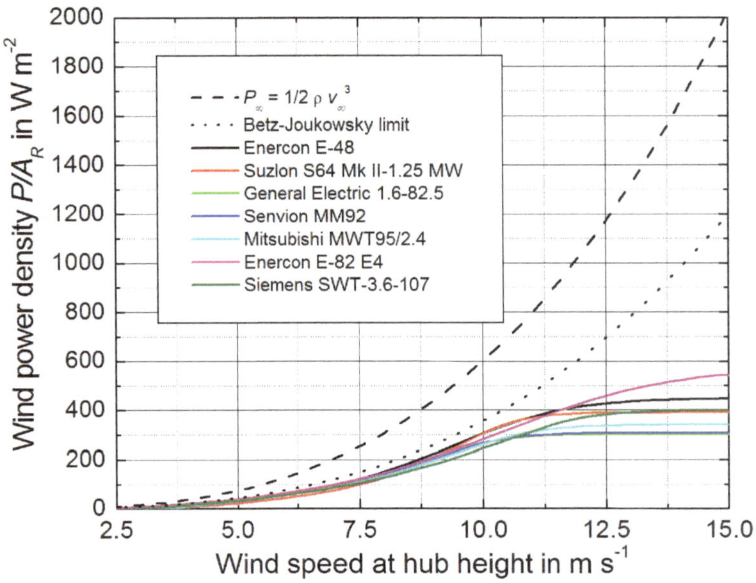

Figure 15. Wind power densities of the seven wind turbines listed in **Table 1**. Also shown are the wind power density $|S_{kin}|$ given by Equation (2.25), and $|S_{kin}|$ weighted by the Betz-Joukowsky limit.

Table 2. Parameters *A*, *K*, *Q*, *B*, *M*, and *u* of the generalized logistic function (3.70) used to model the wind turbines' power curves.

Wind turbine	*A*	*K*	*Q*	*B*	*M*	*u*
Enercon E-48	−24.9	811.2	0.54	1.0	10.9	2.3
Suzlon S64 Mark II-1.25 MW	−56.5	1250.6	3.88	2.0	9.6	4.5
General Electric 1.6 - 82.5	−315.7	1601.3	1.66	2.0	9.8	7.2
Senvion MM92	−267.6	2050.4	19.5	1.9	8.5	6.2
Mitsubishi MWT95/2.4	−270.4	2403.3	12.2	1.5	8.8	4.9
Enercon E-82 E4	−113.8	3038.8	1.49	0.6	10.6	1.7
Siemens SWT-3.6 - 107	−414.3	3599.6	40.0	1.4	9.0	5.4

Joukowsky limit of $C_P = 0.593$, and (c) the power curves of the seven wind turbines shown in **Figure 14**, but normalized by the corresponding swept areas. As shown in **Figure 15**, the wind power densities are lower than the Betz-Joukowsky limit, but follow it up to a wind speed at hub height of $10\ \mathrm{m \cdot s^{-1}}$ or so. Beyond this wind speed, they approach plateau values at the rated wind speeds; these plateau values are ranging from $300\ \mathrm{W \cdot m^{-2}}$ (General Electric 1.6-82.5) to $575\ \mathrm{W \cdot m^{-2}}$ (Enercon E-82 E4). The two other wind power densities continuously increase because of the $\hat{v}^3$-law. Thus, the power efficiencies of all wind turbines considered here are notably higher than the 30-percent limit of Gorban' *et al.* [1] for wind speeds between $6\ \mathrm{m \cdot s^{-1}}$ and $11\ \mathrm{m \cdot s^{-1}}$ (see **Figure 16**). In case of wind speeds higher than $11\ \mathrm{m \cdot s^{-1}}$, the power efficiencies eventually fall below this 30-percent limit and approach very low values in the vicinity of the cut-out wind speed.

4. Summary and Conclusions

We demonstrated that the filtration equation used by Gorban' *et al.* [1] for determining the maximum efficiency of plane propellers at about 30 percent for free fluids plays no role in describing the flows in the ABL and has to be discarded. The ABL is mainly governed by turbulent motion, even though the effect of the turbulence intensity

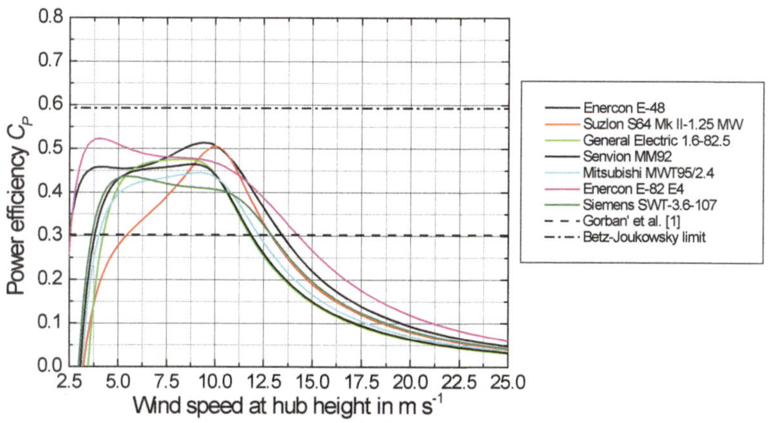

Figure 16. Power efficiencies of the seven wind turbines listed in **Table 1** (in accord with Enercon's product overview updated in September 2012).

is relatively small in the undisturbed wind field over water surfaces. This effect may become more influential in case of aerodynamically rougher landscapes covered, for instance, with vegetation canopies. In case of wind farms the effect by turbulence may considerably increase inside the array of wind turbines. Based on Equation (2.57), we showed that the criticism of van Kuik *et al.* [9] regarding the work of Gorban' *et al.* [1] is quite justified.

We also demonstrate that the stream tube model customarily applied to derive the Rankine-Froude theorem must be corrected in the sense of Glauert to provide an appropriate value for the axial velocity at the rotor area. Including this correction leads to the Betz-Joukowsky limit, namely of a maximum efficiency of 59.3 percent.

We also assessed Joukowsky's constant circulation model that leads to values of the maximum efficiency exceeding the Betz-Jowkowsky limit for very low tip speed ratios. Some of these values, however, have to be rejected because of physical reasons.

Using Glauert's [10] optimum actuator disk, and the results of the blade-element analysis by Okulov and Sørensen [11] [12] we illustrated that the maximum efficiency of propeller-type wind turbines depends on tip-speed ratio and the number of blades.

Finally, we showed that the power efficiencies of seven wind turbines of different rated power are notably higher than 30-percent limit of Gorban' *et al.* [1] for wind speeds between $6\,\mathrm{m}\cdot\mathrm{s}^{-1}$ and $11\,\mathrm{m}\cdot\mathrm{s}^{-1}$. In case of wind speeds higher than $11\,\mathrm{m}\cdot\mathrm{s}^{-1}$, the power efficiencies eventually fall below this 30-percent limit and approach very low values in the vicinity of the cut-out wind speed.

Acknowledgements

We would like to express much gratitude to the Alaska Department of Labor for funding Dr. Gary Sellhorst's project work. We would like to extend gratitude to the National Science Foundation for funding the project work of Hannah K. Ross and John Cooney in summer 2012 through the Research Experience for Undergraduates (REU) Program, grant number AGS1005265. We also express our thanks to the Max Planck Institute for Chemistry for the current financial support for Dr. Dr. habil. Ralph Dlugi.

References

[1] Gorban', A.N., Gorlov, A.M. and Silantyev, V.M. (2001) Limits of the Turbine Efficiency for Free Fluid Flow. *Journal of Energy Resources Technology*, **123**, 311-317. http://dx.doi.org/10.1115/1.1414137

[2] Okulov, V.L. and van Kuik, G.A.M. (2011) The Betz-Joukowsky Limit: On the Contribution to Rotor Aerodynamics by the British, German and Russian Scientific Schools. *Wind Energy*, **15**, 355-344. http://dx.doi.org/10.1002/we.464

[3] Betz, A. (1920) Das Maximum der theoretisch möglichen Ausnützung des Windes durch Windmotoren. *Zeitschrift für das gesamte Turbinenwesen*, **26**, 307-309. (In German)

[4] Joukowsky, N.E. (1920) Windmill of the NEJ Type. *Transactions of the Central Institute for Aero-Hydrodynamics of Moscow*. (In Russian) (as cited by [5])

[5] van Kuik, G.A.M. (2007) The Lanchester-Betz-Joukowsky Limit. *Wind Energy*, **10**, 289-291. http://dx.doi.org/10.1002/we.218

[6] Sørensen, J.N. (2011) Aerodynamic Aspects of Wind Energy Conversion. *Annual Review of Fluid Mechanics*, **43**, 427-448. http://dx.doi.org/10.1146/annurev-fluid-122109-160801

[7] Hartwanger, D. and Horvat, A. (2008) 3D Modelling of Wind Turbine Using CFD. *NAFEMS UK Conference* 2008 "*Engineering Simulation: Effective Use and Best Practice*", Cheltenham, 10-11 June 2008, 14 p.

[8] Blackledge, J., Coyle, E. and Kearney, D. (2011) A Stochastic Model for Wind Turbine Power Quality Using a Levy Index Analysis of Wind Velocity Data. *The Third International Conference on Resource Intensive Applications and Services*, Venice, 22-27 May 2011.

[9] van Kuik, G.A.M., Sørensen, J.N. and Okulov, V.L. (2015) Rotor Theories by Professor Joukowsky: Momentum Theories. *Progress in Aerospace Sciences*, **73**, 1-18. http://dx.doi.org/10.1016/j.paerosci.2014.10.001

[10] Glauert, H. (1935) Airplane Propellers. In: Durand, W.F., Ed., *Aerodynamic Theory, Vol. IV, Division L*, Springer, New York, 169-360. http://dx.doi.org/10.1007/978-3-642-91487-4_3

[11] Okulov, V.L. and Sørensen, J.N. (2008) Refined Betz Limit for Rotors with a Finite Number of Blades. *Wind Energy*, **11**, 415-426. http://dx.doi.org/10.1002/we.274

[12] Okulov, V.L. and Sørensen, J.N. (2010) Maximum Efficiency of Wind Turbine Rotors Using Joukowsky and Betz Approaches. *Journal of Fluid Mechanics*, **649**, 497-508. http://dx.doi.org/10.1017/S0022112010000509

[13] de Groot, S.R. and Mazur, P. (1969) Non-Equilibrium Thermodynamics. North-Holland Publishing Comp., Amsterdam/ London.

[14] Landau, L.D. and Lifschitz, E.M. (1981) Lehrbuch der theoretischen Physik-Hydrodynamik. Akademie-Verlag Berlin. (In German)

[15] Budó, A. (1990) Theoretische Mechanik. In: Rompe, R. and Schmutzer, E., Eds., *Hochschulbücher für Physik*, VEB Deutscher Verlag der Wissenschaften, Berlin, 615. (In German)

[16] Mölders, N. and Kramm, G. (2014) Lectures in Meteorology. Springer International Publishing. http://dx.doi.org/10.1007/978-3-319-02144-7

[17] Prandtl, L. (1905) Über Flüssigkeitsbewegung bei sehr kleiner Reibung. Verhandlungen des III. *Internationalen Mathematiker Kongresses*, Heidelberg, 8-13 August 1904, B. G. Teubner, Leipzig, 485-491. (In German)

[18] Rotta, J.C. (1972) Turbulente Strömungen. B. G. Teubner, Stuttgart. (In German) http://dx.doi.org/10.1007/978-3-322-91206-0

[19] Sanderse, B., van der Pijl, S.P. and Koren, B. (2011) Review of Computational Fluid Dynamics for Wind Turbine Wake Aerodynamics. *Wind Energy*, **14**, 799-819. http://dx.doi.org/10.1002/we.458

[20] Reynolds, O. (1895) On the Dynamical Theory of Incompressible Viscous Fluids and the Determination of the Criterion. *Philosophical Transactions of the Royal Society of London*, **186**, 123-164. http://dx.doi.org/10.1098/rsta.1895.0004

[21] Prandtl, L. (1918) Tragflügeltheorie—I. Mitteilung. Nachrichten der K. Gesellschaft zu Göttingen, Mathematisch-Physikalische Klasse, 451-477.

[22] Kramm, G. and Meixner, F.X. (2000) On the Dispersion of Trace Species in the Atmospheric Boundary Layer: A Re-formulation of the Governing Equations for the Turbulent Flow of the Compressible Atmosphere. *Tellus*, **52A**, 500-522. http://dx.doi.org/10.1034/j.1600-0870.2000.00984.x

[23] Montgomery, R.B. (1954) Convection of Heat. *Archiv für Meteorologie,Geophysik und Bioklimatologie*, **A7**, 125-132. http://dx.doi.org/10.1007/BF02277911

[24] Fortak, H. (1969) Zur Energetik der planetarischen Grenzschicht. *Annalen der Meteorologie* (NF) **4**, 157-162. (In German)

[25] van Mieghem, J. (1949) Les equations générales de la mécanique et de l'énergétique des milieux turbulents en vue des applications à la météorologie. *Inst. R. Météor. Belgique., Mém.*, **XXXIV**, 60. (In French)

[26] van Mieghem, J. (1973)Atmospheric Energetics. Clarendon Press, Oxford, UK.

[27] Herbert, F. (1975) Irreversible Prozesse der Atmosphäre—3. Teil (Phänomenologische Theorie mikroturbulenter Systeme). *Beiträge zur Physik der Atmosphäre*, **48**, 1-29. (In German)

[28] Libby, P.A. and Williams, F.A., Eds. (1980) Turbulent Reacting Flows. Springer-Verlag, Berlin. http://dx.doi.org/10.1007/3-540-10192-6

[29] Pichler, H. (1984) Dynamic der Atmosphäre. Bibliographisches Institut, Zürich. (In German)

[30] Cox, G. (1995) Basic considerations. In: Cox, G., Ed., *Combustion Fundamentals of Fire*, Academic Press, London, San Diego, New York, 3-30.

[31] Kramm, G., Dlugi, R. and Lenschow, D.H. (1995) A Re-evaluation of the Webb-Correction Using Density-Weighted Averages. *Journal of Hydrology*, **166**, 283-292. http://dx.doi.org/10.1016/0022-1694(94)05088-F

[32] Thomson, D. (1995) The Parameterization of the Vertical Dispersion of a Scalar in the Atmospheric Boundary Layer. *Atmospheric Environment*, **29**, 1343. http://dx.doi.org/10.1016/1352-2310(94)00326-G

[33] Venkatram, A. (1998) Response. *Atmospheric Environment*, **32**, 259. http://dx.doi.org/10.1016/S1352-2310(96)00078-7

[34] Kowalski, A.S. (2012) Exact Averaging of Atmospheric State and Flow Variables. *Journal of the Atmospheric Sciences*, **69**, 1750-1757. http://dx.doi.org/10.1175/JAS-D-11-0299.1

[35] Hesselberg, T. (1926) Die Gesetze der ausgeglichenen atmosphärischen Bewegungen. *Beiträge zur Physik der freien Atmosphäre*, **12**, 141-160. (In German)

[36] Lumley, J.L. and Yaglom, A.M. (2001) A Century of Turbulence. *Flow, Turbulence and Combustion*, **66**, 241-286. http://dx.doi.org/10.1023/A:1012437421667

[37] Herbert, F. (1995) A Re-evaluation of the Webb Correction Using Density-Weighted Averages—Comment. *Journal of Hydrology*, **173**, 343-344. http://dx.doi.org/10.1016/0022-1694(95)02867-o

[38] Eliassen, A. and Kleinschmidt Jr., E. (1957) Dynamic Meteorology. In: Flügge, S., Ed., *Handbuch der Physik*, Bd. XLVIII. Springer-Verlag Berlin/Heidelberg/New York, 1-154. http://dx.doi.org/10.1007/978-3-642-45881-1_1

[39] Dutton, J.A. (1995) Dynamics of Atmospheric Motion. Dover, New York.

[40] de Vries, O. (1979) Fluid Dynamic Aspects of Wind Energy Conversion. AGARDograph 243, AGARD, Brussels.

[41] Türk, M. and Emeis, S. (2010) The Dependence of Offshore Turbulence Intensity on Wind Speed. *Journal of Wind Engineering and Industrial Aerodynamics*, **98**, 466-471. http://dx.doi.org/10.1016/j.jweia.2010.02.005

[42] Barthelmie, R.J., Frandsen, S.T., Nielsen, M.N., Pryor, S.C., Rethore, P.-E. and Jørgensen, H.E. (2007) Modelling and Measurements of Power Losses and Turbulence Intensity in Wind Turbine Wakes at Middelgrunden Offshore Wind Farm. *Wind Energy*, **10**, 517-528. http://dx.doi.org/10.1002/we.238

[43] Barthelmie, R.J., Churchfield, M.J., Moriarty, P.J., Lundquist, J.K., Oxley, G.S., Hahn, S. and Pryor, S.C. (2015) The Role of Atmospheric Stability/Turbulence on Wakes at the Egmond aan Zee Offshore Wind Farm. *Journal of Physics: Conference Series*, **625**, conference 1. http://dx.doi.org/10.1088/1742-6596/625/1/012002

[44] Stull, R.B. (1988) An Introduction to Boundary Layer Meteorology. Kluwer Academic Publishers, Dordrecht, Boston, London. http://dx.doi.org/10.1007/978-94-009-3027-8

[45] Garratt, J.R. (1994) The Atmospheric Boundary Layer. Cambridge University Press, Cambridge, New York, Melbourne.

[46] Mellor, G.L. and Yamada, T. (1974) A Hierarchy of Turbulence Closure Models for Planetary Boundary Layers. *Journal of the Atmospheric Sciences*, **31**, 1791-1806. http://dx.doi.org/10.1175/1520-0469(1974)031<1791:AHOTCM>2.0.CO;2

[47] Mellor, G.L. and Yamada, T. (1982) Development of a Turbulence Closure Model for Geophysical Fluid Problems. *Review of Geophysics and Space Physics*, **20**, 851-875. http://dx.doi.org/10.1029/RG020i004p00851

[48] Janjić, Z.I. (2001) Nonsingular Implementation of the Mellor-Yamada Level 2.5 Scheme in the NCEP Meso Model. National Centers for Environmental Prediction (NCEP), Office Note #437.

[49] Lorenz, E.N. (1967) The Nature and Theory of the General Circulation of the Atmosphere. World Meteorological Organization, WMO-No. 218.Tp.115, Geneva.

[50] Bernhardt, K. and Lauter, E.A. (1977) Globale physikalische Prozesse und Umwelt. *Zeitschrift für Meteorologie*, **27**, 1-20. (In German)

[51] Holton, J.R. (1979) An Introduction to Dynamic Meteorology. Academic Press, New York, San Francisco, London.

[52] Peixoto, J.P. and Oort, A.H. (1992) Physics of Climate. Springer-Verlag, New York, Berlin, Heidelberg.

[53] Hilbert, D. and Cohn-Vossen, S. (1952) Geometry and the Imagination. Chelsea Publishing Company, New York.

[54] Lass, H. (1950) Vector and Tensor Analysis. McGraw-Hill, New York, Toronto, London.

[55] Alekseenko, S.V., Kuibin, P.A. and Okulov, V.L. (2007) Theory of Concentrated Vortices. Springer, Berlin, Heidelberg, New York.

[56] Ross, H.K., Cooney, J., Hinzman, M., Smock, S., Sellhorst, G., Dlugi, R, Mölders, N. and Kramm, G. (2014) Wind Power Potential in Interior Alaska from a Micrometeorological Perspective. *Atmospheric and Climate Sciences*, **4**, 100-121. http://dx.doi.org/10.4236/acs.2014.41013

[57] Goorjian, P.M. (1972) An Invalid Equation in the General Momentum Theory of the Actuator Disc. *AIAA Journal*, **10**, 543-544. http://dx.doi.org/10.2514/3.50146

[58] Betz, A. (1926) Wind-Energie und Ihre Ausnutzung durch Windmühlen. Vandenhoeck & Ruprecht, Göttingen, Germany. (In German)

[59] Rankine, W.J.M. (1865) On the Mechanical Principles of the Action of Propellers. *Transaction of the Institute of Naval Architects*, **6**, 13-39.

[60] Froude, W. (1878) On the Elementary Relation between Pitch, Slip and Propulsive Efficiency. *Transaction of the Institute of Naval Architects*, **19**, 22-33.

[61] Froude, R.E. (1889) On the Part Played in Propulsion by Difference in Pressure. *Transaction of the Institute of Naval Architects*, **30**, 390-405.

[62] Wilson, R.E. and Lissaman, P.B.S. (1974) Applied Aerodynamic Performance of Wind Power Machines. Oregon State University, Corvallis.

[63] Snel, H. (1998) Review of the Present Status of Rotor Aerodynamics. *Wind Energy*, **1**, 46-69. http://dx.doi.org/10.1002/(SICI)1099-1824(199804)1:1+<46::AID-WE3>3.3.CO;2-0

[64] Mathew, S. (2006) Wind Energy: Fundamentals, Resource Analysis, and Economics. Springer. http://dx.doi.org/10.1007/3-540-30906-3

[65] Sharpe, D. (2004) A General Momentum Theory Applied to an Energy-Extracting Actuator Disc. *Wind Energy*, **7**, 177-188. http://dx.doi.org/10.1002/we.118

[66] Trefftz, E. (1921) Zur Prandtlschen Tragflächentheorie. *Math. Ann.*, **82**, 306-319. http://dx.doi.org/10.1007/BF01498674

[67] Joukowsky, N.E. (1912) Vortex Theory of Screw Propeller, I. Trudy Otdeleniya Fizicheskikh Nauk Obshchestva Lubitelei Estestvoznaniya **16** (1), 1-31 (In Russian). French translation in: Théorie tourbillonnaire de l'hélice propulsive (Gauthier-Villars, Paris, 1929) 1-47 (as cited by [12]).

[68] Joukowsky, N.E. (1914) Vortex Theory of Screw Propeller, II. Trudy Otdeleniya Fizicheskikh Nauk Obshchestva Lubitelei Estestvoznaniya **17** (1), 1-33 (In Russian). French translation in: Théorie tourbillonnaire de l'hélice propulsive (Gauthier-Villars, Paris, 1929) 48-93 (as cited by [12]).

[69] Joukowsky, N.E. (1915) Vortex Theory of Screw Propeller, III. Trudy Otdeleniya Fizicheskikh Nauk Obshchestva Lubitelei Estestvoznaniya **17** (2), 1-23 (In Russian). French translation in: Théorie tourbillonnaire de l'hélice propulsive (Gauthier-Villars, Paris, 1929) 94-122 (as cited by [12]).

[70] Joukowsky, N.E. (1918) Vortex Theory of Screw Propeller, IV. Trudy Avia Raschetno-Ispytatelnogo Byuro, no 3, 1-97 (In Russian). French translation in: Théorie tourbillonnaire de l'hélice propulsive (Gauthier-Villars, Paris, 1929) 123-198 (as cited by [12]).

[71] Betz, A. (1919) Schraubenpropeller mit geringstem Energieverlust—Mit einem Zusatz von L. Prandtl. Nachrichten d. K. Gesellschaft d. Wissenschaften, Göttingen, Math.-phys. Klasse, 193-217. (In German)

[72] Goldstein, S. (1929) On the Vortex Theory of Screw Propellers. *Proceedings of the Royal Society of London. Series A*, **123**, 440-465. http://dx.doi.org/10.1098/rspa.1929.0078

[73] Okulov, V.L. (2004) On the Stability of Multiple Helical Vortices. *Journal of Fluid Mechanics*, **521**, 319-342. http://dx.doi.org/10.1017/S0022112004001934

[74] Haltiner, G.J. and Martin, F.L. (1957) Dynamical and Physical Meteorology. McGraw-Hill, New York, Toronto, London.

[75] Thomas, T.Y. (1961) Concepts from Tensor Analysis and Differential Geometry. Academic Press, New York, London.

[76] Sokolnikoff, I.S. (1964) Tensor Analysis: Theory and Applications to Geometry and Mechanics of Continua. 2nd Edition, John Wiley & Sons, New York, London, Sydney.

[77] Eisenreich, G. (1971) Vorlesungen über Vektor- und Tensorrechnung. BSB B.G. Teubner Verlagsgesellschaft, Leipzig. (In German)

[78] Teichmann, H. (1973) Physikalische Anwendungen der Vektor-und Tensorrechnung. Bibliographisches Institut Mannheim, Wien, Zürich. (In German)

[79] Fortak, H. (1967) Vorlesungen über theoretische Meteorologie-Kinematik der Atmosphäre. Freie Universität Berlin, Institut für Theoretische Meteorologie. (In German)

Appendix A: Trajectories versus Streamlines

Let us consider a natural coordinate frame with the unit vectors $\boldsymbol{t}_t$, $\boldsymbol{n}_t$, and $\boldsymbol{b}_t$ that form a right-handed rectangular coordinate system at any given point of a curve in space (moving trihedron) like a trajectory (**Figure 3**), where the subscript *t* characterizes the trajectory-related quantities. Such a frame describing the movement of a particle along its trajectory in space is called an intrinsic system because it is closely related to the motion itself (e.g., [16] [54] [74]). As illustrated in **Figure 3**, the unit vector $\boldsymbol{t}_t$ is related to the direction of the instantaneous motion at any point, *i.e.*, it is tangent to the curve of the trajectory, and is, therefore, called the unit tangent. The velocity vector is given by $\boldsymbol{v} = V\boldsymbol{t}_t$, where V is its magnitude. The moving trihedron consisting of $\boldsymbol{t}_t$, $\boldsymbol{n}_t$, and $\boldsymbol{b}_t$ is an orthonormal frame, *i.e.*, the orthonormal conditions

$$\boldsymbol{t}_t \cdot \boldsymbol{t}_t = \boldsymbol{n}_t \cdot \boldsymbol{n}_t = \boldsymbol{b}_t \cdot \boldsymbol{b}_t = 1 \tag{A.1}$$

and

$$\boldsymbol{t}_t \cdot \boldsymbol{n}_t = \boldsymbol{t}_t \cdot \boldsymbol{b}_t = \boldsymbol{n}_t \cdot \boldsymbol{b}_t = 0 . \tag{A.2}$$

are fulfilled. These unit vectors can be considered as either contravariant or covariant basis vectors. We may write

$$\left.\begin{aligned} \boldsymbol{t}_t &= \boldsymbol{n}_t \times \boldsymbol{b}_t \\ \boldsymbol{n}_t &= \boldsymbol{b}_t \times \boldsymbol{t}_t \\ \boldsymbol{b}_t &= \boldsymbol{t}_t \times \boldsymbol{n}_t \end{aligned}\right\} . \tag{A.3}$$

because the triple scalar product $\boldsymbol{t}_t \cdot (\boldsymbol{n}_t \times \boldsymbol{b}_t) = [\boldsymbol{t}_t, \boldsymbol{n}_t, \boldsymbol{b}_t] = \sqrt{g} = 1$. The unit vector $\boldsymbol{n}_t$ is the principal normal. The unit vector $\boldsymbol{b}_t$ is perpendicular to the osculating plane spanned by $\boldsymbol{t}_t$ and $\boldsymbol{n}_t$, and called the binormal. The plane containing $\boldsymbol{n}_t$ and $\boldsymbol{b}_t$ is the normal plane, and the plane determined by $\boldsymbol{t}_t$ and $\boldsymbol{b}_t$ is the rectifying plane (e.g., [53]). As the unit tangent $\boldsymbol{t}_t$ is time-dependent, the other two basis vectors depend on time, too. The unit tangent is defined by

$$\boldsymbol{t}_t = \frac{\mathrm{d}\boldsymbol{r}}{\mathrm{d}s}, \tag{A.4}$$

where $\boldsymbol{r}$ is the position vector, and *s* is the arc length. From Equation (A.4) we can infer that $(\mathrm{d}s)^2 = \mathrm{d}\boldsymbol{r} \cdot \mathrm{d}\boldsymbol{r}$. Thus, the arc length between two points α and β is given by

$$s = \int_\alpha^\beta \left(\frac{\mathrm{d}\boldsymbol{r}}{\mathrm{d}s} \cdot \frac{\mathrm{d}\boldsymbol{r}}{\mathrm{d}s} \right)^{1/2} \mathrm{d}s . \tag{A.5}$$

The principal normal $\boldsymbol{n}_t$ is referred to the change of the unit vector $\boldsymbol{t}_t$ along the space trajectory because

$$\frac{\mathrm{d}}{\mathrm{d}s}(\boldsymbol{t}_t \cdot \boldsymbol{t}_t) = 0 = 2\boldsymbol{t}_t \cdot \frac{\mathrm{d}\boldsymbol{t}_t}{\mathrm{d}s} . \tag{A.6}$$

This means that $\boldsymbol{t}_t$ and $\mathrm{d}\boldsymbol{t}_t/\mathrm{d}s$ are perpendicular to each other, or in other words, $\mathrm{d}\boldsymbol{t}_t/\mathrm{d}s$ is normal to the curve of the trajectory. The magnitude of $\mathrm{d}\boldsymbol{t}_t/\mathrm{d}s$ is defined as the curvature K_t of the trajectory at a given point. Thus, we have

$$\frac{\mathrm{d}\boldsymbol{t}_t}{\mathrm{d}s} = K_t \boldsymbol{n}_t . \tag{A.7}$$

We may also define a radius of curvature by $R_t = K_t^{-1}$. As

$$\frac{\mathrm{d}}{\mathrm{d}s}(\boldsymbol{b}_t \cdot \boldsymbol{b}_t) = 0 = 2\boldsymbol{b}_t \cdot \frac{\mathrm{d}\boldsymbol{b}_t}{\mathrm{d}s} . \tag{A.8}$$

and

$$\frac{\mathrm{d}}{\mathrm{d}s}(\boldsymbol{t}_t \cdot \boldsymbol{b}_t) = 0 = \frac{\mathrm{d}\boldsymbol{t}_t}{\mathrm{d}s} \cdot \boldsymbol{b}_t + \boldsymbol{t}_t \cdot \frac{\mathrm{d}\boldsymbol{b}_t}{\mathrm{d}s} = K_t\, \boldsymbol{n}_t \cdot \boldsymbol{b}_t + \boldsymbol{t}_t \cdot \frac{\mathrm{d}\boldsymbol{b}_t}{\mathrm{d}s} = \boldsymbol{t}_t \cdot \frac{\mathrm{d}\boldsymbol{b}_t}{\mathrm{d}s}, \tag{A.9}$$

we may infer that $\mathrm{d}\boldsymbol{b}_t/\mathrm{d}s$ is perpendicular to both $\boldsymbol{b}_t$ and $\boldsymbol{t}_t$. Thus, $\mathrm{d}\boldsymbol{b}_t/\mathrm{d}s$ can also be expressed by

$$\frac{\mathrm{d}\boldsymbol{b}_t}{\mathrm{d}s} = -T_t \boldsymbol{n}_t, \tag{A.10}$$

where the proportionality constant T_t is the torsion. For the derivative of $\boldsymbol{n}_t$ with respect to s we obtain

$$\frac{\mathrm{d}\boldsymbol{n}_t}{\mathrm{d}s} = \frac{\mathrm{d}}{\mathrm{d}s}(\boldsymbol{b}_t \times \boldsymbol{t}_t) = \frac{\mathrm{d}\boldsymbol{b}_t}{\mathrm{d}s} \times \boldsymbol{t}_t + \boldsymbol{b}_t \times \frac{\mathrm{d}\boldsymbol{t}_t}{\mathrm{d}s} = -T_t \boldsymbol{n}_t \times \boldsymbol{t}_t + K_t \boldsymbol{b}_t \times \boldsymbol{n}_t = T_t \boldsymbol{b}_t - K_t \boldsymbol{t}_t. \tag{A.11}$$

The relations (A.7), (A.10), and (A.11) can be summarized as follows

$$\left.\begin{aligned} \frac{\mathrm{d}\boldsymbol{t}_t}{\mathrm{d}s} &= K_t \boldsymbol{n}_t \\ \frac{\mathrm{d}\boldsymbol{n}_t}{\mathrm{d}s} &= T_t \boldsymbol{b}_t - K_t \boldsymbol{t}_t \\ \frac{\mathrm{d}\boldsymbol{b}_t}{\mathrm{d}s} &= -T_t \boldsymbol{n}_t \end{aligned}\right\}. \tag{A.12}$$

These equations are the central equations in the theory of space curves customarily called the Serret-Frenet formulae (e.g., [74]-[76]). The Serret-Frenet formulae (A.12) allow to determine the change of the vectors of the trihedron, $\boldsymbol{t}_t$, $\boldsymbol{n}_t$, and $\boldsymbol{b}_t$, while it is moving along a given space curve $\boldsymbol{r} = \boldsymbol{r}(s)$ as a function of the arc length s.

Equation (A.11) may also be written as

$$\frac{\mathrm{d}\boldsymbol{n}_t}{\mathrm{d}s} = T_t \boldsymbol{b}_t - K_t \boldsymbol{t}_t = (T_t \boldsymbol{t}_t + K_t \boldsymbol{b}_t) \times \boldsymbol{n}_t. \tag{A.13}$$

The vector $\boldsymbol{\delta}_t = T_t \boldsymbol{t}_t + K_t \boldsymbol{b}_t$ that lies in the rectifying plane (see **Figure 3**) is the Darboux vector of a space curve (e.g., [77] [78]). Using this definition yields

$$\frac{\mathrm{d}}{\mathrm{d}s}\begin{Bmatrix} \boldsymbol{t}_t \\ \boldsymbol{n}_t \\ \boldsymbol{b}_t \end{Bmatrix} = \boldsymbol{\delta}_t \times \begin{Bmatrix} \boldsymbol{t}_t \\ \boldsymbol{n}_t \\ \boldsymbol{b}_t \end{Bmatrix}. \tag{A.14}$$

Obviously, the Darboux vector determines the new orientation (rotation) of the moving trihedron (see **Figure 3**). In other words, the Darboux vector coincides with the instantaneous axis of rotation. If $\boldsymbol{\omega}_t$ denotes the instantaneous angular velocity vector and V the instantaneous speed, we will have

$$\boldsymbol{\omega}_t = V \boldsymbol{\delta}_t. \tag{A.15}$$

The magnitude of the Darboux vector is the total curvature, sometimes also called the Lancret curvature (e.g., [79]).

A trajectory is the actual path of an air particle, *i.e.*, it characterizes the direction of the velocity that such an air particle is taking successively during a certain time interval (e.g., [79]). If the velocity field $\boldsymbol{v}(\boldsymbol{r}, t)$ is always known during that time interval, the trajectory can be calculated by integration of $\mathrm{d}\boldsymbol{r} = \boldsymbol{v}\mathrm{d}t$. A streamline represents a "snapshot" of the directions of the velocity field $\boldsymbol{v}(\boldsymbol{r}, t)$ at various locations at time t_0. As at each location a streamline is parallel to the flow field, we may write

$$\mathrm{d}\boldsymbol{r} \times \boldsymbol{v}(\boldsymbol{r}, t) = \boldsymbol{0}. \tag{A.16}$$

As sketched in **Figure 3**, a trajectory is the envelope of the corresponding streamlines. Thus, at a certain point $P(t)$ the unit tangent $\boldsymbol{t}_t = \boldsymbol{t}_s$ represents both the direction of this trajectory and the direction of the streamline touching the trajectory at this certain point. Even though the unit tangent is the same, the principal normal, $\boldsymbol{n}_t$, and the binormal, $\boldsymbol{b}_t$, of the trajectory generally differ from the principal normal, $\boldsymbol{n}_s$, and the binormal, $\boldsymbol{b}_s$, of the streamline, *i.e.*, $\boldsymbol{n}_t \neq \boldsymbol{n}_s$ and $\boldsymbol{b}_t \neq \boldsymbol{b}_s$. However, we have

$$\boldsymbol{t}_t = \boldsymbol{n}_t \times \boldsymbol{b}_t = \boldsymbol{n}_s \times \boldsymbol{b}_s. \tag{A.17}$$

If we express the substantial derivative of the unit tangent in the Eulerian form, we will obtain

$$\frac{\mathrm{d}\boldsymbol{t}_t}{\mathrm{d}t} = \frac{\partial \boldsymbol{t}_t}{\partial t} + V\frac{\partial \boldsymbol{t}_t}{\partial s} = \frac{\partial \boldsymbol{t}_t}{\partial t} + V K_s \boldsymbol{n}_s. \tag{A.18}$$

Here, the partial derivative of the unit tangent with respect to the arc length was replaced by $\partial \boldsymbol{t}_t / \partial s = K_s \boldsymbol{n}_s$, where $K_s = R_s^{-1}$ is the curvature of the streamline and R_s is the corresponding radius of the curvature. On the other hand, we have

$$\frac{\mathrm{d}\boldsymbol{t}_t}{\mathrm{d}t} = \frac{\mathrm{d}\boldsymbol{t}_t}{\mathrm{d}s}\frac{\mathrm{d}s}{\mathrm{d}t} = VK_t\boldsymbol{n}_t . \tag{A.19}$$

Combining Equations (A.18) and (A.19) yields

$$\frac{\partial \boldsymbol{t}_t}{\partial t} = V\left(K_t\boldsymbol{n}_t - K_s\boldsymbol{n}_s\right) . \tag{A.20}$$

In the two-dimensional case, we have $\boldsymbol{n}_t = \boldsymbol{n}_s$ because $\boldsymbol{b}_t = \boldsymbol{b}_s$. Thus, Equation (A.20) becomes

$$\frac{\partial \boldsymbol{t}_t}{\partial t} = V\left(K_t - K_s\right)\boldsymbol{n}_t . \tag{A.21}$$

With respect to the unit vectors $\boldsymbol{i}$ and $\boldsymbol{j}$ of a Cartesian frame the unit tangent of a plane trajectory can be expressed by

$$\boldsymbol{t}_t = \boldsymbol{i}\cos\vartheta + \boldsymbol{j}\sin\vartheta . \tag{A.22}$$

where ϑ is the angle between $\boldsymbol{i}$ and $\boldsymbol{t}_t$. It is also the angle between $\boldsymbol{j}$ and $\boldsymbol{n}_t$. This angle usually depends on time. Thus, the local rate of change of the unit tangent reads

$$\frac{\partial \boldsymbol{t}_t}{\partial t} = -\frac{\partial \vartheta}{\partial t}\boldsymbol{i}\sin\vartheta + \frac{\partial \vartheta}{\partial t}\boldsymbol{j}\cos\vartheta = \frac{\partial \vartheta}{\partial t}\boldsymbol{n}_t . \tag{A.23}$$

Combining Equations (A.21) and (A.23) yields

$$\frac{\partial \vartheta}{\partial t} = V\left(K_t - K_s\right) . \tag{A.24}$$

The scalar form of the relation (A.24) is Blaton's equation. In the case of steady-state conditions, the left-hand side terms of Equations (A.21) and (A.24) vanish and the curvatures of the trajectories and the streamlines are identical, *i.e.*, the trajectories and the streamlines coincide.

If no tangential acceleration exists, *i.e.*, $\partial \overline{p} / \partial s = 0$, and the horizontal scale is small enough, so that the magnitude of Coriolis acceleration is small in comparison with those of the centripetal acceleration and the acceleration due to the pressure gradient, we will lead to the following conditions for a frictionless flow:

$$\frac{\mathrm{d}V}{\mathrm{d}t} = 0 \tag{A.25}$$

and

$$\frac{V^2}{R_t} = -\frac{1}{\rho}\frac{\partial p}{\partial n_t} . \tag{A.26}$$

The solution of this equation set is then given by

$$V = \pm\sqrt{-\frac{R_t}{\rho}\frac{\partial p}{\partial n_t}} . \tag{A.27}$$

The equation describes the cyclostrophic flow. Obviously, $\partial p / \partial n_t$ and R_t must have opposite signs to ensure that the condition

$$-\frac{R_t}{\rho}\frac{\partial p}{\partial n_t} \geq 0 . \tag{A.28}$$

is fulfilled. Both cyclonic and anti-cyclonic flows are possible. Assuming, for instance, $\rho = 1.2\ \mathrm{kg\cdot m^{-3}}$, $V = 30\ \mathrm{m\cdot s^{-1}}$, and $R_t = 300\ \mathrm{m}$ yields $\partial p / \partial n_t = -3.6\ \mathrm{Pa\cdot m^{-1}}$. The cyclostrophic approximation is acceptable if the Ross by number defined by

$$Ro_b = \left|\frac{V}{fR_t}\right| \quad (A.29)$$

fulfills the condition $Ro_b \gg 1$. **Figure 17** shows the wind speed V as a function of the pressure gradient $\partial p/\partial n_t$ in case of a cyclostrophic flow, where $R_t = 300$ m and $\rho = 1.2$ kg·m^{-3} were assumed. As illustrated, even small values of the pressure gradient like $\partial p/\partial n_t = -0.02$ Pa·m^{-1} can produce a cyclostrophic wind speed of more than 2 m·s^{-1}. In such a case the Rossby number amounts to $Ro_b = 75 \gg 1$ if a typical value of $f = 10^{-4}$ s^{-1} for the Coriolis parameter is considered. In case of $V = 30$ m·s^{-1}, $R_t = 300$ m, and $f = 10^{-4}$ s^{-1}, the Rossby number is $Ro_b = 10^3$. These numbers are typical for an F1 tornado on the Fujita Tornado Damage Scale. The occurrence of a cyclostrophic flow has to be considered if a difference between the static pressure far upstream and far downstream of a wind turbine is presupposed.

Appendix B: The Equations of the General Momentum Theory

In this appendix, we only consider average values. Thus, all symbols that are characterizing average values are ignored here. Let be r the radial distance of any annular element of the propeller disk, v_x and v_r the axial and the radial components of the fluid velocity, respectively. Furthermore, let be p the pressure immediately in front of the propeller and δp the decrease of the pressure behind the propeller associated with the angular velocity ω. Moreover, let p_w be the pressure in the final wake, v_w the corresponding axial velocity, and ω_w the corresponding angular velocity at radial distance r_w from the axis of the slipstream. Since the area of the annular element is $\mathrm{d}A = 2\pi r \mathrm{d}r$, the equation of continuity reads

$$v_w r_w \mathrm{d}r_w = v_x r \mathrm{d}r = \text{const.} \quad (B.1)$$

Generally, the torque is

$$\boldsymbol{Q} = \boldsymbol{r} \times \boldsymbol{F} \quad (B.2)$$

If the force $\boldsymbol{F}$ is only a function of the radial vector $\boldsymbol{r}$ of the form $\boldsymbol{F} = f(r)\boldsymbol{r}$, as considered here, we will obtain $\boldsymbol{Q} = f(r)\boldsymbol{r} \times \boldsymbol{r} = \boldsymbol{0}$. Since the torque is equal to the derivative of the angular momentum, $\boldsymbol{L}$, with respect to time, we may write

$$\boldsymbol{Q} = \frac{\mathrm{d}\boldsymbol{L}}{\mathrm{d}t} = \boldsymbol{0} \quad (B.3)$$

i.e., the angular momentum is invariant with time ($\boldsymbol{L} = \text{const.}$). Equation (B.3) describes the conservation of angular momentum in the central field. The angular momentum can be expressed by

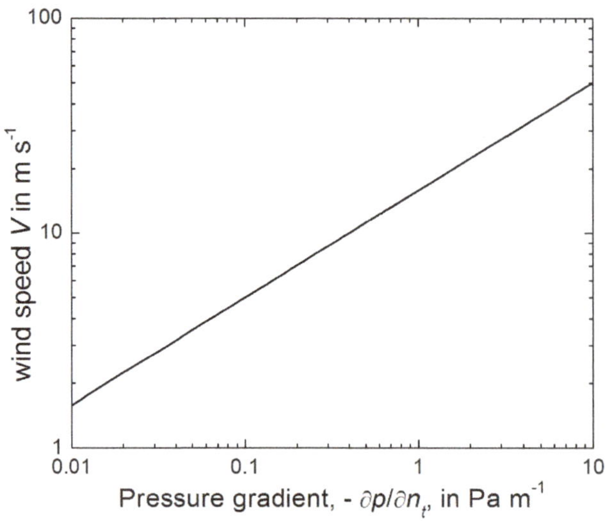

Figure 17. Wind speed versus pressure gradient in case of a cyclostrophic flow.

$$\boldsymbol{L} = m\boldsymbol{r} \times \boldsymbol{v} = \text{const.} \tag{B.4}$$

where $\boldsymbol{v}$ is the velocity vector of a particle with the mass m moving through the central field. As $\boldsymbol{L} \perp \boldsymbol{r}$ and $\boldsymbol{L} \perp \boldsymbol{v}$ throughout the motion, the radius vector of the particle on which the central field is acting lies in the plane spanned by $\boldsymbol{r}$ and $\boldsymbol{v}$ which is perpendicular to $\boldsymbol{L}$. Hence, we may write for convenience

$$L = |\boldsymbol{L}| = m\omega r^2 = \text{const.} \tag{B.5}$$

Thus, the conservation of angular momentum provides

$$\omega_w r_w^2 = \omega r^2 . \tag{B.6}$$

In accord with Equation (2.54), the torque is given by

$$\mathrm{d}Q = \rho v_x \omega r^2 \mathrm{d}A = 2\pi \rho v_x \omega r^3 \mathrm{d}r . \tag{B.7}$$

The Bernoulli equation in its approximated form (2.46) yields

$$H_\infty = p_\infty + \frac{1}{2}\rho v_\infty^2 = p + \frac{1}{2}\rho\left(v_x^2 + v_r^2\right). \tag{B.8}$$

and

$$H_w = p - \delta p + \frac{1}{2}\rho\left(v_x^2 + v_r^2 + \omega^2 r^2\right) = p_w + \frac{1}{2}\rho\left(v_w^2 + \omega_w^2 r_w^2\right). \tag{B.9}$$

Thus, the difference of the total pressure heads, $\Delta H = H_\infty - H_w$, is given by

$$\Delta H = p + \frac{1}{2}\rho\left(v_x^2 + v_r^2\right) - p + \delta p - \frac{1}{2}\rho\left(v_x^2 + v_r^2 + \omega^2 r^2\right) = \delta p - \frac{1}{2}\rho\omega^2 r^2 . \tag{B.10}$$

In addition, the pressure difference $\Delta p = p_\infty - p_w$ reads

$$\begin{aligned} \Delta p &= H_\infty - \frac{1}{2}\rho v_\infty^2 - H_w + \frac{1}{2}\rho\left(v_w^2 + \omega_w^2 r_w^2\right) = \frac{1}{2}\rho\left(v_w^2 - v_\infty^2\right) + \frac{1}{2}\rho\omega_w^2 r_w^2 + \Delta H \\ &= \delta p + \frac{1}{2}\rho\left(v_w^2 - v_\infty^2\right) + \frac{1}{2}\rho\omega_w^2 r_w^2 - \frac{1}{2}\rho\omega^2 r^2 \\ &= \delta p + \frac{1}{2}\rho\left(v_w^2 - v_\infty^2\right) + \frac{1}{2}\rho\left(\omega_w - \omega\right)\omega_w r_w^2 \end{aligned} , \tag{B.11}$$

where the conservation of the angular momentum (see Equation (B.6)) has been used. Rearranging $\left(\omega_w - \omega\right)\omega_w r_w^2$ by

$$\omega_w - \omega = \omega_w + 2\Omega - 2\Omega - \omega = 2\left(\Omega + \frac{1}{2}\omega_w\right) - 2\left(\Omega + \frac{1}{2}\omega\right)$$

leads to

$$\Delta p = \delta p + \frac{1}{2}\rho\left(v_w^2 - v_\infty^2\right) + \rho\left(\Omega + \frac{1}{2}\omega_w\right)\omega_w r_w^2 - \rho\left(\Omega + \frac{1}{2}\omega\right)\omega_w r_w^2 . \tag{B.12}$$

Here, Ω is the angular velocity of the rotor.

In applying Bernoulli's equation to the flow relative to the propeller blades, we have to consider the relative angular velocity of the air that increases from Ω to $\omega + \Omega$ associated with a decrease of the static pressure behind the propeller and given by

$$\delta p = \frac{1}{2}\rho\left(\left(\Omega + \omega\right)^2 - \Omega^2\right)r^2 = \rho\left(\Omega + \frac{1}{2}\omega\right)\omega r^2 = \rho\left(\Omega + \frac{1}{2}\omega\right)\omega_w r_w^2 . \tag{B.13}$$

Thus, we obtain

$$\Delta p = \frac{1}{2}\rho\left(v_w^2 - v_\infty^2\right) + \rho\left(\Omega + \frac{1}{2}\omega_w\right)\omega_w r_w^2 . \tag{B.14}$$

The pressure gradient in the wake balances the centrifugal force on the fluid and is governed by

$$\frac{\mathrm{d}p_w}{\mathrm{d}r_w} = \rho\frac{v_\Theta^2}{r_w} = \rho\omega_w^2 r_w . \tag{B.15}$$

A balance between the pressure gradient force and the centrifugal force in the horizontal direction leads to the cyclostrophic flow well known in meteorology (e.g., [12]). The derivation of Equation (B.14) with respect to r_w leads to

$$\frac{\mathrm{d}}{\mathrm{d}r_w}(\Delta p) = \frac{1}{2}\rho\frac{\mathrm{d}}{\mathrm{d}r_w}\left(v_w^2 - v_\infty^2\right) + \rho\left(\Omega + \frac{1}{2}\omega_w\right)\frac{\mathrm{d}}{\mathrm{d}r_w}\left(\omega_w r_w^2\right) = -\frac{\mathrm{d}p_w}{\mathrm{d}r_w} = -\rho\omega_w^2 r_w \tag{B.16}$$

or

$$\frac{1}{2}\frac{\mathrm{d}}{\mathrm{d}r_w}\left(v_\infty^2 - v_w^2\right) = \left(\Omega + \frac{1}{2}\omega_w\right)\frac{\mathrm{d}}{\mathrm{d}r_w}\left(\omega_w r_w^2\right) + \omega_w^2 r_w , \tag{B.17}$$

where Equation (B.15) was used. Since

$$\omega_w^2 r_w = \frac{1}{2}\omega_w\frac{\mathrm{d}}{\mathrm{d}r_w}\left(\omega_w r_w^2\right), \tag{B.18}$$

Equation (B.17) becomes [62]

$$\frac{1}{2}\frac{\mathrm{d}}{\mathrm{d}r_w}\left(v_\infty^2 - v_w^2\right) = \left(\Omega + \omega_w\right)\frac{\mathrm{d}}{\mathrm{d}r_w}\left(\omega_w r_w^2\right). \tag{B.19}$$

Furthermore, the thrust is given by

$$\begin{aligned}\mathrm{d}T &= \rho v_w\left(v_\infty - v_w\right)\mathrm{d}A_w + \Delta p\mathrm{d}A_w \\ &= -\rho v_w\left(v_w - v_\infty\right)\mathrm{d}A_w + \frac{1}{2}\rho\left(v_w^2 - v_\infty^2\right)\mathrm{d}A_w + \rho\left(\Omega + \frac{1}{2}\omega_w\right)\omega_w r_w^2\mathrm{d}A_w\end{aligned} . \tag{B.20}$$

The pressure increment at the propeller disk is given by [62]

$$\mathrm{d}T = \delta p\mathrm{d}A = \rho\left(\Omega + \frac{1}{2}\omega\right)\omega r^2\mathrm{d}A . \tag{B.21}$$

Combining these two equations leads to

$$-v_w\left(v_w - v_\infty\right)\mathrm{d}A_w + \frac{1}{2}\left(v_w^2 - v_\infty^2\right)\mathrm{d}A_w + \left(\Omega + \frac{1}{2}\omega_w\right)\omega_w r_w^2\mathrm{d}A_w = \left(\Omega + \frac{1}{2}\omega\right)\omega r^2\mathrm{d}A . \tag{B.22}$$

The equation of continuity provides $v_w\mathrm{d}A_w = v_x\mathrm{d}A$ (see Equation (B.1)). Thus, we obtain

$$-v_w\left(v_w - v_\infty\right) + \frac{1}{2}\left(v_w^2 - v_\infty^2\right) = \left(\frac{\Omega + \frac{1}{2}\omega}{v_x} - \frac{\Omega + \frac{1}{2}\omega_w}{v_w}\right)v_w\omega_w r_w^2 . \tag{B.23}$$

Finally, using $v_w\left(v_w - v_\infty\right) - \frac{1}{2}\left(v_w^2 - v_\infty^2\right) = \frac{1}{2}\left(v_w - v_\infty\right)^2$ yields

$$\frac{1}{2}\left(v_w - v_\infty\right)^2 = \left(\frac{\Omega + \frac{1}{2}\omega_w}{v_w} - \frac{\Omega + \frac{1}{2}\omega}{v_x}\right)v_w\omega_w r_w^2 . \tag{B.24}$$

This equation already derived by Wilson and Lissaman [62] suffice to determine the relationship between the thrust and torque of the propeller and the flow in the slipstream. Owing to the complexity of the equations, however, it is customary to adopt certain approximations based on the fact that the rotational velocity in the slipstream is generally very small.

Since Sharpe [65] criticized the work of Glauert [10], Wilson and Lissaman [62]and others because of their dropping of the static pressure in the wake, we compared Sharpe's equation (4), (7), (8), and (11) with our Equations (B.11), (B.13), (B.14), and (B.19). Rearranging Equation (B.11) yields

$$\delta p = -\frac{1}{2}\rho\left(v_w^2 - v_\infty^2\right) - \frac{1}{2}\rho\omega_w^2 r_w^2 + \frac{1}{2}\rho\omega^2 r^2 + \Delta p\,. \tag{B.25}$$

Using $v_r = v_\infty(1-a)$, $v_w = v_\infty(1-b)$, $a' = \omega/(2\Omega)$, and $b' = \omega_w/(2\Omega)$ yields $v_w^2 - v_\infty^2 = v_\infty^2 b(b-2)$ and $-\frac{1}{2}\rho\omega_w^2 r_w^2 + \frac{1}{2}\rho\omega^2 r^2 = -2\Omega^2\rho\left((b'r_w)^2 - (a'r)^2\right)$, so that Equation (B.25) becomes

$$\delta p = \frac{1}{2}\rho\left[v_\infty^2(2-b)b - 4\Omega^2\left((b'r_w)^2 - (a'r)^2\right)\right] + \Delta p\,. \tag{B.26}$$

This is identical with Sharpe's Equation (4). Considering $b'r_w^2 = a'r^2$ (see Equation (B.6)) leads to Sharpe's Equation (5). In a similar manner, we obtain from Equation (B.13)

$$\delta p = \rho\left(\Omega + \frac{1}{2}\omega\right)\omega_w r_w^2 = 2\rho\Omega^2(1+a')b'r_w^2 = 2\rho\Omega^2(1+a')a'r^2\,. \tag{B.27}$$

that is identical with Sharpe's Equation (7). Thus, our Equation (B.14) results in

$$\Delta p = \frac{1}{2}\rho\left(v_w^2 - v_\infty^2\right) + \rho\left(\Omega + \frac{1}{2}\omega_w\right)\omega_w r_w^2 = \frac{1}{2}\rho\left[4\Omega^2(1+b')a'r^2 - v_\infty^2(2-b)b\right]. \tag{B.28}$$

This equation completely agrees with Sharpe's Equation (8). Finally, by rearranging our Equation (B.19)we obtain

$$v_\infty^2(b-1)\frac{db}{dr_w} = 2\Omega^2(1+2b')\frac{d}{dr_w}\left(b'r_w^2\right). \tag{B.29}$$

that is identical with Sharpe's Equation (11). Thus, Equation (B.24) already derived by Wilson and Lissaman [62] and the opposite one derived by Glauert [10] are accurate if an actuator disk is considered. Sharpe's criticism is, therefore, not justified.

Appendix C: Solution of an Irrotational Wake

An exact solution of the general equations of the General Momentum Theory described before can be obtained when the flow in the slipstream is irrotational except along the axis [10] [65]. This condition implies that the rotational momentum ωr^2 has the same value *k* for all radial elements, *i.e.*,

$$\omega_w r_w^2 = \omega r^2 = k\,. \tag{C.1}$$

Here, *r* the radial distance of any annular element of the propeller disk.

On the basis of the equation (see Equation (B.19) of **Appendix B**)

$$\frac{1}{2}\frac{d}{dr_w}\left(v_\infty^2 - v_w^2\right) = \left(\Omega + \omega_w\right)\frac{d}{dr_w}\left(\omega_w r_w^2\right) \tag{C.2}$$

we can deduce that the axial velocity v_w is constant across the wake because $d\left(\omega_w r_w^2\right)/dr_w = 0$ and, hence, $dv_w/dr_w = 0$. Furthermore, it can be shown that Equation (B.24) of **Appendix B** is satisfied by a constant value of the axial velocity v_x across the propeller disk. If v_w and v_x are constant, we will obtain from the equation of continuity (see Equation (B.1) of **Appendix B**),

$$v_w r_w dr_w = v_x r dr = \text{const.}\,, \tag{C.3}$$

and the conservation of angular momentum (see Equation (B.6) of **Appendix B**),

$$\omega_w r_w^2 = \omega r^2\,, \tag{C.4}$$

the following relationship [10]

$$\frac{v_w}{v_x} = \frac{\omega_w}{\omega} = \frac{r^2}{r_w^2} = \frac{R^2}{R_w^2} \tag{C.5}$$

or with respect to Sharpe's [65] notation

$$\frac{r_w \mathrm{d}r_w}{r\mathrm{d}r} = \frac{1-a}{1-b} = \frac{R_w^2}{R^2} = \frac{r_w^2}{r^2} = \frac{a'}{b'}, \tag{C.6}$$

where $v_x = v_\infty(1-a)$, $v_w = v_\infty(1-b)$, R is the radius of the rotor, R_w is the maximum value of r_w,

$$a' = \frac{v_\theta}{2\Omega r} = \frac{\omega}{2\Omega} \tag{C.7}$$

is the azimuthal interference factor, and by analogy with that quantity,

$$b' = \frac{v_{\theta,w}}{2\Omega r_w} = \frac{\omega_w}{2\Omega}, \tag{C.8}$$

considered for the fully developed wake. Thus, Equation (B.24) of **Appendix B** becomes

$$\frac{1}{2}(v_w - v_\infty)^2 = \frac{v_x - v_w}{v_x}\Omega k \tag{C.9}$$

or

$$\frac{1}{2}b^2 v_\infty^2 = \frac{b-a}{1-a}\Omega k . \tag{C.10}$$

In accord with Glauert [10], we define

$$\mathrm{X} = \frac{v_\infty}{\Omega R}, \tag{C.11}$$

$$\mu = \frac{v_x}{\Omega R}, \tag{C.12}$$

$$\mu_w = \frac{v_w}{\Omega R}, \tag{C.13}$$

and

$$q = \frac{v_x k}{\Omega^2 R^3} = \frac{\mu k}{\Omega R^2} . \tag{C.14}$$

Inserting these definitions into Equation (C.9) yields

$$q = \frac{\mu^2}{2}\frac{(\mu_w - \mathrm{X})^2}{\mu - \mu_w} . \tag{C.15}$$

If we assume again, that $\Delta p = \overline{p_\infty} - \overline{p_w} = 0$, we will obtain from Equation (B.14) of **Appendix B**

$$\frac{1}{2}\left(v_w^2 - v_\infty^2\right) = -\left(\Omega + \frac{1}{2}\omega_w\right)k \tag{C.16}$$

or

$$\frac{1}{2}\left(\mathrm{X}^2 - \mu_w^2\right) = \left(1 + \frac{1}{2}\frac{\mu_w q}{\mu^2}\right)\frac{q}{\mu} . \tag{C.17}$$

Inserting Equation (C.11) into this equation provides

$$\mathrm{X}^2 - \mu_w^2 = \mu\left(1 + \frac{1}{4}\mu_w \frac{(\mu_w - \mathrm{X})^2}{\mu - \mu_w}\right)\frac{(\mu_w - \mathrm{X})^2}{\mu - \mu_w} . \tag{C.18}$$

Rearranging this equation yields

$$\mu = \frac{1}{2}(\mu_w + X) - \frac{\mu(\mu_w - X)^3}{8(\mu - \mu_w)}. \tag{C.19}$$

Since $\mu = X(1-a)$ and $\mu_w = X(1-b)$, we obtain

$$a = \frac{1}{2}b\left(1 - \frac{X^2(1-a)b^2}{4(b-a)}\right). \tag{C.20}$$

This formula was already derived by Wilson and Lissaman [62], but they used $X = \lambda^{-1}$, where λ is, again, the tip speed ratio. A similar formula was also deduced by Glauert [10] for an engine-driven propeller.

14

Feasibility Study of Melon Seed Oil as a Source of Biodiesel

Kennedy Izuchukwu Ogunwa[1], Samuel Ofodile[2], Ozioma Achugasim[2*]

[1]World Bank African Centre of Excellence in Oilfield Chemicals, Institute Of Petroleum Studies, University of Port Harcourt, Port Harcourt, Nigeria
[2]Department of Pure and Industrial Chemistry, University of Port Harcourt, Port Harcourt, Nigeria
Email: *ozioma.achugasim@uniport.edu.ng

Abstract

Melon seed oils were extracted at a very high yield of 52.2%. The extracted oil was subjected to oil quality tests and subsequently transesterified to give fatty acid methyl esters or biodiesel. The biodiesel was also subjected to fuel quality tests. The results showed that the extracted oil had specific gravity of 0.91 and moisture content of 0.90% indicating that the oil is a very good energy source, a good candidate for transesterification and will not be easily susceptible to microbial attack and autooxidation. The fuel quality parameters of the produced biodiesel showed that it conforms to standards for biodiesel and compares well with a standard petrodiesel.

Keywords

Biodiesel, Melon, Petrodiesel, Transesterification

1. Introduction

The need for alternatives to fossil fuel sources has become more obvious today than ever and the reasons are not far-fetched. They include increasing cost of fossil fuels, their unrenewable or closed carbon cycle nature, emissions of combustion-generated pollutants, etc [1]. The alternatives are found in biofuels. Among the different biofuels, biodiesel seems to be the most studied and the most promising competitor to the currently used petrodiesel.

Biodiesel has been described as a fuel composed of monoalkyl esters of long-chain fatty acids derived from renewable vegetable oils and fats [2]. Biodiesel production apart from providing a good alternative to petrodiesel, will encourage the market for increased production of vegetable oils and animal fats, reduce global warming

*Corresponding author.

and improve the lubricating properties of diesel engines [3] [4].

Chemically, biodiesels are produced from the transesterification of triglycerides found in vegetables oils and fats to form the monoalkyl esters which are the primary molecules of biodiesel [5] [6].

The formation of the methyl esters (ME) involves a step-wise reaction with intermediate formation of diglycerides (DG) and monoglycerides (MG) from the starting triglyceride (TG) with the production of glycerol (GL) in the final step.

$$TG + ROH \rightleftharpoons DG + R'CO_2R \quad (1)$$

$$DG + ROH \rightleftharpoons MG + R'CO_2R \quad (2)$$

$$MG + ROH \rightleftharpoons GL + R'CO_2R \quad (3)$$

The overall reaction is therefore given as:

$$TG + 3ROH \rightleftharpoons 3R'CO_2R + GL$$

Of course the reaction is catalyzed by either a base or an acid [7]-[9].

The hydrocarbon moiety of the methyl esters are of comparable carbon number with that of the petrodiesel. Also the cetane number, energy content, viscosity and phase changes of biodiesel are akin to that of petrodiesels [10].

The vegetable oils that have been utilized as sources of biodiesel include those from rape seed, cotton seed, soybean seed, fluted pumpkin seed, sunflower seed, palm kernel seed and fruit, corn, linseed etc. [11] [12].

Melon seed (*Citrullus lanatus. thumb. mausf*) oil provides another useful source of biodiesel production. The melon seed (*Citrullus lanatus. thumb. mausf*) popularly known as egusi in Nigeria is from the family cucurbifaceae and mainly cultivated as a soup thickener and snack. The seed is rich in oil, low in cholesterol and contains essential and unsaturated fatty acids [13] [14].

The Local utilization of this seed in biodiesel production will no doubt improve its production/market, generate employment and may earn Nigeria the much needed foreign exchange.

2. Materials and Method

2.1. Oils Extraction

The melon seed samples used for this study were obtained from Ariaria market, Aba in Abia State, Nigeria. The samples were air-dried, weighed, ground and extracted in a soxhlet extractor using hexane as the solvent. The extracted oil was subsequently analyzed for some oil quality parameters as shown in **Table 1**.

2.2. Biodiesel Production

50 ml of methanol was added to 3.0 g of potassium hydroxide KOH in a 100 ml conical flask. The mixture was stirred with gentle heating. The KOH/methanol mixture was subsequently added to 150 ml of the melon seed oil in 1000 ml conical flask and stirred gently for more than 10 hrs. The entire mixture was then transferred to a separatory funnel and allowed to stand overnight to give two distinct layers of lower glycerol and upper methyl esters (biodiesel). The separated biodiesel was subsequently washed severally with water and subjected to fuel

Table 1. Oil quality parameter of oil extracted from melon seed.

PARAMETERS	VALUE
Acid value (mg KOH/g)	4.18
Iodine value (g/100g)	121.65
Peroxide value (meq/kg)	16.07
Specific gravity (g/ml)	0.91
Kinematic viscosity (mm^2/s)	21.65
Moisture content (%)	0.90
Oil yield (%)	52.20
Color	Yellow

quality tests to determine its suitability or otherwise, as an alternative to petro-diesel. The results are presented in **Table 2**.

3. Result and Discussion

The result of the oil quality parameters of the extracted melon seed oil is presented in **Table 1**.

The result of the oil quality parameters shows that melon seed oil could serve as an economic feedstock for biodiesel production given the percentage oil yield of 52.2. The iodine value is not too high showing that bio-diesel obtained from melon seed oil may not be too susceptible to oxidation and quick rancidity.

Given the threshold acid value of 1.0 mg KOH/g for fresh oils, it is obvious that melon seed oil (acid value 4.18) will have catalysis problem during esterification (catalyst deactivation) and formation of soaps as side reaction. These problems can be taken care of with the use of heterogeneous catalysts or the addition of bases like sodium hydroxide.

The specific gravity of 0.91 shows that the oil is a good energy source. The very low moisture content of the melon seed oil of 0.90 is an indication that hydrolysis of the produced esters during esterification and its consequent soap formation is not likely to occur if melon seed oil is used in transesterification reaction. The low moisture content will also prevent microbial attack and autoxidation which leads to oil rancidity.

The result of the fuel quality parameters of the melon seed oil methyl esters are shown in **Table 2**.

It is obvious that transesterification has drastically reduced the moisture content and acid value of melon seed oil giving the produced biodiesel better fuel characteristics. There is also a sharp reduction of the kinematic viscosity of the melon seed oil from 21.65 to about 5.8 mm^2/s in the biodiesel, a value that is comparable to that of the petrodiesel.

The suitability of the produced methyl esters as a biodiesel is better appreciated when these parameters are compared with that of a standard petrodiesel (ASTM D975) and a standard biodiesel (ASTM D6751) as shown in **Table 3**.

The specific gravity of the produced biodiesel compares well with that of the standard biodiesel (D6751) and

Table 2. Fuel quality parameters of methyl esters produced from melon seed oil.

PARAMETERS	VALUE
Specific gravity (g/ml)	0.86
Kinematic viscosity (mm^2/s)	5.80
Flash Point (˚C)	132.00
Diesel Index	46.60
Iodine Value (mg KOH/g)	120.40
Acid Value (mg KOH/g)	0.26
Peroxide value (meq O_2/kg)	51.12
Heat of combustion (mJ/kg)	37.36
Moisture Content (%)	0.04
Color	Light Yellow

Table 3. Fuel quality parameters of a standard biodiesel, a petrodiesel and melon seed oil methyl esters.

PARAMETERS	ASTM D975 (PETRODIESEL)	ASTM D6751 (BIODIESEL)	MELON SEED OIL METHYL ESTERS
Specific gravity (g/ml)	0.85	0.88	0.86
Kinematic Viscosity (mm^2/s)	1.9 - 4.1	1.9 - 6.0	5.8
Flash Point (˚C)	60 - 80	100 - 170	132.0
Diesel Index	40 - 55	47 - 60	46.0
Acid Value		0.80	0.20
Heat of combustion (mJ/kg)	44.80	37.80	37.36
Moisture Content (%)	0.05	0.05	0.04

petrodiesel (D975). The flash point is higher than that of a standard biodiesel making it safer for storage and transportation. The acid value and moisture content is highly reduced and the diesel index similar to that of the standard biodiesel and petrodiesel.

4. Conclusions

The high percentage oil yield of melon seed shows that it is a viable oil source. The oil quality parameters show that the oil from melon seed with a moderate degree of unsaturation is not susceptible to oxidative rancidity. The low moisture content makes it a good candidate for transesterification reaction.

The fuel quality parameters of the methyl esters derived from the melon seed oil show that it conforms to standards for biodiesel and compares very well with a standard petrodiesel. So melon seed oil can serve as a good feedstock for biodiesel production.

Acknowledgements

The authors are grateful to Nigerian national Petroleum Cooperation (NNPC) for assistance in carrying out some of the analysis in the work.

References

[1] Ali, Y. and Hanna, M.A. (1994) Alternative Diesel Fuels from Vegetable Oils. *Bioresource Technology*, **50**, 153-163. http://dx.doi.org/10.1016/0960-8524(94)90068-X

[2] Canaka, M. and Sanli, H. (2008) Biodiesel Production from Various Feedstocks and their Effects on Fuel Properties. *Journal of Industrial Microbiology & Biotechnology*, **35**, 431-441. http://dx.doi.org/10.1007/s10295-008-0337-6

[3] Abayeh, O.J., Ismail, A. and Abayeh, O.M. (2012) Characterization of Pumpkin (Cucurbita Pero) Seed Oil as a Bio-diesels Produced through KOH-Catalzed Transesterification Process. *Journal of the Chemical Society of Nigeria*, **37**, 81-86.

[4] Math, M.C. (2007) Optimization of Restaurant Waste Oil-Methyl Ester Yield. *Journal of scientific and Industrial Research*, **66**, 772-776.

[5] Romanski, J., Nowak, P., Kosinski, K. and Jurczak, J. (2012) High Pressure Transesterification of Sterically Hindered Esters. *Tetrahedron Letters*, **53**, 5287-5289. http://dx.doi.org/10.1016/j.tetlet.2012.07.094

[6] Sidhara, R. and Mathai, I.M. (1974) Transesterification Reaction. *Journal of Scientific and Industrial Research*, **33**, 178-187.

[7] Damoko, D. and Munir, C. (2000) Kinetics of Palm Oil Transesterification in a Batch Reactor. *Journal of American Oil Chemical Society*, **77**, 1263-1267. http://dx.doi.org/10.1007/s11746-000-0198-y

[8] Freedom, B., Buterfield, R.O. and Pryde, E.H. (1986) Transesterification Kinetics of Soyabean Oil. *Journal of the American Oil Chemists' Society*, **63**, 1375-1380. http://dx.doi.org/10.1007/BF02679606

[9] Noureddini, H., and Zhn, D. (1997) Kinetics of Transesterification of Soyabean Oil. *Journal of the American Oil Chemists' Society*, **74**, 1457-1463. http://dx.doi.org/10.1007/s11746-997-0254-2

[10] Muniyappa, P.R., Brammer, S.C. and Noureddini, H. (1996) Improved Conversation of Plant Oils and Animal Fat into Biodiesel and Co-Product. *Bioresource Technology*, **56**, 19-24. http://dx.doi.org/10.1016/0960-8524(95)00178-6

[11] Eckey, E.W. (1954) Vegetable Fats and Oils. Reinhold, New York, 75-79. http://dx.doi.org/10.1097/00010694-195407000-00026

[12] Akintayo, E.T. (1997) Chemical Composition, Physiochemical Properties of Fluted Pumpkin Seed and Seed Oils. *Rivista Italiana Sostanze Grasse*, **74**, 13-16.

[13] Lazos, E.S. (1986) Nutritional Fatty Acid and Oil Characterization of Pumpkin and Melon Seeds. *Journal of Food Science*, **51**, 1382-1384. http://dx.doi.org/10.1111/j.1365-2621.1986.tb13133.x

[14] Gushini, G., Wehner, T.C. and Jarret, R.L. (2004) Inheritance of Egusi Seed Type in Water Melon. *Journal of Heredity*, **94**, 268-270. http://dx.doi.org/10.1093/jhered/esh031

15

The Challenges of the Engineering Inspection of Pipelines and Heat Engines by Temperature Deformation and Stress

Eduard A. Mikaelian[1], Saif A. Mouhammad[2*]

[1]Gubkin Russian State University of Oil and Gas, Moscow, Russia
[2]Physics Department, Faculty of Science, Taif University, Taif, Saudi Arabia
Email: *saifnet70@hotmail.com

Abstract

Pipelines and engines are studied under the thermal stress during operation. With the calculated ratios obtained, it is possible to settle practical issues in operation in order to increase the reliability and safety of equipment. To this end, it is first necessary to investigate the thermal mode of the pipeline, the engine for the planned period by the operations control centre. At the same time, the equipment with the average temperature at a certain time of the year exceeding the design value should be identified. In addition, the equipment concerning the nature of jamming should be studied at the sections specified in such a way. In case of identifying the jammed sections based on the above calculated ratios by thermal deformation and stressing the necessary value of movement of the sections, it should be determined that the actual voltage does not exceed the permissible level.

Keywords

Thermal Stress, Deformation, Jamming in Pipes, Equipment, Thermal Stress Characteristics

1. Introduction

In the operation of one of the main types of machines in the oil, gas and energy sectors pipelines are known pipelines and engines associated with a reduction of reliability and safety [1] [2]. One of the reasons that create unsafe operational conditions for machines in that case is thermal stress, deformation, jams in the machines (rods, pipes and engines) [3] [4].

*Corresponding author.

In connection with this, the article addresses the issue based on the thermodynamic ratios and thermal conductivity to build design ratios by determining the occurring thermal stresses that cause abnormal and unsafe conditions for pipelines, rods and heat engines. The obtained solutions are made available for use in the inspecting, diagnosing or during the technical examination of machines in question [5] [6].

2. Settlement Ratios of Thermodynamics and Heat Transfer for the Thermal Stress of the Equipment

Thermal stress occurs in engineering. In case of gas pipeline systems, thermal stress can occur in a pipeline particularly in the jammed sections, gas compressor units, pistons, turbines etc. [1] [3]. Since the gas temperature in the pipeline section is substantially above the temperature of the soil, an intensive heat exchange between the soil and pipe takes effect. The temperature along the pipeline drops but due to the high coefficient of heat transfer from gas to the wall, the wall temperature is practically the same as the gas temperature (t_g). Since the pipeline is laid at a different temperature (t_{la}), thermal stresses and deformations will inevitably arise due to relative elongations in the linear part of the gas pipeline:

$$\Delta L/L = \alpha_w (t_g - t_{la}). \tag{1}$$

where ΔL—change of length of the pipeline L; α_w—coefficient of thermal, linear expansion of the pipeline and of the rode.

Due to pipeline jamming (soil or dead spots) the linear expansion can be completely destroyed. Compression stress takes place:

$$\sigma = E \times \Delta L/L = \alpha_w (t_g - t_{la}) \times E. \tag{2}$$

where E—the elasticity modulus of the rod material and of the pipe material.

These additional stresses are the greatest value in the initial section of the gas pipeline, at the output of the compressor station CS where gas has the highest temperature. A rational way to combat the thermal stress is to pre-cool gas before it enters the pipeline [2] [4]. Under certain conditions, such as permafrost gas pre-cooling to low temperatures may become the only means of gas transport. In the case of the use of special coolers, unlike air coolers, for example, the compression- or absorption-type chillers, it is expedient to install them before blowers prior to the compression process at the compression station input. It is possible for air coolers during the cold season, when there is a considerable difference in temperature between gas and ambient air temperature while the cooled gas temperature with the help of air coolers does not reach the dew point. In this case, it is possible to either gain in the power supplied on compressing the gas, or increase the capacity of the compressor station reaching 10% or more.

Changing the length of the rod (ΔL) is subject to the following law:

$$\Delta L = \alpha_w \times L(T_1 - T_2) = \alpha_w \times L \times \Delta T, \tag{3}$$

where L—rod length; $\Delta T = T_1 - T_2$—temperature change.

If the elements of bodies, structures freely expand or constrict, *i.e.* are not fixed, the temperature changes do not cause stress. The change in length caused by temperature change (ΔT), is eliminated by the influence of stress σ according to the equation:

$$\Delta L = \sigma \times L/E, \tag{4}$$

On the basis of the last two equations it follows:

$$\alpha_w \times L(T_1 - T_2) = \sigma \times L/E, \tag{5}$$

from where:

$$\sigma = -\alpha_w \times E(T_1 - T_2). \tag{6}$$

If lengthening or shortening the rod is limited only in part, the last equation can be written as follows:

$$\sigma = -k \times \alpha_w \times E(T_1 - T_2), \tag{7}$$

where k—factor depending on the nature of consolidation.

The calculated ratios presented above can be used to conduct the technical inspection of pipeline sections concerning thermal strains and stresses and determine the nature of jamming. In case of clamped areas are identified on the basis of the above calculated ratios in terms of thermal strains and stresses, it may be possible to determine the necessary amount of transfer of the sections based on the condition that the actual voltage does not exceed the permissible one.

3. Examples of the Calculated Temperature Stresses in Pipelines and Heat Engines

3.1. Example 1

A steel rod 50 cm long is fitted between the two walls without any preload at 32˚C. What the tension in the rod at 35˚C is going to be. How this tension will change, if the wall moves by ΔL = 0.25 mm from one another. The thermal expansion coefficient and the modulus of elasticity of the material of the rod are respectively equal to $\alpha_w = 12.5 \times 10^{-6}$ °C^{-1} and $E = 2.04 \times 10^6$ bar.

We first define the tension in the rod without taking into account the nature of fixing:

$$\sigma = \alpha_w \times E(T_2 - T_1) = 12.5 \times 10^{-6} \times 2.04 \times 10^6 (35 - 32) = 76.5 \text{ bar} = 7.65 \text{ Mpa}. \quad (8)$$

Given the nature of fixing the tension in the rod becomes equal to:

$$\sigma = E \times \alpha_w (T_1 - T_2) - \Delta L \times E/L = 76.5 - 0.025 \times 2.04 \times 10^6 / 50 = 943.5 \text{ bar} = 94.35 \text{ Mpa}. \quad (9)$$

The coefficient, which depends on the nature of fixing the rod, is determined by the following formula:

$$k = 1 - \Delta L / [\alpha \times L(T_2 - T_1)] = 1 - 0.025/12.5 \times 10^{-6} \times 50 \times 3 = -12.3. \quad (10)$$

In view of the obtained values of the coefficient, the tension in the rod will fall sharply.

3.2. Example 2

A steel pipe is clamped at a section with the length L = 1000 m without any preload at an average temperature T_1 = 23˚C. What voltage will be in the pipe with an increase in its temperature to T_2 = 26˚C, which is due to its increased load. How this tension will change in the pipeline, if its clamped ends will at a distance of 0.5 m. The thermal expansion coefficient and modulus of elasticity of the pipe material are respectively equal to $\alpha_w = 12.5 \times 10^{-6}$ °C^{-1} and $E = 2.04 \times 10^6$ bar.

First we determine the voltage at a given section of the pipeline when its temperature changes taking into account his clamping:

$$\sigma = \alpha_w \times E(T_2 - T_1) = 12.5 \times 10^{-6} \times 2.04 \times 10^6 (26 - 23) = 76.5 \text{ bar} = 7.65 \text{ Mpa}. \quad (11)$$

In view of the possibility of moving the clamped ends of the pipe by 0.5 m the tension at the considered section will be determined from the following equations:

$$\alpha_w \times L(T_2 - T_1) - \sigma \times L/E - \Delta L = 0; \quad (12)$$

from where

$$\sigma = E \times \alpha_w (T_2 - T_1) - \Delta L \times E/L = 76.5 - 0.5 \times 2.04 \times 10^6 / 1000 = -943.5 \text{ bar} = -94.35 \text{ Mpa}. \quad (13)$$

The coefficient, which depends on the nature of the pipeline's clamping, will be determined by the following formula:

$$k = 1 - \Delta L / [\alpha_w \times L(T_2 - T_1)] = 1 - 0.5/12.5 \times 10^{-6} \times 1000(26 - 23) = -12.3. \quad (14)$$

Since the clamped ends of the pipe can be moved, it is possible to reduce the tension at the considered section to a value that does not exceed the allowable stress of the pipe material. Thus, the possibility of moving the clamped ends of the pipe to the desired value makes it possible to prevent rupture and destruction of the pipeline [6] [7].

The objectives of the technical inspection of pipeline sections by the thermal deformation and stress the nature of clamping. If there are any clamped areas on the basis of the above calculated ratios of thermal deformations and stresses, it may be necessary to determine the amount of movement of the sections based on the condition that the actual stress does not exceed the permissible one.

3.3. Example 3

According to the pipeline temperature data, for the coming period is set exceeding the mean pipe temperature by 5 degrees at the section with the length of 10,000 km. To determine the value by which it is necessary to shift the considered section by its ends, so that the actual stress by the thermal deformations in the pipeline does not exceed the permissible σ = 720 Mpa. The thermal expansion coefficient and modulus of elasticity of the pipe material are taken to be equal, respectively $\alpha_w = 12.5 \times 10^{-6}$ °C^{-1} and $E = 2.04 \times 10^6$ bar = 0.204×10^6 Mpa.

The above problem is an example of this kind of problems of technical inspection of pipelines by thermal deformation and stress, which are offered for use in the inspection, diagnosis, or during the technical examination of pipelines [8] [9].

3.4. Example 4

Determination of the Actual Thermal Stresses in the Cylinder Group Induced by Abnormalities in the Engine

The presented problem is associated with the definition of thermal deformations that can cause scratching in the moving parts of the engine cylinder. The initial data set as follows: *D*—diameter of the cylinder; *b*—the gap between the piston and cylinder; the piston material and its characteristics: α_w—the coefficient of the linear thermal expansion; *E*—modulus of elasticity.

Calculations are made for a diesel engine with the compression ratio equal to $\varepsilon = v_1/v_2 = 12.5$. The rest of the original data and additional conditions of the problem are given in the course of solving. The solution of the problem begins by determining the maximum temperature of the operating procedure for the engine in the following sequence. The molecular weight of the working mixture is determined by the composition of the combustion products: oxygen-O_2 = 0.08; nitrogen-N_2 = 0.72; carbon dioxide-CO_2 = 0.14; water vapour-H_2O = 0.06:

$$\mu = 1/(0.08/32 + 0.72/28 + 0.14/44 + 0.06/18) = 28.98 \text{ kg/mol}. \quad (15)$$

The characteristic constant of the gas mixture:

$$R = 8314/\mu = 8314/28.98 = 286.9 \text{ J/(kg}\cdot\text{K)}. \quad (16)$$

The first average mass isobaric and isochoric heat capacities and adiabatic indices:

$$\begin{aligned} c_{pm} &= \sum_{i=1}^{n} m_i \times c_{pm_i} = 0.72 \times 1.051 + 0.08 \times 0.929 + 0.14 \times 0.859 + 0.06 \times 1.881 \\ &= 1.064 \text{ kJ/(kg}\cdot\text{K)} = 1064 \text{ J/kg}\cdot\text{K}; \end{aligned} \quad (17)$$

$$c_{vm} = c_{pm} - R = 1064 - 286.9 = 777 \text{ J/(kg}\cdot\text{K)}; \quad (18)$$

$$k = c_{pm}/c_{vm} = 1064/777 = 1.37. \quad (19)$$

Then the thermodynamic parameters at each point of the operating procedure are calculated (points: 1) the beginning of the compression process; 2) the beginning of the supply of heat to the working medium—the beginning of the combustion process; 3) and 4) beginning and the end of the enlargement process of the working medium in the cylinder) [5] [10]. Point 1: the initial data-p_1 = 105 Pa; t_1 = 30°C; from the equation of state of this point we can obtain the specific volume:

$$v = R \times T/P = 286.9 \times 303/10^5 = 0.869 \text{ m}^3/\text{kg}. \quad (20)$$

Point 2: the initial data- $\varepsilon = v_1/v_2 = 12.5$, from where v_2 = 0.869/12.5 = 0.070 m^3/kg; rate of expansion n = 1.32; from the equation of:

$$p_1/p_2 = (v_2/v_1)^{n_1}, \quad (21)$$

we derive the pressure at the end of the compression process:

$$p_2 = p_1 \times (v_1/v_2)^{n_1} = 10^5 \times 12.5^{1.32} = 2.767 \times 10^6 \text{ Pa} = 2.767 \text{ Mpa}. \quad (22)$$

From the equation of state for point 2 we obtain the value of the absolute temperature:

$$T_2 = p_2 \times v_2/R = 2.767 \times 10^6 \times 0.07/286.9 = 675 \text{ K} = 402^\circ\text{C}. \quad (23)$$

Point 3: the initial data- $v_3/v_2 = 1.9$; from where $v_3 = 1.9 \times 0.07 = 0.133 \text{ m}^3$; $p_2 = p_3 = 2.767$ Mpa .

From the corresponding equation of state we evaluate the temperature of the working medium of the engine at the beginning of the expansion process:

$$T_3 = p_3 \times v_3/R = 2.767 \times 10^6 \times 0.133/286.9 = 1283 \text{ K} = 1010^\circ\text{C}. \quad (24)$$

Point 4: at $v_4 = v_1 = 0.869 \text{ m}^3$ from the equation of the expansion process calculated is the temperature at the end of the process:

$$p_4 = p_3 \times (v_3/v_4)^{n_2} = 2.767 \times 10^6 \times (0.133/0.869)^{1.25} = 0.265 \times 10^6 \text{ Mpa}. \quad (25)$$

The gas temperature at the end of the expansion process is determined from the equation of state:

$$T_4 = p_4 \times v_4/R = 0.265 \times 10^6 \times 0.869/286.9 = 801 \text{ K} = 528^\circ\text{C}. \quad (26)$$

Thus, based on thermodynamic calculations the maximum temperature of the working medium of the engine will be equal to the temperature at the end of expansion: $T_3 = 1283 \text{ K} = 1010^\circ\text{C}$.

Further, the structural and technological parameters are included as the initial data in the calculations. Cast iron is taken as piston material, for which the thermal expansion coefficient and modulus of elasticity are respectively as follows: $\alpha = 1.07 \times 10^{-6}\,{}^\circ\text{C}^{-1}$ and E = 10 bar; the piston diameter D = 40 cm = 0.4 m; maximum and minimum clearances between the piston and cylinder, respectively, in the upper and lower parts of the piston are recommended equal:

$$b_1 = (1/175 - 1/200) \times D; \text{ take } b_1 = 2.2 \text{ mm}; \quad (27)$$

$$b_2 = (1/900 - 1/1000) \times D; \text{ take } b_2 = 0.4 \text{ mm}; \quad (28)$$

In the operation, the real and maximum temperature of the engine exceeds the maximum and design cycle temperature calculated using the formula (24). The growth of the real maximum temperature may occur due to all kinds of losses occurring in the running engine, high loads, deteriorating technical condition, disturbance of the cooling mode etc. [11] [12]. In this regard, the following values of the possible, actual and maximum temperatures in the engine are set:

$$T^1 = 1333 \text{ K}; T^2 = 1383 \text{ K}; T^3 = 1433 \text{ K}. \quad (29)$$

The thermal elongations are calculated in the case of extensions of the minimum clearance b_2 = 0.4 mm = 0.04 cm; piston diameter $d = D - b_2 = 400 - 0.4 = 399.6 \text{ mm}$.

The phenomenon of scuffing in the cylinder occurs under the condition when the temperature elongation becomes equal to or greater than the minimum clearance between the piston and cylinder.

A change in the diameter of the piston caused by an increase in temperature of the combustion products is computed for the above-taken three maximum temperature values based on the known formula:

$$\begin{cases} \Delta d^1 = \alpha \times d(T^1 - T_1) = 1.07 \times 10^{-5} \times 399.6 \times (1333 - 1283) = 0.214 \text{ mm}; \\ \Delta d^2 = \alpha \times d(T^2 - T_1) = 1.07 \times 10^{-5} \times 399.6 \times (1383 - 1283) = 0.428 \text{ mm}; \\ \Delta d^3 = \alpha \times d(T^3 - T_1) = 1.07 \times 10^{-5} \times 399.6 \times (1433 - 1283) = 0.641 \text{ mm}. \end{cases} \quad (30)$$

Then, calculated are the real gaps formed for three adopted and maximum temperatures caused by different operating conditions:

$$\begin{cases} \Delta^1 = \Delta_2 - \Delta d^1 = 0.420 - 0.214 = 0.206 \text{ mm}; \\ \Delta^2 = \Delta_2 - \Delta d^2 = 0.420 - 0.428 = -0.008 \text{ mm}; \\ \Delta^3 = \Delta_2 - \Delta d^3 = 0.420 - 0.641 = -0.221 \text{ mm}. \end{cases} \quad (31)$$

The obtained results indicate that in the risen actual temperature in the engine cylinder with respect to the maximum design temperature of the cycle by 100 degrees, the conditions for scuffing of the cylinder face, which correspond to the second value of the adopted maximum temperature value in the engine cylinder, are created. If this difference is exceeded above 100 degrees, dangerous failures in the engine are possible. Thus, in the third case, when the difference between the actual engine temperature and maximum design temperature of the cycle is equal to 150 degrees, a serious engine breakdown associated with piston seizure in the cylinder may occur.

Let us determine the temperature stress in the considered three cases of the maximum temperature in the engine by the previously given formula:

$$\begin{cases} \sigma_1 = \alpha \times E\left(T^1 - T_2\right) = 1.07 \times 10^{-5} \times 10^6 \times 50 = 535\,\text{bar}; \\ \sigma_2 = \alpha \times E\left(T^2 - T_2\right) = 1.07 \times 10^{-5} \times 10^6 \times 100 = 1070\,\text{bar}; \\ \sigma_3 = \alpha \times E\left(T^3 - T_2\right) = 1.07 \times 10^{-5} \times 10^6 \times 150 = 1605\,\text{bar}. \end{cases} \quad (32)$$

The derived results indicate that when the temperature inside the engine cylinder is exceeded relative to the maximum design cycle temperature by 100 degrees (whether in °C or °K), there is a tension in the engine exceeding the permissible temperature equal to 800 - 900 bar for cast iron.

Similar calculations can be made for internal combustion engines made of other materials. For example, if a piston is made of aluminium, the thermal expansion coefficient and elastic modulus of the material will be equal respectively to: 2.2×10^{-5} °C^{-1} and 7.5×10^5 bar; the maximum and minimum clearances between the piston and cylinder of the engine are recommended within the following limits: $b_1 = \left(1/100 - 1/125\right) D$ and $b_2 = \left(1/350 - 1/500\right) D$.

4. Conclusions

The Chapter "Settlement ratio of thermodynamics and heat transfer for the thermal stress of equipment" provides the corresponding formulas to carry out the task of the technical inspection of pipeline sections and heat engines by thermal deformation and stress and to determine the nature of clamping.

The Chapter "Examples of the Calculated Temperature Stresses in Pipelines and Heat Engines" considers two groups of tasks to control the temperature mode of pipelines and heat engines.

Example 1 concerns the thermal stresses and deformation in the rods. In case of any clamped areas based on the above calculated ratios by thermal strains and stresses determined is the required value of movement of these areas based on the condition that the actual stress does not exceed the permissible one.

Examples 2 and 3 address the issues related to the technical inspection of pipelines for thermal strains and stresses based on which practical recommendations for use during the inspection, diagnosing or technical examination of pipelines and ensuring the safe operation of pipelines are presented.

Example 4 addresses some issues devoted to control the temperature mode of the cylinder group to provide a reliable and safe operation of internal combustion engines trouble-free.

References

[1] Mikaelian, E.A. (1994) Operation of Gas Turbine Pumping Units at Compressor Stations of Gas Pipelines (Ekspluatatsiya gazoturbinnykh perekachivayushchikh agregatov kompressornykh stantsii gazoprovodov). NEDRA, Moscow, 304. http://ogbus.ru/authors/Vanchin/Vanchin_3.pdf

[2] Mikaelian, E.A. (2000) Maintenance of the Power Engineering Equipment, Gas-Turbine and Gas-Compressor Units of the Gas Collection and Transfer Systems (Tekhnecheskoe obuslogivani energotekhnologicheskovo oborudivani, gazoturbinnykh perekachivayushchikh agregatov sestemi cobra e transporta gasa). Fuel and Energy, Moscow, 304. http://www.dobi.oglib.ru/bgl/5076.html

[3] Mikaelian, E.A. (2014) Research into Gas-Turbine and Gas-Pumping Units of Gas Pipelines under the Low-Pressure Gas-Transfer Technology (Esledovanye effectivnosti rabote gazoturbinnykh gazoperekachivaushikh agregatov gazoprovodov v uslovyakh nizkonapornoi tekhnalogy transporta gaza). *Gas Turbo Technology*, **7**, 28-33. http://www.gtt.ru/content/view/1387/123/

[4] Mikaelian, E.A. (2015) Evaluation of the Effective Performance and Technical Condition of the Gas-Turbine and Gas-Compressor Units in Operation (Otsenka effectivnoi rabote e tekhnicheskovo sostoyani gazoturbinnykh GPA v oslovyakh Ekspluatatsiy). *Gas Turbo Technology, Designated Research and Information Journal*, **2**, 18-22.

http://oil-book.narod.ru/door/door2/521.htm

[5] Mikaelian, E.A. (2015) Evaluation of Mechanical Losses of Power of the Gas-Transfer Units (Otsenka mekhanecheskikh poter moshnosty gazoperekachivaushikh agregatov). *Gas Turbo Technology, Designated Research and Information Journal*, **3**, 28-32. http://www.gtt.ru/content/view/1441/1/

[6] Mikaelian E.A. (2015) Proposals for the Harmonisation of the Characteristics of Gas-Turbine Units and Gas Centrifugal Compressors Given the Power Lost for the Mechanical Losses during Power Transmission from the Drive (Soglasovani kharakteristik gazotorbennekh ostanofak e gazovekh tsentrobegnekh kompresorof s ochotom moshnosti, teryaemoi na mekhanicheskie poteri pri peredache energy ot prevoda). *Quality Management in Oil and Gas Industry*, **1**, 15-18. http://instoilgas.ru/d/988947/d/microsoftword-annotatsii1-15sokrrus.pdf

[7] Mikaelian, E.A. (2001) Improving the Quality, Ensuring Reliability and Safety of the Gas and Oil Mains to Improve the Serviceability (Poveshene kachestva, obesbecheny nadognosty e bezopasnosty magestralnekh gazonefteprovodof dla covershenstvovaneya ekspluatatsionnoi prigodnosty). Fuel and Energy, Moscow, 640.

[8] Mikaelian, E.A. and Mouhammad, S.A. (2015) Survey of the Gas-Turbine Gas-Compressor Units of Compressor Plants Based on the Modified Thermodynamics Model. *International Journal of Alternative Fuels*, **17**, 1125-1131. http://recentscience.org/article/myarticles.php

[9] Vladimirov, A.I. and Kershenbaum, Y.V. (2008) Industrial Safety Compressor Stations. Management of Safety and Reliability. Inter-Sector Foundation "National Institute of Oil and Gas", Moscow, 640. http://www.mdk-arbat.ru/bookcard?book_id=3304125

[10] Mikaelian, E.A. (2015) Methods for Determining the Characteristics and Condition of Pipelines (Sposobe opredelenie kharakteristik e tekhnicheskovo coctoyanie gazoprovodof). *Energetik*, **5**, 19-22. http://elibrary.ru/item.asp?id=23618568

[11] Mikaelian, E.A. (2014) From the Energy-Saving Technology to the Resource-Saving One, from Energy Audit to Resource Audit (Ot energosberegaushy tekhnalogy k resorsosberegaushy, ot energoaodeta k resorsoaodetu). *Gas Industry*, **4**, 70-73. http://chemtech.ru/ktp/kompressornaya-tekhnika-i-pnevmatika-22015

[12] Mikaelian, E.A. (2014) Survey of Equipment of the Gas and Oil Transport Systems (Obsledovanie oborodivaniya gazoneftetransportnikh sestem). *Heavy Engineering Industry*, **4**, 18-25. http://www.tiajmash.ru/

16

Harmonic Contributions of Utility and Customer Based on Load Model Using Field Measurements

Ömer Gül, Tayfun Gündoğdu

Department of Electrical Engineering, Istanbul Technical University, Istanbul, Turkey
Email: gulomer@itu.edu.tr, tgundogdu@itu.edu.tr

Abstract

In recent years, sinusoidal waveform of the current and voltage disturbs in the electrical distribution system because of the due to the increasing number of non-linear loads. Many standards of IEC and IEEE standards have been published in order to limit the voltage and current waveform distortion. The operators of the electricity distribution network widely use the power quality monitoring systems at the point of common connection (PCC). It has been identified that there are substantial number of harmonic currents excess of the standards transferred to the grid according to the data obtained from power quality monitoring systems. In case of exceeding the limits specified in the standards, there is a need to determine the network and customer responsibilities for the implementation of required sanctions. In this study, using recorded data at the PCC of a medium voltage electrical distribution system, voltage and current harmonic distortion responsibilities of the network and customer are determined by the improved harmonic current vector method. Up-to-date load model based on field measurement which provides more accurate results has been used instead of the constant load impedance in the proposed method.

Keywords

Harmonic Contributions, Harmonic Distortion Load Model, Norton Equivalent

1. Introduction

In addition to the known harmonics sources, the wide spread use of power electronic devices cause to injection of more harmonic currents in the electrical distribution network. Therefore, voltage waveform distortion is increasing with each passing day. Consisting harmonic currents as a result of connecting of a non-linear load to the network causes the network's voltage to become harmonic as a result of the harmonic voltage drop at the network impedance. In this case, even if the linear load is connected to the network, harmonic currents will continue to flow reference measurement point [1]. To avoid an increase in waveform distortion of the electricity distribution network, some harmonic currents restrictions have been imposed on the production stage of the electrical devices according to standards.

In case of to be exceeded of the limit values which are determined by the standards, it is required to take precautions in order to reduce the waveform distortion. Some responsibilities in the distortion of the waveform should be determined to realize who is responsible for this distortion. The source side of the PCC is under the network responsibility while the load side of the PCC is under the customer responsibility. Therefore, the field measurements are of great importance for determining the responsibilities of the customer or/and the network. In the case of the determination of responsibilities, it will be able to punish to customer and/or the network according to their responsibility rate at the PCC.

The harmonic analysis studies of three-phase power distribution networks, modeling of loads [2]-[8], and the identification of customer contributions [9] are still up-to-date. Efforts to seek solutions for harmonics injected by the nonlinear load and under the condition of the load is not fully specified have begun a few decades ago [10]. A virtual instrument dedicated for the analysis of power signals, in order to determine the existence and characteristics of harmonic distortion is proposed. The processing algorithm are based on estimate harmonic emission level and obtain the real-time estimates of regression coefficients, and then calculate the level of the harmonic impedance and emission estimates in real time [11].

In a study using the value of the reference impedance, changes in the harmonic sources are included by converting the impedance changes into equivalent current sources [9]. The method of harmonic current vector, which shows the network and customer as their Norton equivalent, seems to be theoretically correct [10]. Harmonics caused by the customer and network are calculated by using this method. The most important shortcoming of this method is that it requires knowledge of customer and consumer impedances. The load impedance varies quite a lot according to the network impedance. Therefore, it is not clear how to use reference impedance approach for the loads [10]. One other study clarifies the harmonic current vector method to determine the reference impedance. By selection of load resistance as reference impedance calculations are not affected under the resonance situation [12]. Reactive power flow direction at PCC has been based on by another study [14]. Many methods have been applied in order to keep the harmonic distributions within certain limits between the client and distribution network. Since the measurements taken from different points in a network at the same time is quite difficult and expensive [14]-[18], to be able to perform the measurements at the PCC more practical approaches should be developed [19]. Despite using various measurement methods in the determination of the total harmonic distribution at the PCC, none of them is practical. Power flow direction method, which is a well-known method [20], has been considered as not a feasible method in the determination of the harmonic source [21]. Several different electrical quantities have been discussed as a criterion for harmonic cost distortion, however, only estimated harmonic cost distortion could be calculated [22].

Harmonic Current Method, which is modeled by Norton equivalent circuit of the customer and the network for a specific harmonic degree, is considered theoretically correct, although it shows some practical shortcomings [12]. This method does not take into account any resonance condition; it just requires knowledge of impedance of the customer and the network at any harmonic degree. Reference impedance methods were found in order to determine the customer and the network impedances, but the studies are still continuing [12].

While the load impedance is constantly changing, to use the reference impedance method which assumes the load impedance as constant does not provide sufficiently accurate results. Because, switching positions of the filter and compensation capacitors can cause very large changes in the load impedance. Instead of the reference impedance method, use of harmonic current method with the approach to updating the load impedances online by using the field measurements will provide more accurate results. Furthermore, switching positions of the filters and capacitors have been already taken in to account in the proposed algorithm by means of the long-term measurements at the PCC.

This study proposes a new method, which is based on measurement of current and voltage at the PCC, in order to determine the responsibilities of the network and costumer. Moreover, it is provided that the current vector method has been improvement by using constantly changing impedance. In the proposed method, up-to-date load model based on field measurement has been used instead of the constant load impedance.

In order to determine the current and voltage contributions to the PCC, Norton and Thevenin equivalent circuits of the network and costumer have been used with an algorithm. The proposed algorithm is verified in

MATLAB by processing of the measurements. After the processing of the data collected from the PCC for a long-time period, the responsibility of the customer and the network at the PCC and current and voltage THD values have been obtained.

2. Harmonic Vector Method with Reference Impedance

Impedance of the customer and the network are of great importance in determining the compliance of the harmonic standards. The contract between the customer and the network is determined by value of the reference impedance [15]. Usually, the reference impedance value is not exactly equal to the actual value of the impedance. These reference impedance values are taken as basis in determining whether the customer violates or not the harmonic limits.

Harmonic Vector Method is the representation of the circuit of the customer and network sides with Thevenin/Norton equivalent circuits. Harmonic current or voltage at the PCC are separated into two components (customer and network) in order to identify the individual harmonic distribution. The equivalent circuit is composed as h order harmonic current or voltage source for both customer and network circuits.

Although, characteristics of the customer impedance and especially frequency are known rarely, the impedance of the network side is usually particular. Active harmonic impedance of the customer side is very difficult to determine due to the constantly changing characteristics of the customer side. Therefore, equivalent impedances are determined as reference impedance. The main idea is to describe either the customer or the network side impedances as a reference to an additional source of harmonic. In general, resistive component of the customer side is determined as customer reference impedance. This situation depends on the distribution of the current or voltage harmonics [23].

3. Harmonic Voltage Contributions

To determine the contribution of the harmonic voltage, the Thevenin equivalent circuit shown in **Figure 1** is used. This equivalent circuit consists of network (U_{Uh}) and customer (U_{Ch}) voltage source, and network $(Z_{Uh-\text{ref}})$ and customer $(Z_{Ch-\text{ref}})$ reference impedance which include h order harmonic.

As can be seen in **Figure 2** where R_{Cls} is the customer reference impedance, an additional reactance voltage drop is converted to a harmonic source. Customer reference impedance is modeled by a resistor and reactance which are connected in series as seen in **Figure 2**.

$$Z^*_{Ch-\text{ref}} = R_{Cls} = \frac{P}{|I_{P1}|^2} = \frac{|U_{P1}| \cdot \cos\phi_{P1}}{|I_{P1}|} \tag{1}$$

$$U_{Ch} = U_{Ph} - I_{Ph} \cdot R_{Cls} \tag{2}$$

$$U_{Uh} = U_{Ph} + I_{Ph} \cdot Z_{Uh-\text{ref}} \tag{3}$$

The customer reference impedance is calculated as Equation (1) using the active power P measured at connection point and current at the fundamental frequency I_{P1}. Where, ϕ_{P1} is the phase angle between the current

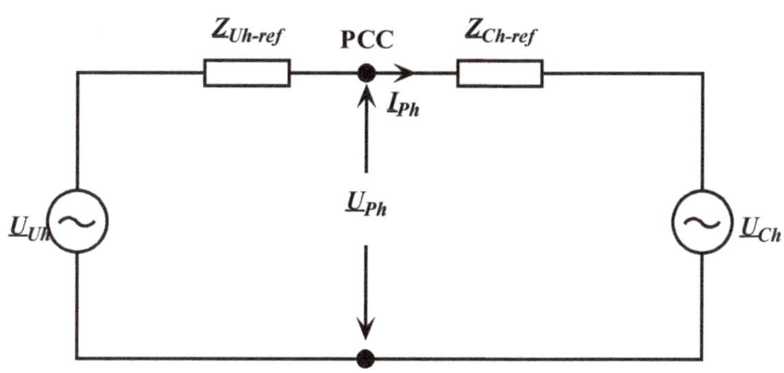

Figure 1. Thevenin equivalent circuit of network and customer at any order of the harmonic.

and voltage at the fundamental frequency. Customer harmonic voltage (U_{Ch}) and network harmonic voltage (U_{Uh}) are obtained from the Equations (2) and (3) according to **Figure 3**.

Generally, customers are connected to the network via transformer. So, reference impedance of the network is determined as the sum of the impedance of the last transformer before connected to the customer and the short-circuit impedance of the network. This impedance composes the basic part of the network impedance and value of this impedance can be found easily. However, networks should provide an impedance close to the current impedance and should avoid from the unnecessary impedances which cause the harmonic distortion.

Additives to voltage value at the customer and network connection point are given Equations (4) and (5) applying the superposition principle as shown in **Figure 3**.

$$\underline{U}_{U-Ph} = \frac{R_{Cls}}{\underline{Z}_{Uh-\text{ref}} + R_{Cls}} \cdot \underline{U}_{Uh} \tag{4}$$

$$\underline{U}_{C-Ph} = \frac{\underline{Z}_{Uh-\text{ref}}}{\underline{Z}_{Uh-\text{ref}} + R_{Cls}} \cdot \underline{U}_{Ch} \tag{5}$$

Harmonic voltage phasor measured at the connection point (U_{Ph}), is the vectoral summation of the harmonic voltage contributions of the customer and network. Taking account the projection of contribution of the network voltage (U_{U-Ph}) and customer voltage (U_{C-Ph}) to the measured voltage of the connection point (U_{Ph}), scalar network voltage harmonic contribution (U_{Uhs}) and customer scalar harmonic contribution (U_{Chs}) are calculated as Equations (6) and (7).

$$U_{Uhs} = \left|\underline{U}_{U-Ph}\right| \cdot \cos\left(\delta_{U-Ph} - \delta_{Ph}\right) \tag{6}$$

$$U_{Chs} = \left|\underline{U}_{C-Ph}\right| \cdot \cos\left(\delta_{C-Ph} - \delta_{Ph}\right) \tag{7}$$

where, δ_{U-Ph} is angle between contribution of the network harmonic voltage and connection point voltage, and δ_{C-Ph} is the angle between contribution of the customer harmonic voltage and connection point voltage. Total harmonic voltage of the connection point is equal to sum of the scalar harmonic components of the network and the customer. This scalar quantities can be positive or negative values, and thus current can increase or decrease at the PCC.

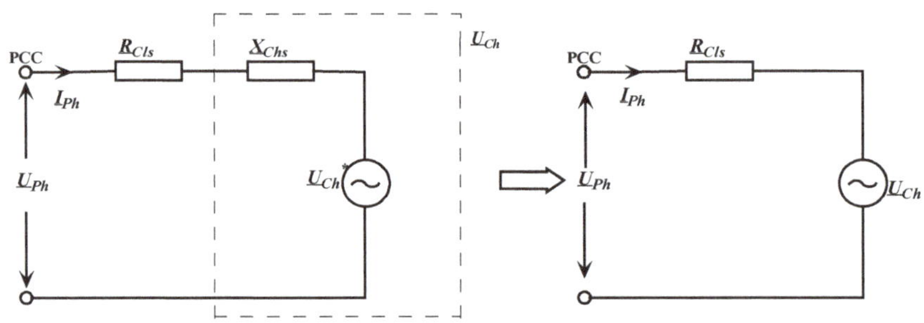

Figure 2. Conversion of reactance voltage drop to an additional source.

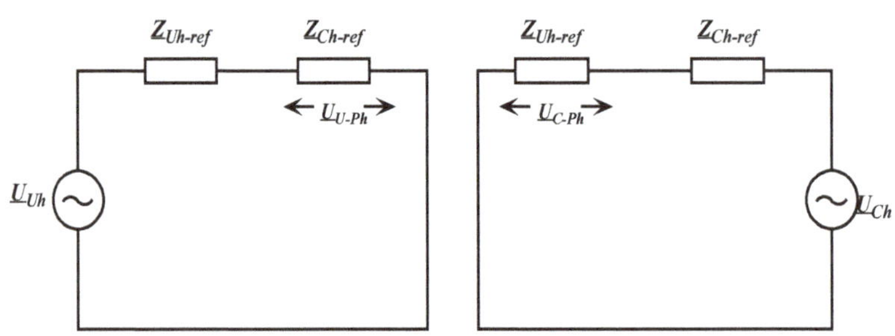

Figure 3. Implementation of superposition principle for the detection of customer and network voltage harmonics.

4. Harmonic Current Contributions

Resonance between the load and the network status is a common occurrence. Limits defined in the standards are usually exceeded due to the increasing harmonic distortion in case of resonance. Responsibilities should be determined in order to put leverage on the customer and network sides when the limits are exceeded. Norton equivalent circuit in which network and customer reference impedances are used shown in **Figure 4**. Where I_{Uh} and $Z_{Uh\text{-ref}}$ indicate the network harmonic current and reference impedance at the *h*. harmonic order, respectively, and I_{Ch} and $Z_{Ch\text{-ref}}$ indicate the customer harmonic current and reference impedance at the *h*. harmonic order, respectively.

Reference impedance calculation is quite difficult due to the continuously changing state of the load. Reference impedances may be determined due to the measured values at the reference connection point (PCC) according to the general impressions among the researchers. Reference impedance of the network is the sum of the short-circuit impedances and the impedance of the first transformer before the customer connection point. Resistive part of the load impedance of the customer (R_{Clp}) may be found by the measurements at the connection point. Customer impedance can be expressed by a resistance and an impedance parallely connected while dealing with the calculations of the harmonic current contributions.

Customer impedance can be calculated as in Equation (8) by using the active power (P) and basic component of the voltage (U_{P1}) in absence of customer side of active load production.

$$Z_{Ch\text{-ref}} = R_{Clp} = \frac{|U_{P1}|^2}{P} = \frac{|U_{P1}|}{|I_{P1}| \cdot \cos\phi_{P1}} \tag{8}$$

In the Equation (8), is the angle between measured current and voltage at the fundamental frequency, and $|U_{P1}|$ and $|I_{P1}|$ are the voltage and current magnitudes are the fundamental frequency. Customer and network harmonic currents are obtained by Equations (9) and (10) according to **Figure 5**.

$$I_{Ch} = \frac{\underline{U}_{Ph}}{\underline{R}_{Clp}} - I_{Ph} \tag{9}$$

$$I_{Uh} = \frac{\underline{U}_{Ph}}{\underline{Z}_{Uh\text{-ref}}} + I_{Ph} \tag{10}$$

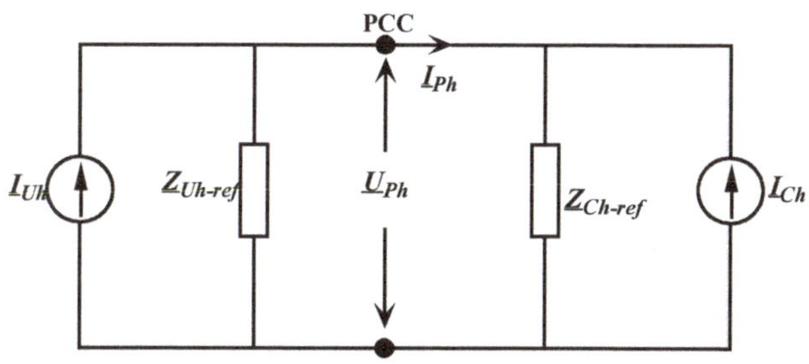

Figure 4. Norton equivalent circuit for the analysis of harmonic current in a specific-grade.

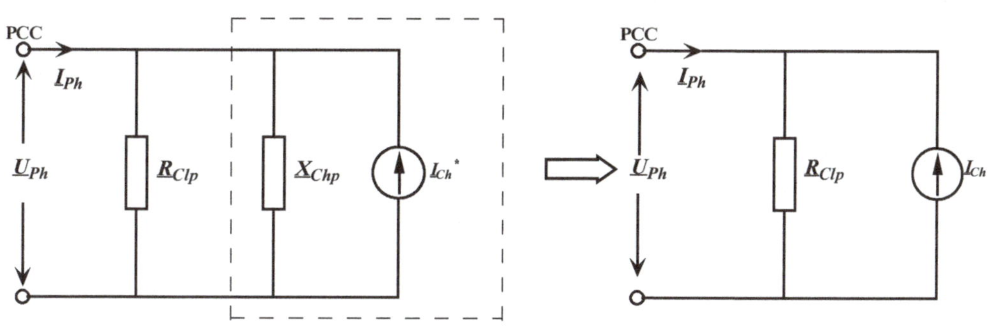

Figure 5. Transformation of the reactance current to an additional harmonic current source.

Harmonic current contributions of the network and customer to the I_{Ph} connection point are determined by principle of Superposition. Customer and network harmonic current contributions are determined by using Equations (11) and (12) according to **Figure 6**.

$$\underline{I}_{U-Ph} = \frac{\underline{Z}_{Uh-\text{ref}}}{\underline{Z}_{Uh-\text{ref}} + R_{Clp}} \cdot \underline{I}_{Uh} \tag{11}$$

$$I_{Uh} = \frac{\underline{U}_{Ph}}{\underline{Z}_{Uh-\text{ref}}} + I_{Ph} \tag{12}$$

Total Harmonic current (I_{Ph}) at the connection point can be separated into two scalar components as seen in **Figure 7**.

Scalar harmonic contributions of the Customer and Network are calculated as in Equations (13) and (14) by using harmonic currents of customer (I_{C-Ph}) and network (I_{U-Ph}) using the projection of these currents to total harmonic current (I_{Ph}).

$$I_{Uhs} = \left|\underline{I}_{U-Ph}\right| \cdot \cos\left(\phi_{U-Ph} - \phi_{Ph}\right) \tag{13}$$

$$I_{Chs} = \left|\underline{I}_{C-Ph}\right| \cdot \cos\left(\phi_{C-Ph} - \phi_{Ph}\right) \tag{14}$$

where, ϕ_{U-Ph} is the phase angle between the I_{U-Ph} and I_{Ph}, and ϕ_{C-Ph} is the phase angle between the I_{C-Ph} and I_{Ph}. As seen in **Figure 7**, I_{Ph} is equal to sum of the scalar harmonic current of the network (I_{Uhs}) and customer (I_{Chs}). If there any negative signed component in the summation, it compensates a specific-grade positive signed component and reduces the amplitude of related harmonic.

5. Load Model Using Field Measurements

A distribution system comprises a number of loads conveniently supplied by circuits from the nearest distribution point. The distribution circuits' configuration depends on the particular load requirements. A considerable number of loads (linear or nonlinear) are located so close together and supplied from the main distribution point, that they can be considered as whole. The aggregate form of various individual loads called general loads. Residential and commercial buildings can be called as general loads since they have a number of linear and nonlinear load combinations rated at low power values. Owing to the fact that some of the loads in buildings switch on and switch off continuously, the buildings as a general load have varying characteristics and mostly operate unbalanced. Moreover, due to harmonic currents drawn by nonlinear loads, the waveform of voltage in the

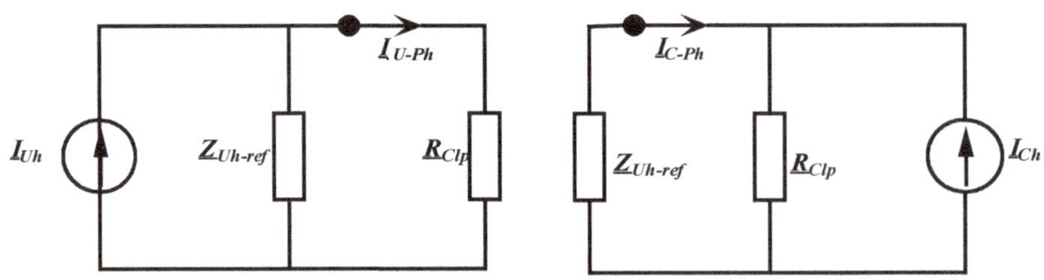

Figure 6. Determination of harmonic current contributions by principle of Superposition.

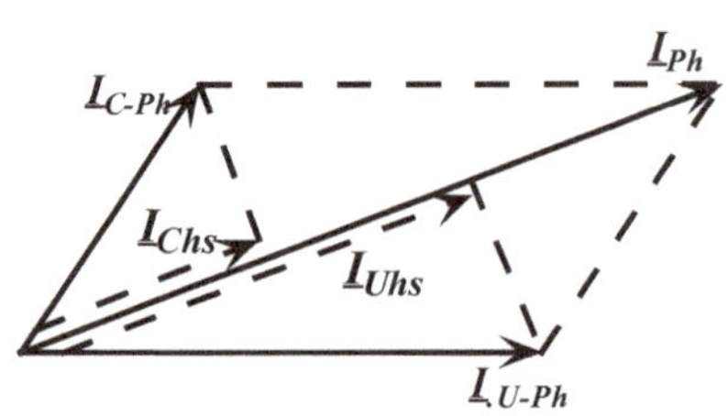

Figure 7. Harmonic current contributions of the customer and network to the harmonic current of the connection point.

buildings becomes distorted. It is therefore that the buildings are very sophisticated loads in terms of modeling. Consequently, using the measured voltage and current data to model these types of loads can be an alternative solution.

The generalized time domain load model is based on actual recorded data and is an extension of the ideas proposed in [24]. The total load current $(i_T(t))$ is split into components that are orthogonal to the voltage. That part of the total current that corresponds to uncommon harmonics with the voltage is called the generated component (i_g) [18]. The current (i_p) is the component in phase with the source voltage and is modeled as the current through a resistor. The quadrature component of the current (i_q) given in (15) is split up into an inductive reactive component (i_l), a capacitive reactive component (i_c), and a residual component (i_r). The component (i_g) exists only when there are uncommon harmonics between the load voltage and the current. The residual component $i_r(t)$ exists when complete extraction of orthogonal components cannot be done. This component is orthogonal to the voltage and is shown in **Figure 8** as the current flowing through a "Black Box". For load modeling applications however, this "Black Box" may be replaced by a current source of appropriate magnitude and phase angle at each common harmonic frequency or as an equivalent impedance at each harmonic frequency.

$$i_q(t) = i_T(t) - i_g(t) - i_p(t) = i_c(t) + i_l(t) + i_r(t) \tag{15}$$

Given the measured voltage $v(t)$ and current $i_T(t)$ as

$$\begin{aligned} v(t) &= \sqrt{2}\sum_{n=1}^{N} V_n \cos(n\omega t + \alpha_n), \\ i_T(t) &= \sqrt{2}\sum_{n=1}^{M} I_n \cos(n\omega t + \beta_n). \end{aligned} \tag{16}$$

where N and M are the number of harmonics in the voltage and current respectively. V_n and I_n represent the RMS values of the nth harmonic voltage and current respectively. α_n and β_n represent the phase angles of the nth harmonic voltage and current respectively. The load parameters *R*, *L*, and *C* are calculated based on the voltage $V(t)$ and current $i(t)$ as follows:

$$R = \frac{(v(t))^2}{\frac{1}{T}\int_0^T v(t)\cdot i(t)\,\mathrm{d}t}, \quad L = \frac{\frac{1}{T}\int_0^T (w(t))^2\,\mathrm{d}t}{\frac{1}{T}\int_0^T i(t)\cdot w(t)\,\mathrm{d}t}, \quad C = \frac{\frac{1}{T}\int_0^T i(t)\cdot\frac{\mathrm{d}v(t)}{\mathrm{d}t}\mathrm{d}t}{\frac{1}{T}\int_0^T \left(\frac{\mathrm{d}v(t)}{\mathrm{d}t}\right)^2 \mathrm{d}t} \tag{17}$$

where T denotes the period, $w(t)$ is defined as the alternating part of $\int_0^T v(t)\mathrm{d}t$. In the calculation of the required costumer parameters, mentioned load model has been used. And the calculated parameters are given in Section 5.

6. Proposed Harmonic Contribution Method

In order to determine the responsibilities of the network and costumer, real impedance value of the customer which is calculated by measuring current and voltage at the PCC has been used in proposed model instead of the constant load impedance. To be able to use the proposed method, equivalent model of the grid system should be

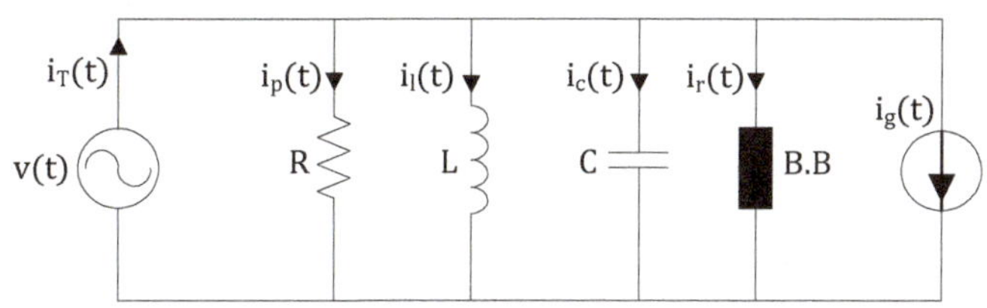

Figure 8. General equivalent network and current components.

prepared after the measurements at the PCC. After the establishment of Norton and Thevenin equivalent, contributions at the PCC can be found precisely.

In this study, superposition method is applied to the equivalent circuit of this grid system by using MATLAB after modeling of a real grid system. Norton equivalent circuit is used for harmonic current contributions while Thevenin equivalent circuit is used for harmonic voltage contributions. It is very important to determine the network and customer sides in a real grid system. Actually, there are lots of customers connected in parallel to each other on a real grid system. In **Figure 9** it shown that from where the field measurements should be performed in the case of there are multiple customer on the system.

MATLAB program has been prepared to calculate the fundamental, 5th, 7th, 11th, 13th and 17th harmonic levels of the current and voltage of customer and network sides, individually. Harmonics that are the 3th and multiples of three and that are not found on symmetrical and balanced systems are not taken into account. The circuit, whose field measurements are performed, has been modelled as given in **Figure 10**.

Decomposition of the network and customer current and the voltage are done at the PCC by using **Figure 11** and **Figure 12**. In order to determine percentage of the contributions of the network and customer currents at the PCC, Equation (18) is used. $I_r(t)$ and $I_g(t)$ are modeled as a single current source (I_C), and RLC are modeled as a single impedance (Z_C) at the customer side of the grid model as shown in **Figure 11** and **Figure 12**. Load impedance has been updated with repeated measurements.

$$I_u = \frac{V_{pcc}}{Z_u} + I_{pcc}, \quad I_c = \frac{V_{pcc}}{Z_c} - I_{pcc}, \qquad (18)$$

$$I_{upcc} = \frac{Z_u}{Z_c + Z_u} I_u, \quad I_{cpcc} = \frac{Z_c}{Z_c + Z_u} I_c.$$

where, I_{upcc} indicates the contribution of the PCC current to the network current and I_{cpcc} indicates the contribution of the PCC current to the customer current. I_u and I_c indicate the modelled currents of the network

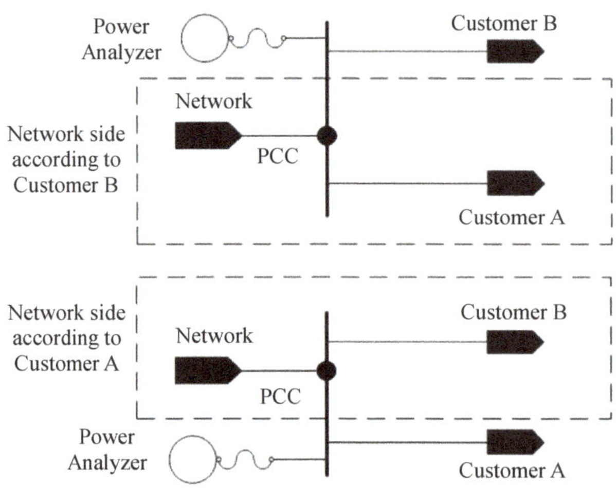

Figure 9. View of the PCC and determination of the grid according to customers.

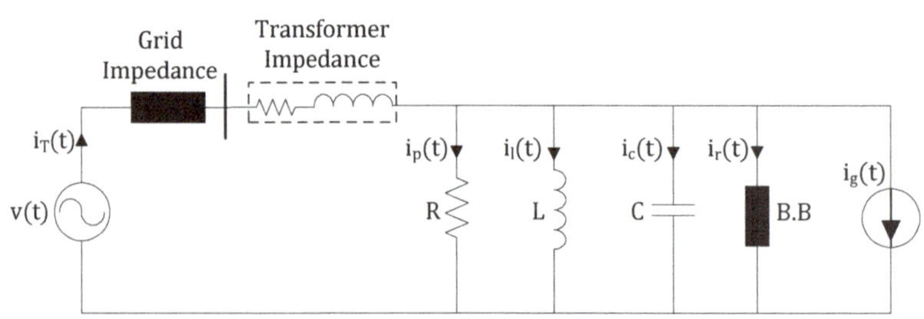

Figure 10. Model of customer and network.

and customer, respectively. The accuracy of the calculated network and customer currents can be proven with the condition of being equal to each other of the result of Equation (19) and measured value of the PCC current [9] [14].

$$\left|I_{pcc}\right| = I_{upcc} + \left(-I_{cpcc}\right) \tag{19}$$

In order to realize the contribution of the network and customer voltages to the voltage of PCC in what rate Equation (20) is used.

$$V_{upcc} = \frac{Z_u * Z_c}{Z_u + Z_c} I_u, \quad V_{cpcc} = \frac{Z_u * Z_c}{Z_u + Z_c} I_c \tag{20}$$

where, V_{upcc} indicates the contribution of the PCC current to the network voltage and V_{cpcc} indicates the contribution of the PCC current to the customer voltage. The accuracy of the calculated network and customer voltages can be proven with the condition of being equal to each other of the result of Equation (21) and measured value of the PCC voltage [9] [13].

$$\left|V_{pcc}\right| = V_{upcc} + V_{cpcc} \tag{21}$$

7. Results of the Proposed Method with Field Measurements

One-line diagram of the test system where the measurements were recorded with permission of the Bosporus Electricity Distribution company (BEDAŞ) is shown in **Figure 13**. As seen in one-line diagram of the test system, there is only one customer connected to the network. Analyzed energy system given in **Figure 13** has (2.5 + 12i) mΩ grid impedance and (42.5 + 108.75i) mΩ transformer impedance per phase.

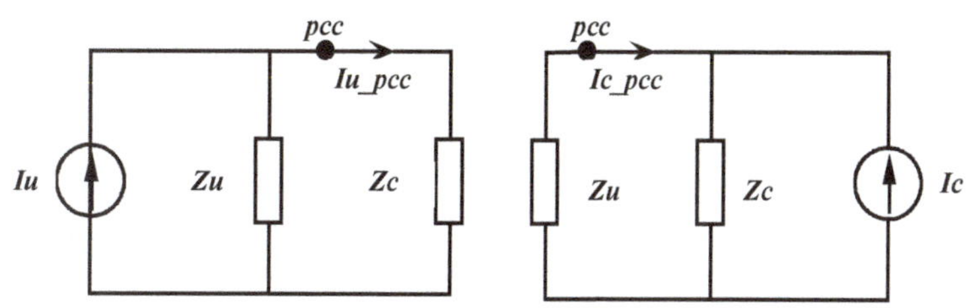

Figure 11. Modelled Norton equivalent circuit for the determination of currents on the PCC.

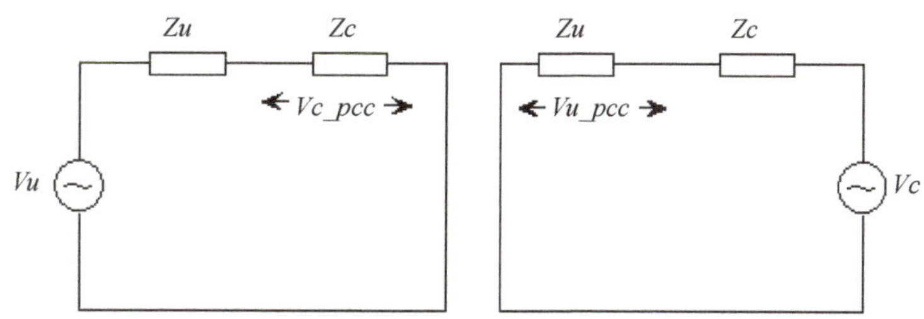

Figure 12. Modelled Thevenin equivalent circuit for the determination of voltages on the PCC.

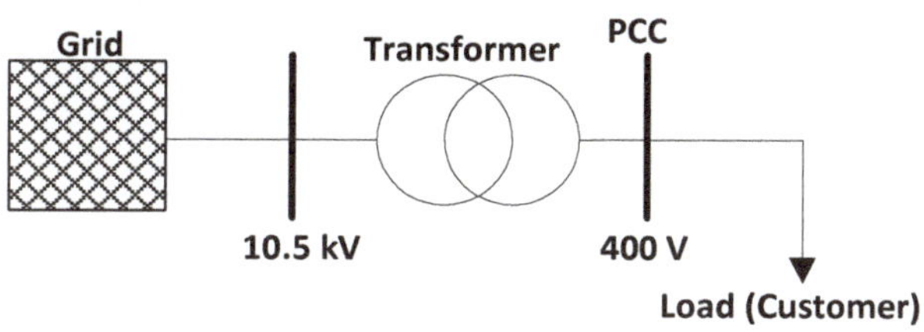

Figure 13. One-line diagram of the test system.

Load model were determined with the field measurements. Measurements had been taken and recorded during the 40 hours in order to determine the contributions of the network and customer current and voltage to the PCC. Sampling time of the measurement is 5 minutes. Instant measurement taken from the PCC were processed via MATLAB according to method described Section 3. According to this method the customer load impedance varies between (0.14 + 0.1i) Ω and (0.6 + 1.1i) Ω.

As a result of processing of the measurement data at the PCC shown in **Figure 13** voltage and currents contributions of the network and customer to the PCC at the fundamental frequency are shown in **Figure 14** and **Figure 15**. The indices "*c*", "*u*" and "pcc" used the graphics indicate costumer, utility (network), and point of common coupling, respectively. As shown in the **Figure 14** and **Figure 15**, power drawn from the network varies according to the time.

After the measurements of the current and voltage, load impedance was calculated by using (8) and given in **Figure 16**. As seen in the **Figure 16**, magnitude of the load impedance varies between 0.172 Ω and 1.25 Ω with time. Even if the type of the load does not known specifically, their affect to the total load impedance have already been taken into account by means of the constantly updated voltage and current data.

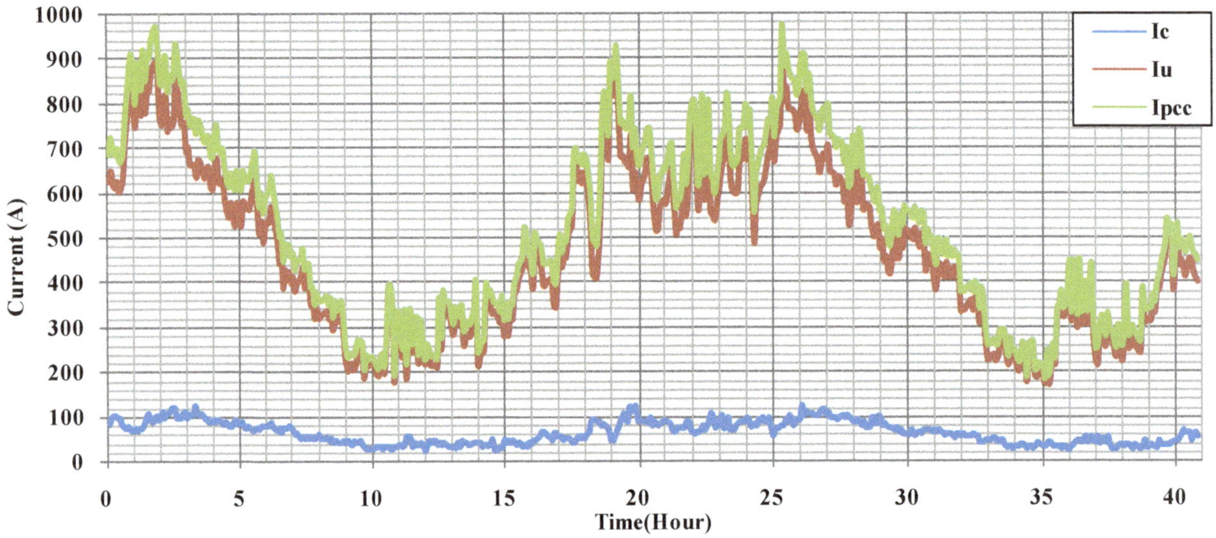

Figure 14. Customer and network current contributions to the PCC at the fundamental frequency.

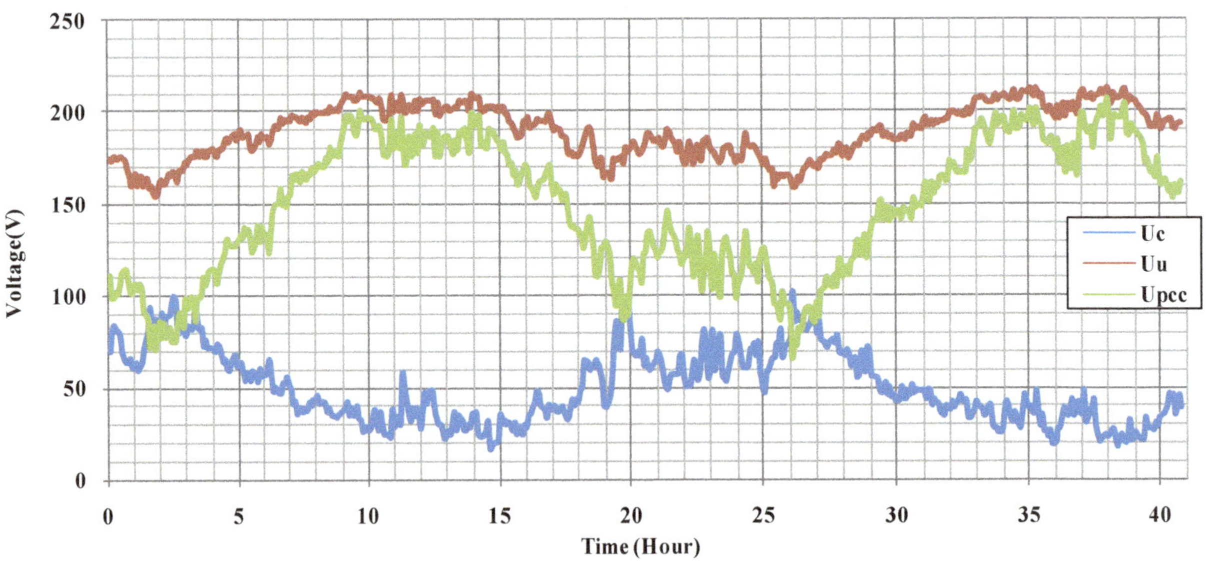

Figure 15. Customer and network voltage contributions to the PCC at the fundamental frequency.

Harmonic voltage contribution is calculated by adapting the impedance value of the customer and network according to rank of the harmonic analysis. As a result of processing of the measured data, voltage harmonic contribution levels between the 5th and 17th of the customer and network to the PCC is given between **Figures 17-21**, respectively.

As can be seen from the referred graphs, reason for disturbance in voltage quality as well as current is results from the customer. That is, largest share in the disturbance in the waveform of the voltage at the PCC belongs to the customer. It is understood that the 5th and 7th voltage harmonic levels are higher than the other harmonic levels as expected from a medium power energy system.

Harmonic current contribution is calculated by adapting the impedance value of the customer and network according to rank of the harmonic analysis. As a result of processing of the measured voltage data, voltage harmonic contribution levels between the 5th and 17th of the customer and network to the PCC is given between **Figures 22-26** respectively. As can be seen from the referred graphs, largest share in the disturbance in the waveform of the current at the PCC belongs to the customer. As shown in **Figure 22** and **Figure 23**, 5th and 7th current harmonic levels are higher than the other harmonic levels as a result of the voltage harmonics.

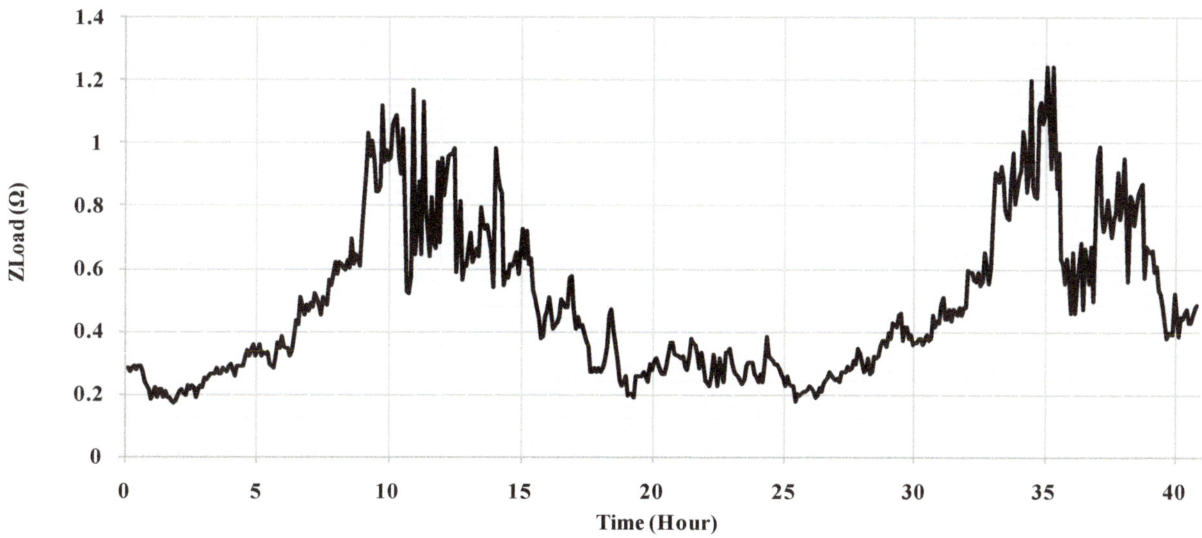

Figure 16. Magnitude of the load impedance.

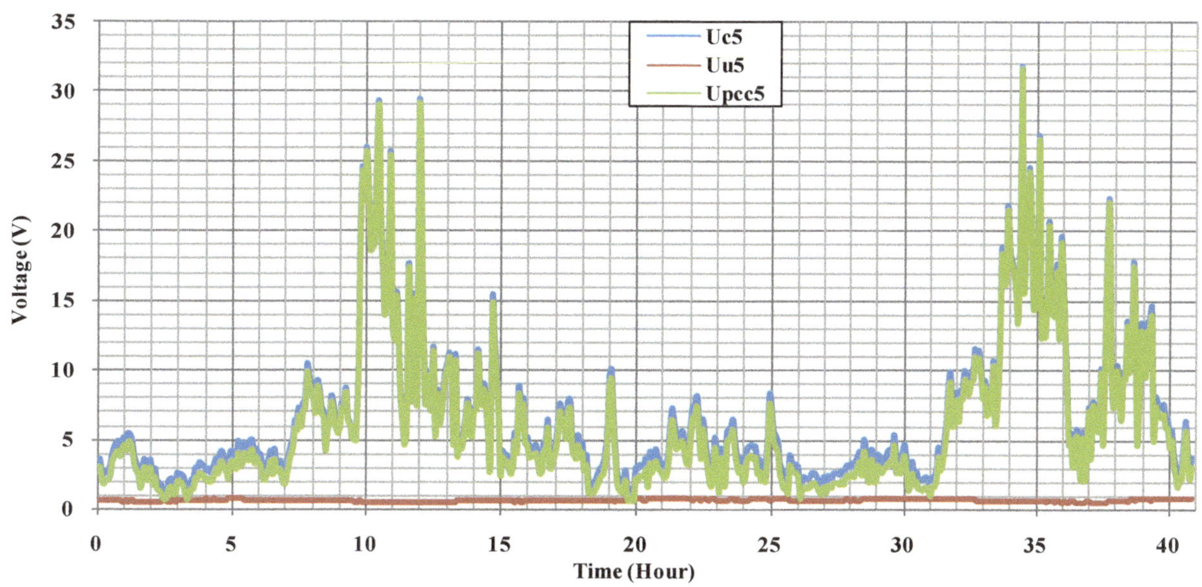

Figure 17. 5th harmonics of the voltage components at the PCC.

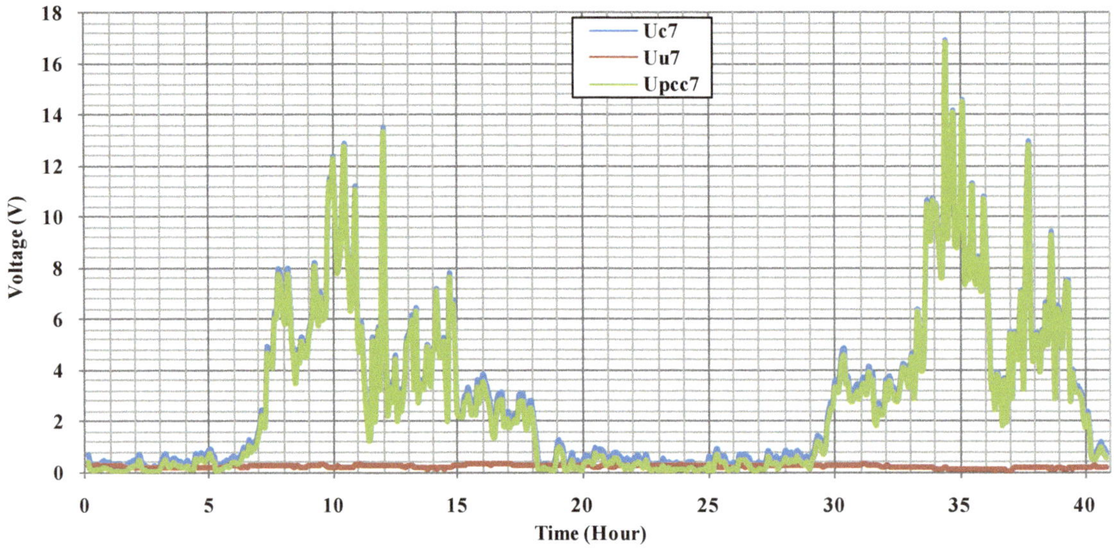

Figure 18. 7th harmonics of voltage components at the PCC.

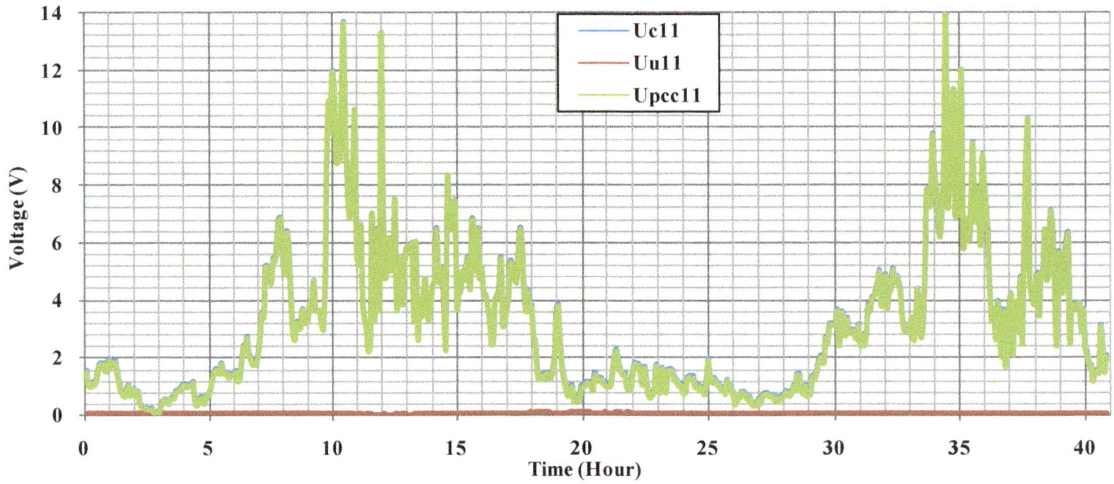

Figure 19. 11th harmonics of voltage components at the PCC.

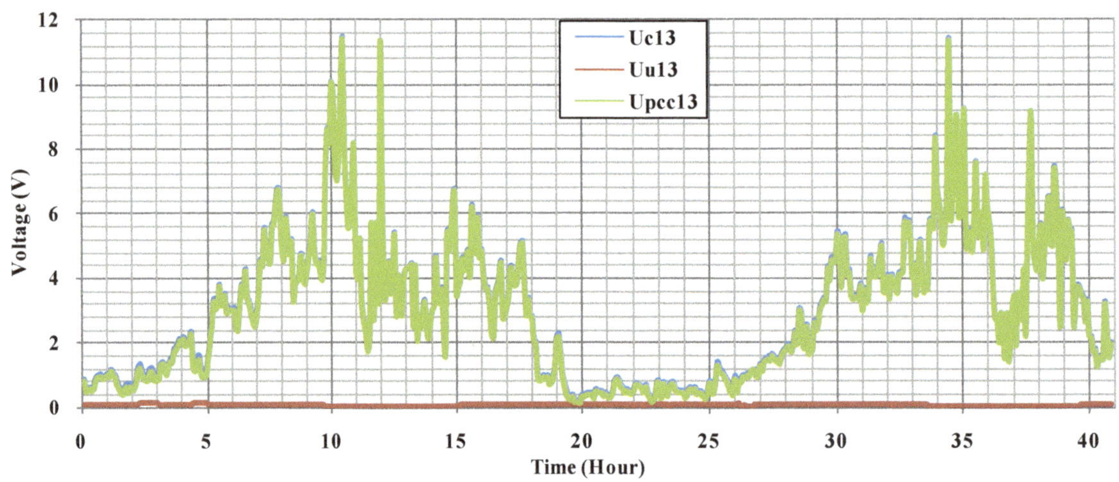

Figure 20. 13th harmonics of voltage components at the PCC.

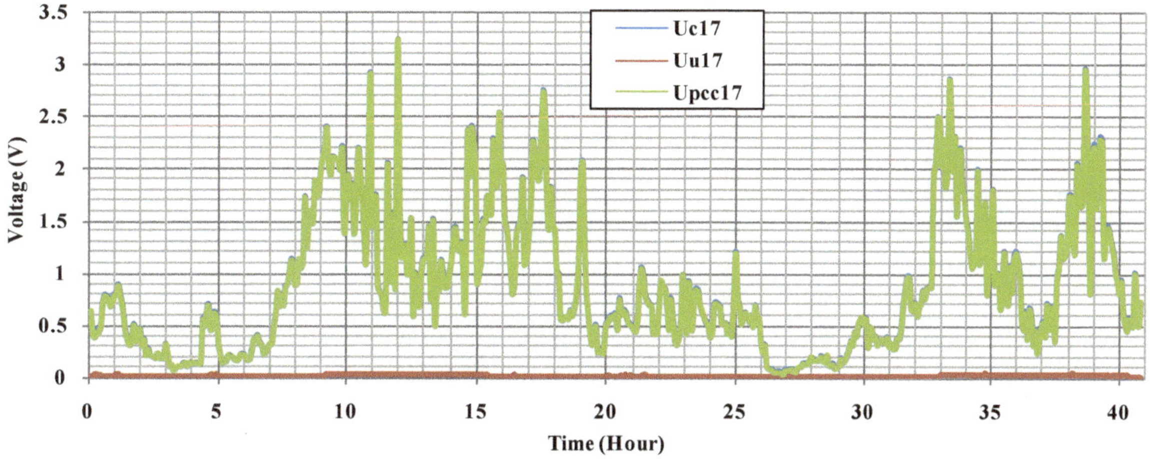

Figure 21. 17th harmonics of voltage components at the PCC.

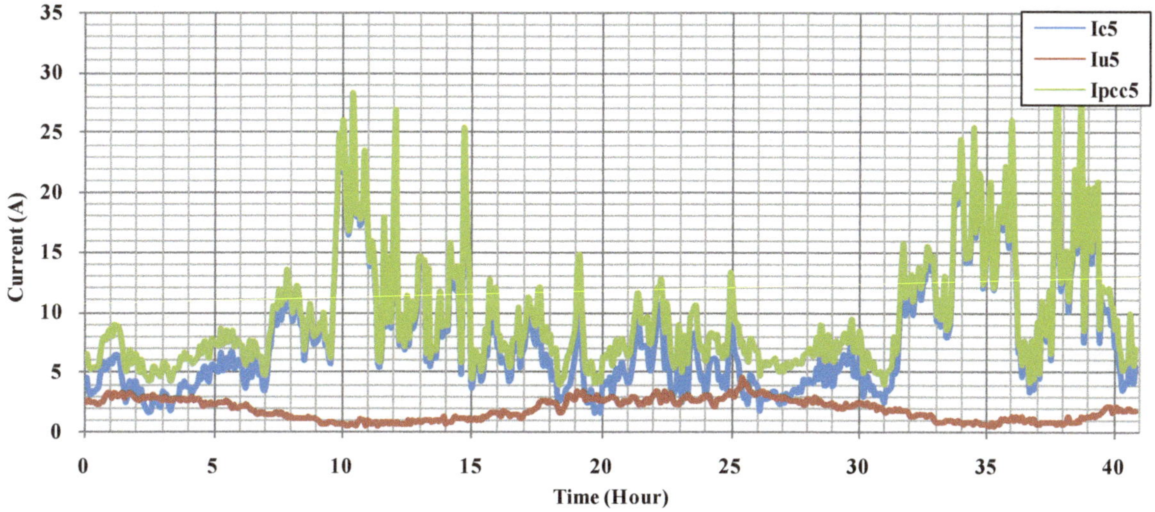

Figure 22. 5th harmonics of current components at the PCC.

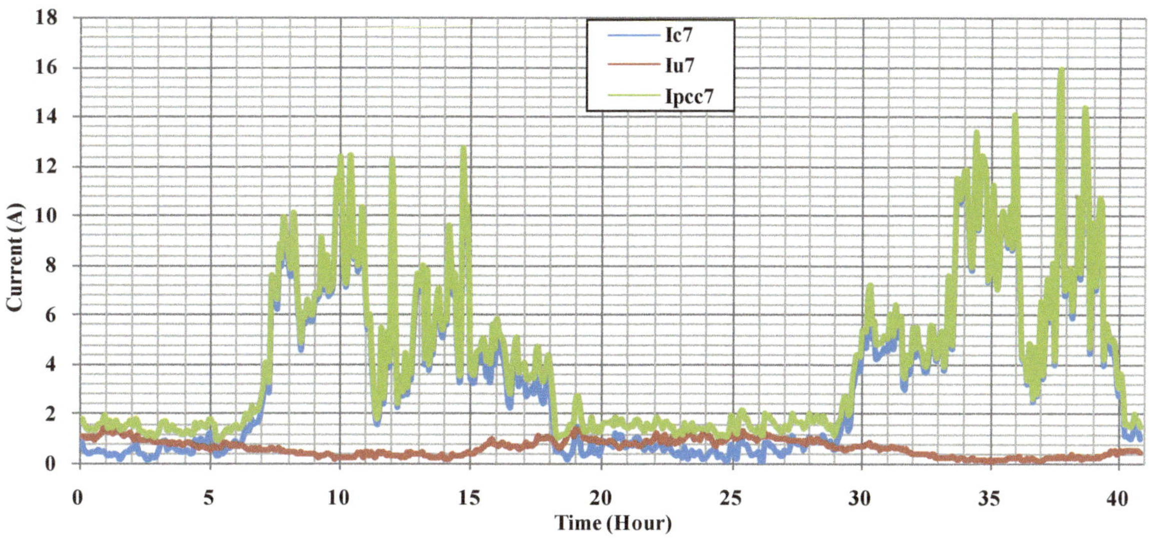

Figure 23. 7th harmonics of current components at the PCC.

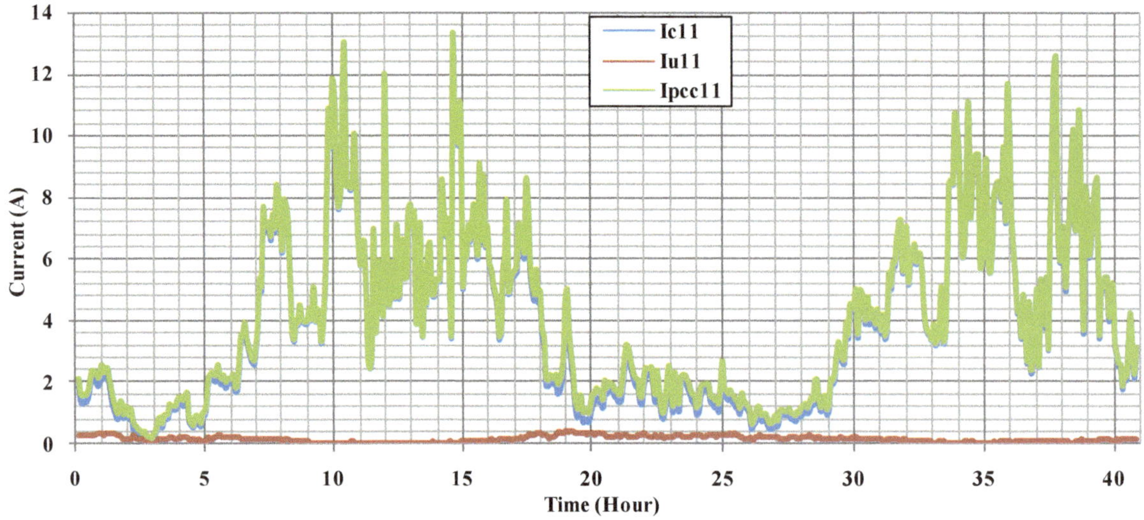

Figure 24. 11th harmonics of current components at the PCC.

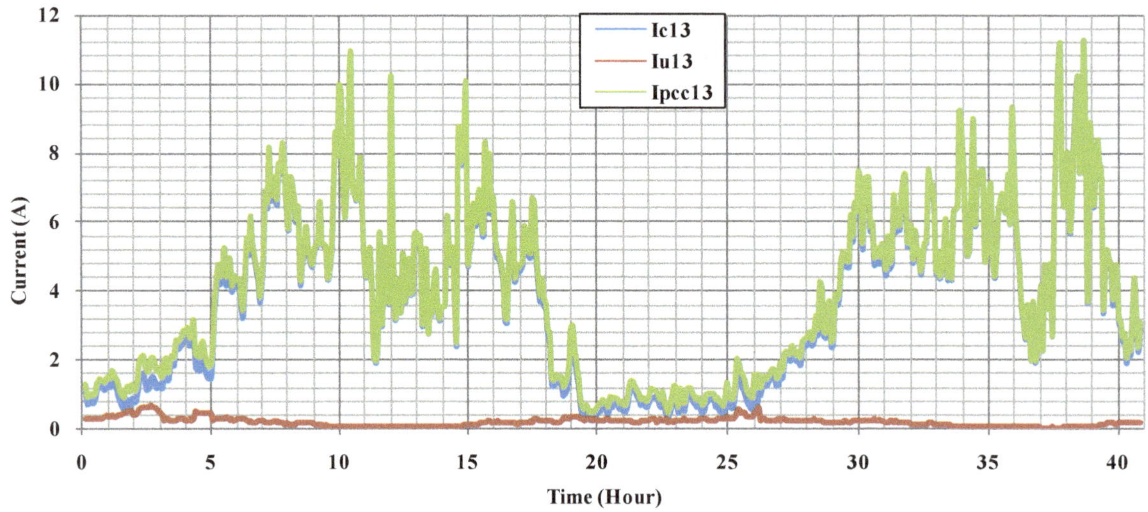

Figure 25. 13th harmonics of current components at the PCC.

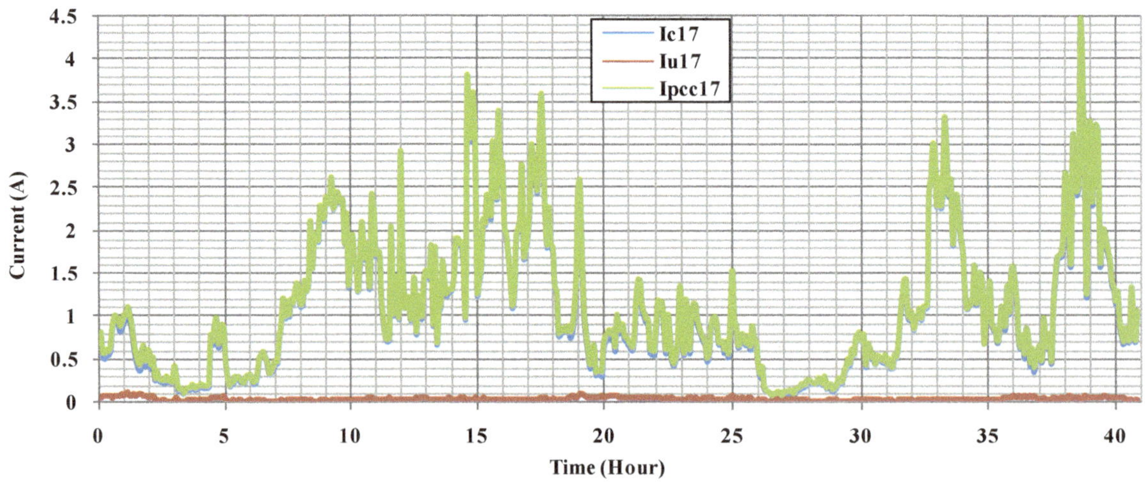

Figure 26. 17th harmonics of current components at the PCC.

Separately calculated average (Avr.) and maximum (instantaneous value) harmonic level values of the current and voltage are given in **Figure 27** and **Figure 28.** Actually, standards of voltage and current distortion limits are determined according to peak value of the measured current or voltage data. In order to make comparison with the max harmonic levels, average harmonic levels of the current and voltage are also presented. As seen in these figures, a large part in the responsibility for distortion at the quality of the current and voltage belongs to customer. On the other hand, the network has at least some contribution on the distortion on the waveform of the current. These values should be taken into account while calculating the reactive power consumption of the customer. As seen in the figures the 5th and 7th harmonic levels of the network is higher than the others. And his is very common phenomenon in the power systems. On the other hand, not only 5th and 7th harmonic levels of the costumer are high but also other harmonic levels are very high according to network.

Calculated voltage and current THDs according the measurement given in **Table 1**. Distortion at the quality of the voltage is higher than the current because of the non-sinusoidal load characteristics of the customer. As seen in the harmonic levels and THDs, there are large differences between maximum and average values of the voltage and current. In order to check the suitability of the current and voltage THDs of the network and costumer, voltage and current distortion limits are provided in **Table 2** and **Table 3**, respectively.

Limits of allowable voltage and current distortion set by IEEE 519 [25]. IEEE 519 provides limits for harmonic current distortion in general distribution and sub-transmissJion systems based on the load compared to the system's short-circuit capacity (the ratio I_{SC}/I_L) as given in **Table 3**. As seen in **Table 1**, average voltage

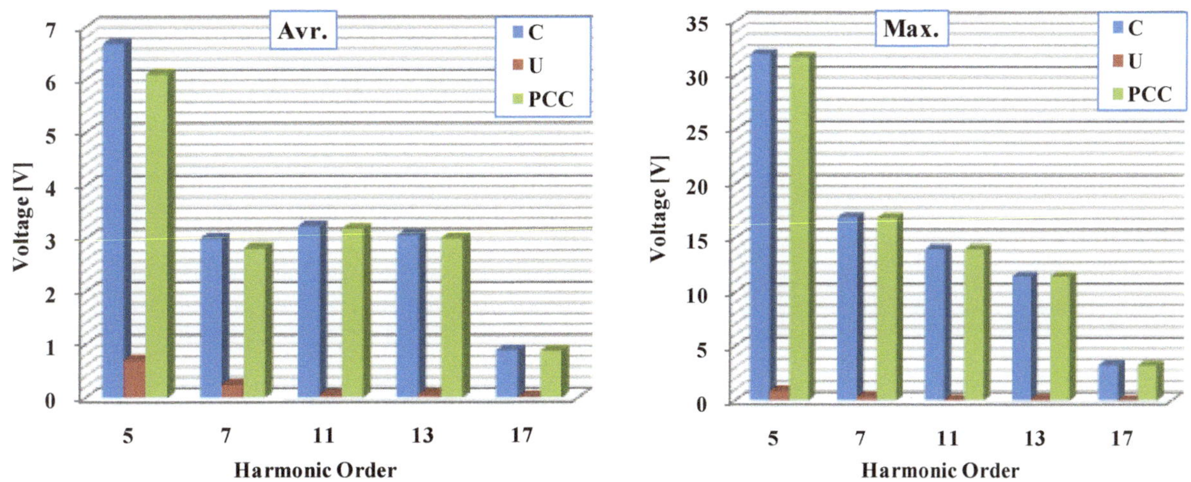

Figure 27. Harmonic level and amplitudes of the voltage components at the PCC.

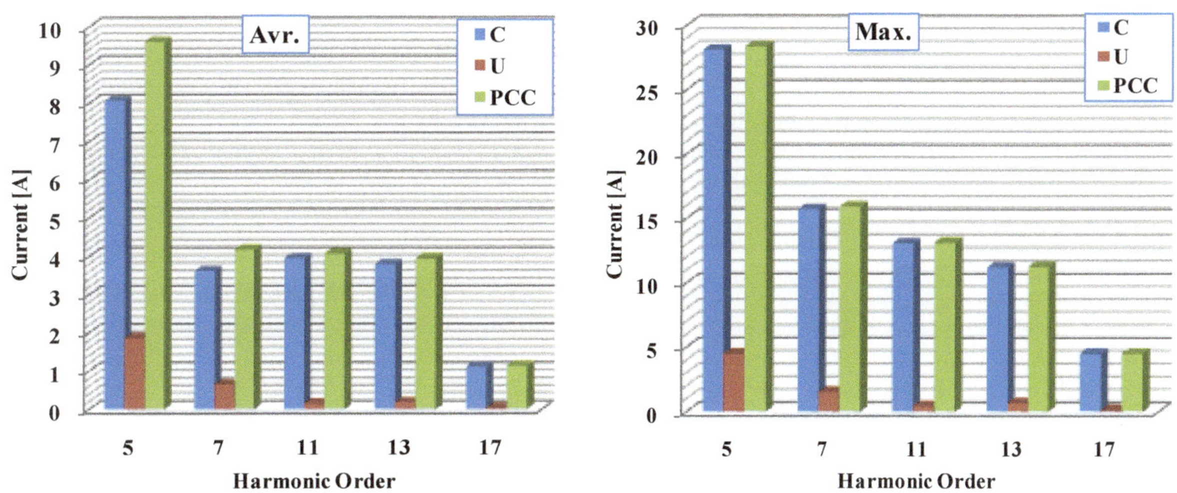

Figure 28. Harmonic level and amplitudes of current components at the PCC.

Table 1. Current and voltage THDs.

THD	Current		Voltage	
	Avr.	Max.	Avr.	Max.
C	0.165	0.296	0.172	0.399
U	0.005	0.006	0.004	0.005
PCC	0.023	0.038	0.055	0.195

Table 2. IEEE 519 voltage distortion limits [24].

Bus voltage at PCC	Vh, %	THDv, %
$V < 69$ kV	3.0	5.0
$69 \text{ kV} \leq V < 161$ kV	1.5	2.5
$V \geq 161$ kV	1.0	1.5

Table 3. IEEE 519 current distortion limits [24].

I_{SC}/I_L	(120 V - 69 kV)					%
	$H < 11$	$11 \leq h < 17$	$17 \leq h < 23$	$23 \leq h < 35$	$35 \leq h$	THD
<20	4.0	2.0	1.5	0.6	0.3	5
20 - 50	7.0	3.5	2.5	1.0	0.5	8
50 - 100	10	4.5	4.0	1.5	0.7	12
100 - 1000	12	5.5	5.0	2.0	1.0	15
>1000	15	7.0	6.0	2.5	1.4	20

THD of the customer is 17.2% while maximum voltage THD is 39.9%. Both of these values exceed the voltage distortion limits as seen in **Table 2**.

On the other hand, neither maximum nor average voltage THD of the network exceeds the limits. Therefore, penalty sanctions should be imposed to the customer for exceeding the voltage distortion limits.

I_{SC}/I_L of the test system is approximately 83. According to **Table 3**, current distortion limit for the customer is 12% while for the network is 6%. Considering these current distortion limits, both of the maximum and average values of the customer exceed the limits. Therefore, another penalty sanctions should be imposed to the customer for exceeding the voltage distortion limits. Penalty rates are determined according to difference between the distortion limit and distortion of the investigated system. Considering the maximum THD values, costumer has 34.9% more voltage and 4.5% more current THD in the test system. Considering the average THD values, costumer has 12.2% more voltage and 4.5% more current THD in the test system. Therefore, penalty rate of the customer will be proportional with these percentages.

8. Conclusions

In this study, a medium voltage electrical distribution system has been investigated by using field measurement in order to determine the customer and network responsibilities. To be able to determine these responsibilities, a new algorithm based on the current vector method with the reference impedance method modified via constantly updated data has been proposed. The proposed algorithm is verified in MATLAB by processing of the measurements and the responsibilities of the customer and the network have been determined, successfully. It is understood that analysis performed with harmonic vector method using updated reference impedance is a method that is very practical and easy to apply in determining the components of network and customer at the PCC. Customer and network's individual current and voltages and also customer load impedance have been calculated by processing current and voltage data at the PCC with the proposed algorithm. Average and maximum value of the current and voltage THDs has been also calculated.

According to the analysis results of test system, it is determined that overall responsibilities in the current and voltage waveform distortion at the PCC only belong to the customer. Therefore, penalty sanctions should be imposed to the customer in the amount of it exceeds allowable limits. To be able to reduce to impact of the harmonics to the minimum and determine the harmonic contribution rates accurately and also determine the share of responsibility, some of the arrangements can be made as follows:

- The necessary arrangements about the harmonic contribution of the network and the customer should be performed and determined financial penalties should be applied when related limits given in the standards were exceeded.
- To take into account the average values instead of instantaneous values will be satisfy in terms of feasibility and justice while calculating the effect of the network's and the customer's harmonic distortion.
- Variable load condition should be taken into account in order to obtain more accurate results. Constantly changing load and also switching positions of the harmonic filters and compensator capacitors have been already taken into account because of the calculated load impedance by using the voltage and current data updated constantly during the measurement period.

In the future, it is planned to use the same algorithm for determining the customer and network distortion contributions in an unbalanced three phase systems. Still, billing is not made for the harmonic contributions in practice. In this regard, a digital recorder including a harmonic contribution algorithm would be useful to determine the harmonic contribution of the each customer in the allowable limits or not.

Acknowledgements

This study has been financially supported by TUBITAK with 110E134 project number. The authors would like to express their gratitude to BEDAŞ DistCo for facilitating and helping with the measurements at the substation.

References

[1] Dugan, R.C., *et al.* (2002) Electrical Power Systems Quality. 2nd Edition, McGraw-Hill, New York.

[2] Gul, O., Kaypmaz, A. and Tanrioven, M. (2004) A Novel Three-Phase Time Domain Load Modelling Technique Using Instantaneous Symmetrical Components. *European Transactions on Electrical Power*, **14**, 361-375. http://dx.doi.org/10.1002/etep.28

[3] Gül, Ö. (2004) Identification of Three Phase Generic Load Model Parameters Using Field Measurements. *IEEE MELECON*2004, Dubrovnik.

[4] Katić, V.A., Mujović, S.V., Radulović, V.M. and Radović, J.S. (2011) The Impact of the Load Side Parameters on PC Cluster's Harmonics Emission. *Advances in Electrical and Computer Engineering*, **11**, 103-110. http://dx.doi.org/10.4316/AECE.2011.01017

[5] Gül, Ö. and Milanovic, J. (2000) Sensitivity of Harmonic Load Model Parameters to Voltage and Current Waveforms. *The International Conference on Harmonics and Quality of Power*, Orlando.

[6] Qian, K., Zhou, C., Allan, M. and Yuan, Y. (2011) Effect of Load Models on Assessment of Energy Losses in Distributed Generation Planning. *IJEPES*, **33**, 1243-1250. http://dx.doi.org/10.1016/j.ijepes.2011.04.003

[7] Stojanovic, D.P., Korunovic, L.M. and Milanovic, J.V. (2008) Dynamic Load Modelling Based on Measurements in Medium Voltage Distribution Network. *EPSR*, **78**, 228-238. http://dx.doi.org/10.1016/j.epsr.2007.02.003

[8] Liang, X., *et al.* (2012) Dynamic Load Models for Industrial Facilities. *IEEE Transactions on Power Systems*, **27**, 69-80. http://dx.doi.org/10.1109/TPWRS.2011.2161781

[9] Farhoodnea, M., Mohamed, A., Shareef, H. and Zayandehroodi, H. (2012) An Enhanced Method for Contribution Assessment of Utility and Customer Harmonic Distortions in Radial and Weakly Meshed Distribution Systems. *IJEPES*, **43**, 222-229. http://dx.doi.org/10.1016/j.ijepes.2012.05.013

[10] Thunberg, E. and Soder, L. (1999) A Norton Approach to Distribution Network Modeling for Harmonic Studies. *IEEE Transactions on Power Delivery*, **14**, 272-277. http://dx.doi.org/10.1109/61.736738

[11] Fan, R.R., *et al.* (2013) A Method for Assessing Customer Harmonic Emission Level Based on the Iterative Algorithm for Least Square Estimation, *Engineering*, **5**, 6-13. http://dx.doi.org/10.4236/eng.2013.59B002

[12] Srinivasan, K. (1996) On Separating Customer and Supply Side Harmonic Contributions. *IEEE Transactions on Power Delivery*, **11**, 1003-1012. http://dx.doi.org/10.1109/61.489362

[13] IEEE Std 519-1992, IEEE Recommended Pradice and Requirements for Harmonic Control in Electrical Power Systems.

[14] Xu, W. and Liu, Y. (2000) A Method for Determining Customer and Utility Harmonic Contributions at the Point of Common Coupling. *IEEE Transactions on Power Delivery*, **15**, 804-811.

[15] Bergeron, R. and Slimani, K. (1999) A Method for the Determination of the Customer Share of the Contribution to the Level of Harmonic Voltage on an Electric Network. *Proceedings of IEEE Power Engineering Society Summer Meeting*, **1**, 354-360.

[16] Chen, C.Y., Liu, X.L., Koval, D., Xu, W. and Tayjasanant, T. (2004) Critical Impedance Method—A New Detecting Harmonic Sources Method in Distribution Systems. *IEEE Transactions on Power Delivery*, **19**, 288-297. http://dx.doi.org/10.1109/TPWRD.2003.820424

[17] Hamzah, N., Mohamed, A. and Hussain, A. (2005) Identification of Harmonic Source at the Point of Common Coupling Based on Voltage Indices. *Jurnal Teknologi*, **43**, 11-32.

[18] Li, C. and Xu, W. (2002) On Defining Harmonic Contributions at the Point of Common Coupling. *IEEE Power Engineering Review*, 45-46.

[19] Bergeron, R. and Slimani, K. (1999) Method for an Equitable Allocation of the Cost of Harmonics in an Electrical Network. *Proceedings of IEEE Power Engineering Society Summer Meeting*, **1**, 347-353.

[20] Cristaldi, L. and Ferrero, A. (1995) Harmonic Power Flow Analysis for the Measurement of the Electric Power Quality. *IEEE Transactions on Instrumentation and Measurement*, **44**, 683-685. http://dx.doi.org/10.1109/19.387308

[21] Xu, W., Liu, X. and Liu, Y. (2003) An Investigation on the Validity of Power Direction Method for Harmonic Source Determination. *IEEE Transactions on Power Delivery*, **18**, 214-219. http://dx.doi.org/10.1109/TPWRD.2002.803842

[22] Davis, E.J., Emanuel, A.E. and Pileggi, D.J. (1996) Evaluation of Single Point Measurements Method for Harmonic Pollution Cost Allocation. *IEEE Transactions on Power Delivery*, **11**, 1003-1008.

[23] Pfajfar, T., Blažič, B. and Papič, I. (2008) Harmonic Contributions Evaluation with the Harmonic Current Vector Method. *IEEE Transactions on Power Delivery*, **23**, 425-433. http://dx.doi.org/10.1109/TPWRD.2007.911165

[24] Varadan, S. and Makram, E.B. (1993) Generalized Load Modeling in the Presence of Hamonics and Distorion. *The 25th Southeastern Symposium on System Theory* (*SSST*'93), 124-128.

[25] ANSI/IEEE (1993) IEEE Recommended Practice and Requirements for Harmonic Control in Electrical Power Systems. *Standard*, *ANSI/IEEE* 519, IEEE, New York.

17

Thermodynamics of Efflux Process of Liquids and Gases

E. A. Mikaelian[1], Saif A. Mouhammad[2*]

[1]Gubkin Russian State University of Oil and Gas, Moscow, Russia
[2]Physics Department, Faculty of Science, Taif University, Taif, Kingdom of Saudi Arabia
Email: [*]saifnet70@hotmail.com

Abstract

The main objective of this work is to obtain the calculated ratio of efflux processes for liquids, vapors, gases on the basis of the developed mathematical model, which allows to determine the characteristics of the channel profiles nozzles and diffusers, to solve a number of subsequent applications for analysis modes. On the basis of the calculated ratios are equations of the first law of thermodynamics for the flow of liquids and gases. The obtained calculated ratios are extended for the case of the efflux of compressible liquids, vapors and gases and as a special case, for incompressible liquids. The characteristics of the critical efflux regime liquids, which allows to determine the linear and the mass efflux rate of the critical regime and the calculated characteristics of the channel profiles nozzles and diffusers, Laval nozzles for different modes of operation are obtained.

Keywords

Thermodynamics, Efflux, Compressible, Incompressible, Liquids, Diffusers, Nozzles

1. Introduction

The efflux processes are quite common in various technological processes performed with the power technology equipment in the gas and oil industry, in heat engines, pumps, compressor machines, mas-and-heat exchange units, pipelines, in separate elements of machines and devices: nozzles, diffusers, convergent nozzles, mud guns, fittings, locking devices, gate valves, valves, various calibration holes etc. It is worth emphasising a special role in studying the processes of the gas and liquid efflux through various sorts of leakiness and gap [1] [2].

The efflux process can be considered as a special case of the occurrence and distribution of potential work. Effective work in the efflux process is distributed on the work, directly transmitted to the bodies of external sys-

[*]Corresponding author.

tem (in our case in the efflux process this work is absent: $\delta W_{cz}^{*}=0$) and to a change in the energy of external position of working medium itself (de_{cz}). The last term, in turn, consists of the kinetic energy $d\left(c^2/2\right)$ and the potential energy (gdz).

Thus, the initial equation of the theoretical efflux process has the following form:

$$\delta W=-VdP=d\left(c^2/2\right)+gdz. \qquad (1)$$

Switching to the real efflux processes then is carried out by introducing correction factors: velocity rates (φ) and flow rates (ϕ). The integral of the initial equation of efflux for the expression of potential flow work from the initial 1 to the final section 2 of a flow has the form as follows:

$$W_{12}=c_2^2/2-c_1^2/2+g\left(z_2-z_1\right); \qquad (2)$$

$$W_{12}=\left[1-\left(P_2/P_1\right)^{(n-1)/n}\right]P_1V_1n/(n-1). \qquad (3)$$

The rate of gas efflux in the initial section can be considered as a result of the efflux of a conditional initial state 0 - 0 at zero velocity $c_0=0$; with the graded level $z_0=z_2$ and pressure P_0.

Then the calculated expression for the potential work and linear velocity of the efflux of the final section of the flow is determined by the following equations:

$$W_{02}=W_{12}+W_{01}=\left[1-\left(P_2/P_0\right)^{(n-1)/n}\right]P_0V_0n/(n-1); \qquad (4)$$

$$c_2=\left(2W_{02}\right)^{0.5}=\left[2W_{12}+c_1^2+2g\left(z_1+z_2\right)\right]^{0.5}. \qquad (5)$$

The theoretical efflux process is regarded as an adiabatic one, then based on the first law of thermodynamics for the flow the potential flow work is determined as the specific heat drop of the flow equal to the difference between the heat content (enthalpy) of it, taken with the opposite sign [3] [4]:

$$W_{12}=h_1-h_2; \quad q_{12}=0. \qquad (6)$$

Further the mass rate of the efflux is entered in calculations:

$$u=G/f=V\rho/f=\rho c, \qquad (7)$$

where f—the cross section of the flow under consideration; G and V are the mass flow rate and volumetric flow rate; ρ—liquid density; c—the linear velocity of the liquid in the direction of movement (the average velocity in the section f in the direction of a normal to this section).

The concept of the mass flow rate in research is most essential. The concept of linear velocity characterises only the kinetic energy of the flow, averaging of such a velocity depends on the flow mode (laminar, transitional, turbulent) and is not identical with the mass flow rate.

The calculated expression of the theoretical mass flow rate at the outlet section is obtained according to the last equation depending on the linear velocity (4) and (5) and the equation of the efflux process:

$$u_2=\left\{\left[1-\left(P_2/P_0\right)^{(n-1)/n}\right]\left(P_2/P_0\right)^{2/n}\left(P_0/V_0\right)2n/(n-1)\right\}^{0.5}. \qquad (8)$$

For transition to the real characteristics of a flow we input correction factors in our calculations:

$$c=\varphi c_2; \quad u=\phi u_2; \quad G=uf=\phi u_2 f=\phi c_2\rho_2 f. \qquad (9)$$

In the formula (9) the velocity and flow-rate factors are determined as the ratio of theoretical and actual velocities:

$$\varphi=c/c_2=V/fc_2; \quad \phi=u/u_2=G/fu_2. \qquad (10)$$

The work of irreversible energy losses associated with the real efflux process:

$$W^{**}=\left(c_2^2-c^2\right)/2=\left(1-\varphi^2\right)c_2^2/2=\xi c_2^2/2, \qquad (11)$$

where ξ—the factor of energy losses in the real process.

To calculate the velocity and flow rate, as follows from the formulas (10) it is necessary to arrange for mass (volume) measurements of the liquid flow rates [5] [6].

2. Efflux of Incompressible Liquids

The initial condition ($\rho_1 = \rho_2 = \rho = 1/v$ = idem):

$$W_{12} = (P_1 - P_2)/\rho;\ W_{02} = (P_0 - P_2)/\rho. \tag{12}$$

Further, by using the initial general ratios (5), (7), (9) and Equation (12), we will obtain the calculated ratios for a particular case of the efflux of incompressible liquids:

$$\begin{aligned} c_2 &= (2W_{02})^{0.5} = \left[2W_{12} + c_1^2 + 2g(z_1 + z_2)\right]^{0.5} = \left[2(P_0 - P_2)/\rho\right]^{0.5} \\ &= \left[2(P_1 - P_2)/\rho + c_1^2 + 2g(z_1 + z_2)\right]^{0.5}; \end{aligned} \tag{13}$$

$$u_2 = G/f = V\rho/f = \rho c_2 = \left[2(P_0 - P_2)\rho\right]^{0.5}; \tag{14}$$

$$G = \phi u_2 f. \tag{15}$$

The obtained ratios can be applied to the efflux of compressible liquids (gases) with the condition of insignificant fluctuations of the densities. In this case, in Formulae (12)-(15) there should be introduced the average density value, for example, as arithmetic mean:

$$(\rho_1 + \rho_2)/2 = \rho_m$$

3. Efflux of Compressible Liquids (Gases)

The general solution of problems concerning the efflux of compressible liquids is obtained by a corresponding development of the previously obtained initial relationships.

From a consideration of the original ratio (8) it follows that the mass velocity becomes zero for the following values of pressure ratios: 1) $P_2/P_0 = 1$, this takes place at the beginning of the efflux $c = 0$; $u = c\rho = 0$ due to the initial rate; 2) $P_2/P_0 = 0$—in the efflux to vacuum at the outlet section $\rho = 0$; $u = c\rho = 0$ due to density. Within this range, the mass flow rate passes through the maximum (Rolle's theorem). This means the variable factor of the radicand (8) passes through the maximum:

$$\Psi = \left[1 - (P_2/P_0)^{(n-1)/n}\right](P_2/P_0)^{2/n}. \tag{16}$$

Let us introduce the following designations:

$$(P_2/P_0) = \tau^{n/(n-1)};\ (P_2/P_0)^{2/n} = \tau^{2/(n-1)}, \tag{17}$$

using the Rolle's theorem for investigating the function to the maximum, we obtain the parameters of the critical mode of efflux for compressible liquids

$$\tau_{cr} = (P_2/P_0)_{cr}^{(n-1)/n} = 2/(n+1), \tag{18}$$

$$\beta = (P_2/P_0)_{cr} = \tau_{cr}^{n/(n-1)} = \left[2/(n+1)\right]^{n/(n-1)}, \tag{19}$$

$$\Psi_{cr} = (1 - \tau_{cr})\tau_{cr}^{2/(n-1)}. \tag{20}$$

Depending on parameters of the critical efflux mode, the linear and mass efflux rates of the critical mode are determined:

$$c_{cr} = \left[n(PV)_{cr}\right]^{0.5}, \tag{21}$$

Table 1. Characteristic values of the discharge critical mode.

n	1.1	1.2	1.3	1.4	
$\tau_{cr} = 2/(n+1)$	0.953	0.909	0.870	0.833	
$\beta = \tau_{cr}^{n/(n-1)}$	0.5847	0.5645	0.5457	0.5283	~0.55
Ψ_{cr}	1.9677	2.0309	2.0896	2.1443	~2.05

$$u_{cr} = \left[2P_0\rho_0\Psi_{cr}\, n/(n-1)\right]^{0.5}. \tag{22}$$

Table 1 shows the values of the critical discharge characteristics depending on the performance of the efflux process.

4. Particular Cases of Efflux

The ideal gas ($PV = RT$):

$$c_{cr} = \left[nRT_{cr}\right]^{0.5};\quad u_{cr} = P_0\left[2\Psi_{cr} n/(n-1)RT_{cr}\right]^{0.5}, \tag{23}$$

The incompressible liquids (V = idem; $n = \infty$): $c_{cr} = \infty$.

This means that the critical mode for incompressible fluids is unattainable. The critical linear velocity of the adiabatic efflux ($n = k$) is the velocity of sound:

$$a^* = \left[k(PV)_{cr}\right]^{0.5},$$

for the ideal gas:

$$a^* = \left[nRT_{cr}\right]^{0.5}. \tag{24}$$

5. Conclusion

According to the energy conservation law, the equation of distribution and occurrence of the potential work of any thermodynamic systems is obtained. Taken as a basis for the theory of the efflux of gases, compressible and incompressible liquids, the characteristic features of the critical mode of the liquid efflux are obtained. The derived calculated ratios will further determine the calculated characteristics of the channel profiles of nozzles and diffusers, Laval nozzles for a range of modes of operation.

References

[1] Mikaelian, E.A. (2000) Maintenance Energotechnological Equipment, Gas Turbine Gas Compressor Units of Gas Gathering and Transportation. *Methodology, Research, Analysis and Practice*, Fuel and Energy, Moscow, 304. http://www.dobi.oglib.ru/bgl/5076.html

[2] Mikaelian, E.A. (2001) Improving the Quality, to Ensure Reliability and Safety of the Main Pipelines. In: Margulov, G.D., Ed., *Series: Sustainable Energy and Society*, Fuel and Energy, Moscow, 640. http://www.dobi.oglib.ru/bgl/4625.html

[3] Vladimirov, A.I. and Kershenbaum, Y.V. (2008) Industrial Safety Compressor Stations. Management of Safety and Reliability. Inter-Sector Foundation "National Institute of Oil and Gas", Moscow, 640. http://www.mdk-arbat.ru/bookcard?book_id=3304125

[4] Mikaelian, E.A. (2008) Diagnosis Energotechnological Equipment GGPA Based on Various Diagnostic Features. *Gas Industry*, **4**, 59-63.

[5] Mikaelian, E.A. (2014) Determination of the Characteristic Features and Technical Condition of the Gas-Turbine and Gas-Compressor Units of Compressor Stations Based on a Simplified Thermodynamic Model. *Quality Management in Oil and Gas Industry*, **1**, 44-48. http://instoilgas.ru/ukang

[6] Mikaelian, E.A. and Mouhammed, S.A. (2014) Survey Equipment Gas Transmission Systems. *Quality Management in Oil and Gas Industry*, **4**, 29-36. http://instoilgas.ru/ukang

18

Short Term Stability Using Phase Plane

Luis Aromataris[1], Sergio Preidikman[1,2], Marcos Galetto[1], Fabian Rinaudo[1], Eduardo Toledo[3]

[1]Grupo de Análisis de Sistemas Eléctricos de Potencia, Facultad de Ingeniería, UNRC, Córdoba, Argentina
[2]Departamento de Estructuras, Facultad de Ciencias Exactas, Físicas y Naturales, Universidad Nacional de Córdoba, Córdoba, Argentina
[3]Centro de Estudios de Energía para el Desarrollo (CEED), Facultad de Ingeniería, UNAM, Oberá, Argentina
Email: laromata@ing.unrc.edu.ar

Abstract

One of the most important problems in the study of transient stability of power systems is the determination of perturbation's maximum time of permanence without losing the synchronism of the generators that feed the network. The problem is generally solved by either the application of the equal-area criterion or through numerical integration methods. In the present work, the phase-plane is proposed as an alternative tool to solve the above-mentioned problem with greater efficiency.

Keywords

Short Term Stability, Phase Plane, Swing Equation

1. Introduction

Transient stability is the power system ability to maintain synchronism after being subject to a severe perturbation, such as: line faults, generation losses or large load losses. When analyzing power systems, the short-term transient stability study is usually presented by using an array consisting on a generator connected to an infinite bus through a transmission network. The generator is modeled as an internal voltage source behind the transient reactance. Generally, in such studies, excitation and velocity control systems are not taken into account.

The equal-area criterion is used to examine the stability of a system composed by two machines or one machine connected to an infinite bus. By applying this criterion, there is no need to solve the so-called swing equation [1].

Currently, numerical integration methods are mostly used to make simulations of the transient stability phenomenon in the time domain. The use of these methods implies the numerical integration of the differential equations describing this phenomenon [1].

In this paper, we revisit an alternative technique that exploits the topology of the phase plane [2]. The previously mentioned technique is developed in references given in this work [3]-[7]. This reference does not provide direct comparisons among the proposed technique, the equal-area criterion, and numerical integration methods.

2. Conventional Calculation Methods

The rotational motion of a synchronous machine is governed by a second order, ordinary, non-lineal differential equation. This equation is the well-known swing equation. When damping is neglected, this equation has the following form:

$$\begin{aligned} P_m - P_e &= \frac{2H}{\omega_s}\frac{\mathrm{d}^2\delta(t)}{\mathrm{d}t^2} \\ P_e &= \frac{E'E\sin\delta(t)}{X} \end{aligned} \tag{1}$$

where:

P_m = mechanical power input to the synchronous machine [pu].

P_e = electric power output from the synchronous machine [pu].

H = inertia constant [MWs/MVA].

δ = rotor angle measured with respect to a reference axis that rotates at the synchronous velocity [electric radians].

ω_s = synchronous angular velocity [rad/s].

t = time [s].

E' = voltage measures following the transient reactance [pu].

E = infinite bus voltage [pu].

X = reactance between E' and E [pu].

Solving the differential Equation (1), both the rotor angle time history, $\delta(t)$, and the rotor generator deviation speed, with respect to the synchronous speed, $\Delta\omega(t)$, can be obtained. The time histories of the variables $\delta(t)$ and $\Delta\omega(t)$ allow the study of the transient stability of a power system consisting on a generator connected to an infinite bus through a transmission network. In order to achieve that, it can be assumed that the system is initially working on a stable state. That is, the generator rotor angle keeps a constant value, indicated by δ_0. If for a time t_f we apply a three-phase fault in any network node and immediately after, at $t_f + \Delta t$, the protection devices act eliminating the fault, then the curve δ *vs.* t, shown in **Figure 1** can be obtained.

Depending on the fault extent, in this case Δt, the following cases may occur:

- Stable case: the generator rotor oscillates around a new equilibrium position; this means that there is not synchronism loss.
- Unstable case: the generator rotor angle grows indefinitely; this means that the synchronism is lost.

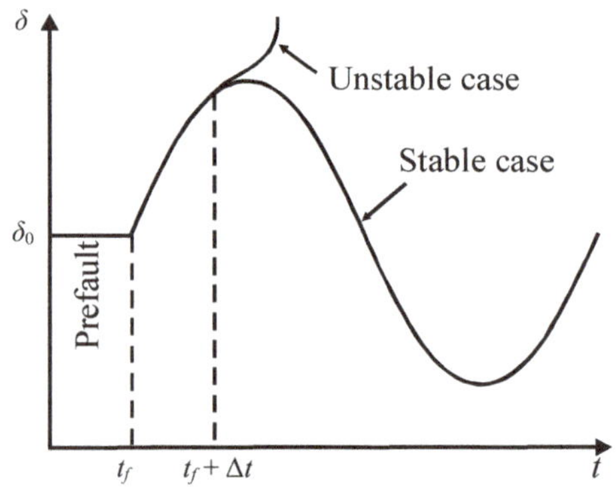

Figure 1. Response of $\delta(t)$ vs. t for different values of fault-clearing times.

The fault maximum extent without losing the generator synchronism is called critical time t_c. Such period of time is generally determined through trial and error procedures.

A similar methodology to be used to determine the transient stability is the one resulting from the application of the equal-area criterion. Using this method, there is no need to integrate the swing equation. As an alternative, the diagram shown in **Figure 2** is built. In such diagram, the pre-fault and post-fault curves show the relation between the electric power output from the synchronous machine, P, and the rotor angle, δ. In this method, the mechanical power input to the synchronous machine, P_m, is considered constant during the study. The angle δ_0 represents the rotor position for the system in stable state. After the fault, the angle δ increases to its δ_c value, which represents the rotor angle when the protection devices eliminate the fault. Once the fault is eliminated, the angle δ continues growing to its maximum value δ_{max}. Taking this into consideration, the following cases may occur:

- Stable case: the angle δ starts decreasing. Consequently, the generator rotor will oscillate around a new equilibrium position. That is, synchronism is maintained.
- Unstable case: the angle δ exceeds its maximum value and grows indefinitely. Then, the generator synchronism is lost.

The objective of this method is to determine the critical time. This can be achieved by obtaining the maximum value of the angle δ_c, which makes the areas A_1 and A_2 be equal.

3. Phase Plane

In order to develop the advantages of the phase plane topology, as a tool to study the solutions of the differential Equation (1), the transformation of the second-order differential equation into a pair of first-order differential equations is necessary. For this purpose, the state variable $\Delta\omega(t) = \dot{\delta}(t)$ is introduced. In terms of the state variables, the Equation (1) may be restated as follows:

$$\begin{cases} \dot{\delta}(t) = \Delta\omega \\ \Delta\dot{\omega}(t) = \frac{\omega_r}{2H} P_m \frac{E'E \sin\delta}{X} \end{cases} \tag{2}$$

The equilibrium points of the dynamical system governed by Equations (2) are determined by imposing the following conditions:

$$\dot{\delta}(t) = 0 \quad y \quad \Delta\dot{\omega}(t) = 0 \tag{3}$$

which leads to:

$$\begin{aligned} &\Delta\omega_e = 0 \\ &\frac{\omega_r}{2H} P_m \frac{E'E \sin\delta_e}{X} = 0 \end{aligned} \tag{4}$$

where, the subscript e indicates an equilibrium state. The first equation in (4) shows that every possible equilibrium state correspond to null deviations of the generator rotor speed with respect to the synchronous speed. Graphically, this means that every equilibrium state will be located on the axis $\Delta\omega = 0$ of the phase plane. For

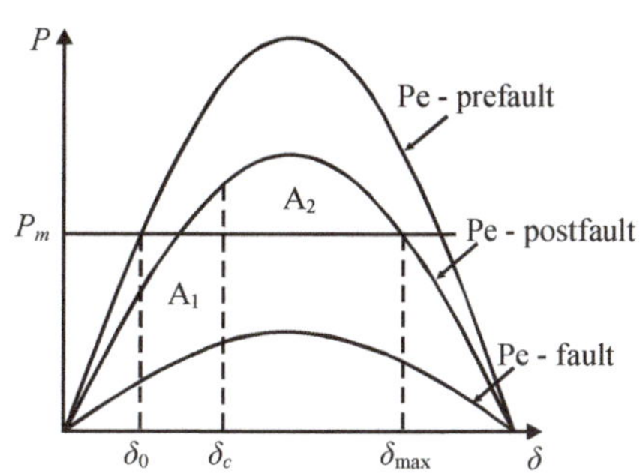

Figure 2. Equal-area method.

the determination of the δ_e possible values, it is necessary to take into consideration the second equation in (4). This equation has infinite solutions and they are represented by:

$$\delta_e = \sin^{1}\left(\frac{\omega_r P_m}{2H}\frac{X}{E'E}\right) \tag{5}$$

From the infinite solutions of the Equation (5), only those included between $0 \le \delta_e \le \pi$ have physical meaning. Then, by imposing this restriction to Equation (5), only 2 equilibrium points P_1 and P_2 can be obtained, as it can be seen in **Figure 3**. The point $P_1 = (\delta_0, 0)$, corresponds to a Poincaré-stable center. On the other hand, P_2 is a hyperbolic-equilibrium point known as saddle point, which is inherently unstable.

In **Figure 3**, different trajectories of the phase plane corresponding to a determined configuration of the power system and for different initial conditions can be observed. The arrowheads in **Figure 3** indicate the directions in which we must proceed along the phase paths for increasing time. In case A, non-perturbed case, the trajectory reduces to the point P_1. If a perturbation is introduced, producing a deviation of the angle δ from its equilibrium point and therefore, a change in the initial values required to solve the swing equation, we may find different cases.

For example, in case B the angle value has been diminished in relation to its equilibrium value, whereas in case C this value has been increased. As it can be observed, in both cases the generator rotor stays indefinitely oscillating round its equilibrium position. In case D, the angle has been increased near the point P2 and the solution corresponds (in the phase plane) to a homoclinic trajectory [2]. For the case in which the angle deviation exceeds the point P_2, the trajectory shows the same aspect as case E, where the rotor angle and speed grow indefinitely. In case F, where the angle has been diminished to the intersection point between the homoclinic trajectory and the axis $\Delta\omega = 0$, the trajectory shows an oscillation tendency round the equilibrium point P_1 at the beginning. But, nearby the point P_2, the angle and speed start growing indefinitely. The analyzed cases show that for initial conditions given within the region of the phase plane bounded by the homoclinic orbit, the system is stable from a transient-stability point of view.

Supposing that such power system is initially in its stable-equilibrium position, which is $\delta(t) = \delta_0$ and $\Delta\omega(t) = 0$. Supposing also that the network configuration is ultimately changed through the application of a three-phase fault in some node, and that after a while the protection devices act eliminating it.

The superimposed phase planes corresponding to three different topologies of the network: pre-fault, fault and post-fault can be observed in **Figure 4**. Curve D′ corresponds to the post-fault homoclinic trajectory and curve G corresponds to a fault trajectory. Point P_1 represents the above-mentioned initial condition. After the fault, the rotor angle and speed values increase, as it can be seen in the fault G trajectory. When the fault is not eliminated, the values δ and $\Delta\omega$ grow indefinitely and therefore the generator loses its synchronism. When the fault is eliminated, the power system configuration changes. Under these post-fault conditions, the equilibrium points are P_1' and P_2'. As for point P_1, point P_1' represents a stable equilibrium point. Point P_2' represents the maximum angle (δ_{max}) the rotor may adopt for not losing synchronism under a post-fault configuration. For fault elimination before the intersection of G and D′ trajectories, the system is stable from a transient-stability point of view. This occurs because δ and $\Delta\omega$ values during fault removal represent the initial conditions needed to solve the equation system (2) under post-fault conditions. As it was mentioned before, the generator maintains its

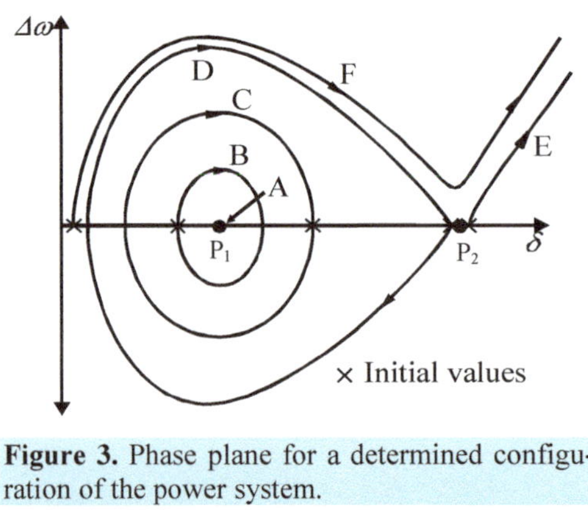

Figure 3. Phase plane for a determined configuration of the power system.

synchronism as well as the initial values are kept within the region limited by the homoclinic trajectory. Therefore, the critical angle δ_c is determined by the intersection of fault G and the homoclinic post-fault D′ trajectories. The time necessary for the evolution of the system from $(\delta, \Delta\omega) = (\delta_0, 0)$ state to $(\delta, \Delta\omega) = (\delta_c, \Delta\omega_c)$ state is the critical time, t_c.

The developed method may be extended to multiple-generator systems. For these cases, a phase plane for each machine of the system should be built.

4. Numerical Results

As an application we analyze the example 13.1 of the book "*Power System Stability and Control*" [1]. In such example, the transient stability of a thermo-generation station consisting on four 555 MVA, 24 Kv, 60 Hz generators, supplying power to an infinite bus through two transmission circuit, is analyzed (**Figure 5**). For $t_f = 0$, a three-phase fault is applied at the end of one of the lines. After a while, the protection devices separate the perturbed line from the system. The objective of this method is to determine the fault permanence maximum value without losing synchronism.

The network reactance shown in **Figure 5** is in p.u. on 2200 MVA and 24 Kv of base. The resistances are negligible.

The initial-operation conditions, expressed in p.u. on the previous base, are:

$P = 0.9$ [pu].

$Q = 0.436$ [pu] (overexcited).

$H = 3.5$ [MWs/MVA].

$E_t = 1.0$ ang 28,34°.

$E = 0.90081$ [pu].

$X'_d = 0.3$ [pu].

The X value for each state is:

Pre-fault:

$$X = X'_d + X_t + \frac{X_1 X_2}{X_1 + X_2} = 0.7752 \text{ [pu]}$$

Fault:

$$X = X'_d + X_t = 0.45 \text{ [pu]}$$

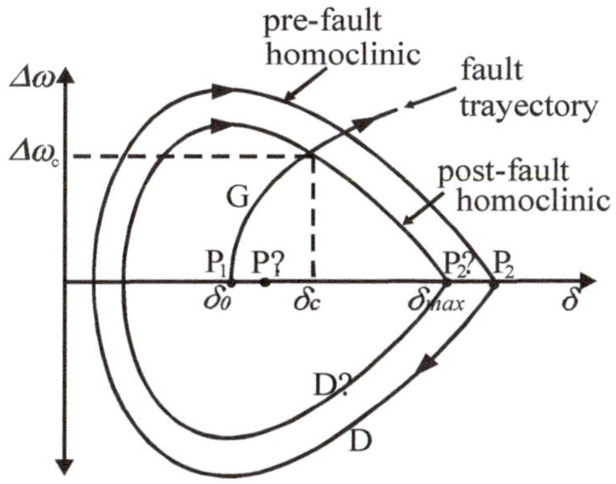

Figure 4. Phase plane for conditions of pre-fault, fault and post-fault.

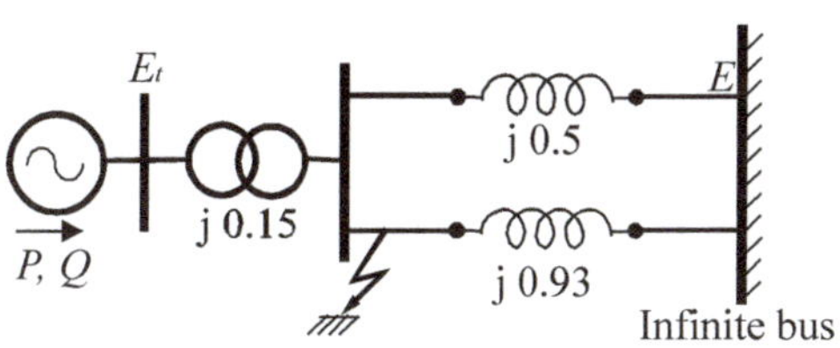

Figure 5. Example 13.1.

Post-fault:

$X = X'_d + X_t + X_1 = 0.95$ [pu]

The E' value is obtained as follows:

$E' = E_t + jX'_d \cdot I$

$E' = 1.1626 \angle 41.77^\circ$

Considering these values and using the Equal-Area Criterion (**Figure 6**), the following results are obtained:

$\delta_0 = 41.77^\circ$

$\delta_{max} = 125.27^\circ$

$\delta_c = 52.29^\circ$

$t_c = 0.0870$ [s].

The following results are obtained by using the conventional numerical-integration method (**Figure 7**):

$\delta_0 = 41.77^\circ$

$\delta_{max} = 125.27^\circ$

$\delta_c = 52.35^\circ$

$t_c = 0.0875$ [s].

Using the phase plane (**Figure 8**), we obtain:

Therefore, the following values result:

$\delta_0 = 41.77^\circ$

$\delta_{max} = 125.27^\circ$

$\delta_c = 52.52^\circ$

$t_c = 0.0879$ [s].

5. Conclusions

A method that uses the phase plane $\delta - \Delta\omega$ to solve the problem of transient stability in a machine connected to an infinite bus through an interconnection network, has been presented. The proposed method combines the

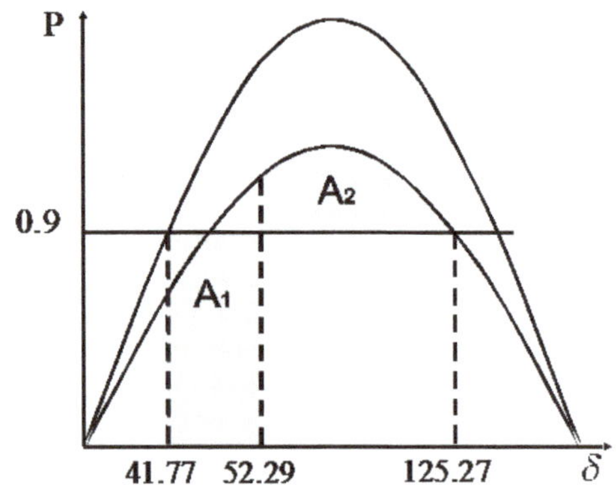

Figure 6. Equal-area criterion.

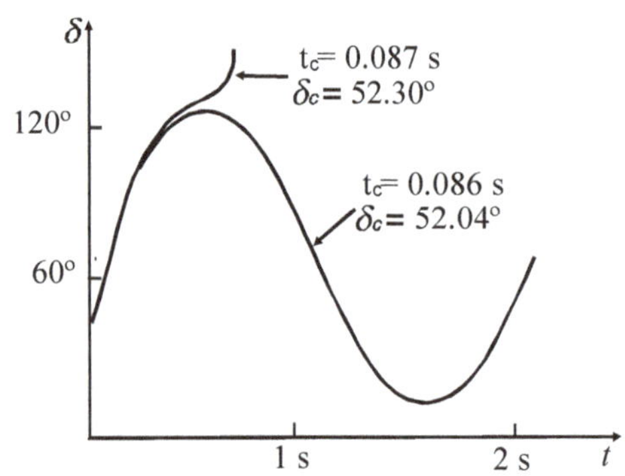

Figure 7. Numeric integration method.

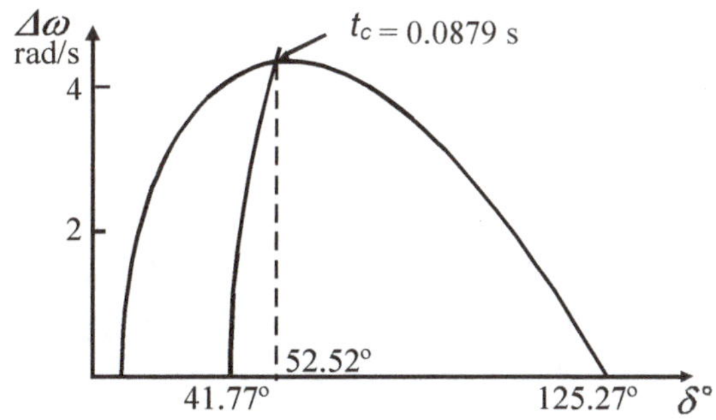

Figure 8. Phase plane.

strength of the conventional numerical-integration method and the simplicity and graphic representation of the equal-area criterion-based method. It also has additional properties in relation to both methods. For instance, it presents the following advantages associated to the equal-area criterion-based method:

- It offers the same integral vision of the problem. In addition, it shows the time-domain evolution of generator rotor speed variation.
- It allows extending the problem to multiple-generator electrical systems, with no need of reducing the network to an infinite bus.

On the other hand, it presents the following advantages in relation to conventional numerical-integration methods:

- It provides an integral vision of the problem.
- It allows the visualization in the same graphic of the angle variations and the rotor speed.
- It allows the direct determination of the critical time for fault elimination, avoiding thus the trial and error method.

References

[1] Kundur, P. (1994) Power System Stability and Control. McGraw-Hill, Inc., New York.

[2] Sansone, G. and Conti, R. (2014) Non-Linear Differential Equations. Elsevier, Pergamon.

[3] Sun, K., Lee, S.T. and Zhang, P. (2011) An Adaptive Power System Equivalent for Real-Time Estimation of Stability Margin Using Phase-Plane Trajectories. *IEEE Transactions on Power Systems*, **26**, 915-923. http://dx.doi.org/10.1109/TPWRS.2010.2055900

[4] Zhou, J. (2011) Stabilisation of Individual Generators with SVC Designed via Phase Plane Partitioning. *IET Control Theory and Applications*, **26**, 915-923.

[5] Pai, M.A. (1989) Energy Function Analysis for Power System Stability. Kluwer, Norwell.

[6] Yan, J., Liu, C.-C. and Vaidya, U. (2011) PMU-Based Monitoring of Rotor Angle Dynamics. *IEEE Transactions on Power Systems*, **26**, 2125-2133. http://dx.doi.org/10.1109/TPWRS.2011.2111465

[7] Harikrishna, D. and Srikanth, N.V. (2011) Phase Plane Method Stability Analysis of Fuzzy Logic PSS. *Journal of Electrical Engineering*, **11**, 1-7.

19

Measuring Porosity of Anodes in Solid Oxide Fuel Cell (SOFC) through Water Archimedeans Porosimetry

Syed Mubashar Hassan[1], Syed Ali Hasnain[2]

[1]School of Engineering and Built Environment, Edinburgh Napier University, Edinburgh, UK
[2]Centre of Excellence in Sustainable Building Design, School of Energy, Geosciences, Infrastructure & Society, Heriot-Watt University, Edinburgh, UK
Email: mubashar_naqvi@live.co.uk

Abstract

The present research is aimed to measure the porosity of anodes in solid oxide fuel cell through water Archimedeans method. There are various alternatives available to replace fossil fuel cells like nuclear power, wind energy, solar energy, bio fuel, and geothermal and fuel cells. Among all the alternatives of fossil fuel, one form of energy production that stands out from the rest and promises a sustainable future energy is fuel cell. Moreover, it offers many advantages in contrast to other forms of energy generation. An Archimedean approach for water immersion porosimetry is carried out. Some of the results are beyond rational limits, and given negative and sometime above 100 percent porosity. The reasons for these unacceptable results are either due to water ingress into the sample or the sample turns into buoyant due to air in the cling film. The results from Archimedean porosimetry should only be used qualitatively due to errors associated with the results. It is also noted that Archimedean porosimetry is not the ideal technique for measuring the porosity of coated samples. It is suggested that larger samples should be analyzed that will help to minimize the weighing errors.

Keywords

Solid Oxide, Fuel Cell, Porosity, Anodes, Archimedeans

1. Introduction

Increasing energy prices is a testimony to the fact that as we already are facing energy shortages, and the worst is still to come because according to a UN report the world population will be 10.5 billion in 2050 [1] [2] compared with 6.5 billion now and in 2040 we will be requiring double the energy compared with what we are using today, and our need for electricity will double by 2030 [3]. Living conditions and life standards are improving throughout the world especially in the developing and underdeveloped countries. Therefore, if the entire world has same living conditions as Britain then we will be requiring energy resources (non-renewable) equivalent to three earths [4]. There are various alternatives available to replace fossil fuel cells like nuclear power, wind energy, solar energy, bio fuel, and geothermal and fuel cells. Among all the alternatives of fossil fuel, one form of energy production that stands out from the rest and promises a sustainable energy future is fuel cell. Moreover, it offers many advantages in contrast to other forms of energy resource [5].

Solid Oxide Fuel Cells (SOFC) in Relation to Porosity of Anodes

SOFC is high temperature fuel cell which uses solid material as electrolyte for their energy production. The operating temperature for the SOFC is 800˚C - 1000˚C. The high temperature is advantageous as it generates high quality exhaust heat which is used for cogeneration, but on the other hand, high temperature also limits the use of most of the materials [6]. So, scientist have tried and succeeded in bringing down the temperature to 600˚C to 800˚C, thus classifying them as intermediate temperature solid oxide fuel cells (ITSOFC) and high temperature solid oxide fuel cells (HTSOFC) [7] [8]. Though low temperature brought some material flexibility, it also decreased the electrode kinetics and electrolyte conductivity which limited their use. So, HTSOFC still remains more prevalent [9]. SOFC is the most likely candidate for the commercialization of fuel cells because of the advantages they offer.

- SOFC is highly efficient, the efficiency of fuel cell is not dependent on the Carnot cycle of heat engine and their combined heat and electric efficiency is higher than all other types of fuel cells.
- SOFC is tolerant to impurities where CO is used without damaging the working of SOFC, unlike other fuel cells where carbon monoxide (CO) acts as a poison.
- Fuel flexibility is the single biggest advantage of SOFC because of such a high temperature; a great number of hydrogen containing compounds can be used as fuels which are internally reformed by the unit.
- As SOFC uses solid electrolyte, there are no moving parts involved for the circulation of electrolyte. So, it is noise pollution free.
- Design simplicity and ability to work without expensive catalysts are the few of the other advantages which make SOFC most probable commercial reality [10] [11].

2. Material & Methods

In this section details of the experimental work carried out during the project is discussed. This includes information about the sample, calculations, instruments and chemical used.

Health and safety issues involved and problem faced during experimentation are also discussed briefly. In the first part sample naming along with sample preparation for the experiments are presented. Archimedean approach for water immersion porosimetry are carried out, details along with calculations are discussed.

2.1. Sample Information

The samples to be analyzed are the alumina coated Ni/YSZ anodes which are manufactured by electro less deposition by varying four variables (**Table 1**).

Each variable has two variations (level).The purpose of all of this is to analyze the effect of these variables on the porosity of the anode. One relatively simple way of checking the effect of variables on the final output is one variable at time technique (OVAT), but according to (Jiju, 2003) OVAT technique is not a reliable one, because some time combination of variables give different results from the ones when they are individually analyzed. Let's suppose that by keeping all the other three variables constant increase in particle size increases the porosity of the anodes, but it may happen that by increasing pH, increase in particle size actually decreased the porosity of anodes. so while manufacturing anodes the knowledge of design of experiment (DOE) was kept in mind and OVAT technique is avoided and all the sixteen possible combinations from these four variable are manufactured, so that a broader and clear picture can be achieved and more conclusive results can be drawn [12].

Three sets of these sixteen samples are analysed (**Table 2**). The term replicate is used for each set. A total of 48 samples are analysed. Each sample is identified by two numbers separated by a dot First number (1-3)

Table 1. Variables and their different levels.

Variable	Symbol	Level	
		1	2
Particle size	A	2	10
Agitation	B	Air	Stir
pH	C	4.9	5.4
Surface treatment	D	HF	MB

Table 2. All sixteen possible combinations of the sample.

Number	Combination	Number	Combination
1	A1B1C1D1	9	A1B2C2D1
2	A2B1C1D1	10	A1B2C1D2
3	A1B2C1D1	11	A1B1C2D2
4	A1B1C2D1	12	A2B2C2D1
5	A1B1C1D2	13	A1B2C2D2
6	A2B2C1D1	14	A2B1C2D2
7	A2B1C2D1	15	A2B2C1D2
8	A2B1C1D2	16	A2B2C2D2

represents which replicate sample is from, while the second number (1-16) represent the combination, for example Sample 1.15 means sample is from replicate 1 and combination 15 (A2B2C1D2).

2.2. Cutting of the Samples

Samples are in the form of squares of dimensions 25 × 25 mm. So they are cut in to 10 × 10 mm dimensions with the help of pencil type Diamond scriber. initially tungsten carbide scriber was used but the Alumina tile was too hard for that, because both has hardness of 9.0 on Moh's hardness scale [13], so even harder material *i.e.* diamond is used as scriber which has hardness 10 on Moh's scale [13]. While samples are cut, special care is taken that the scriber draw a straight line and it don't scratch the surface of the sample, because if a scratch is present on the surface then that will act as a pore and result in the increase in the porosity of the material (**Figure 1(a)** and **Figure 1(b)**).

2.3. Grinding

Once all the samples are cut they are then grinded on Metaserve hard grinder using grit of 120,400,800 and 1200 microns respectively. The purpose of grinding is to balance the edges and give the samples more uniform dimensions and reduce the possibility of error because of irregular dimensions.

2.4. Heating

Samples are then heated in oven at 100°C for three hours to remove the water which has come in contact with the sample during grinding and also the moisture which is absorbed from atmosphere. Sanyo OMT oven is used for this purpose. They are then wrapped in paper and kept in the desiccator. It is essential to remove the moisture or any other impurity from the surface of the sample because if the pores are already occupied by the impurities then intruding liquids used in porosimetry cannot enter the pores, which will gives lower porosity then the actual one. All the identification markings of the samples are also removed using acetone, because ink of the writing

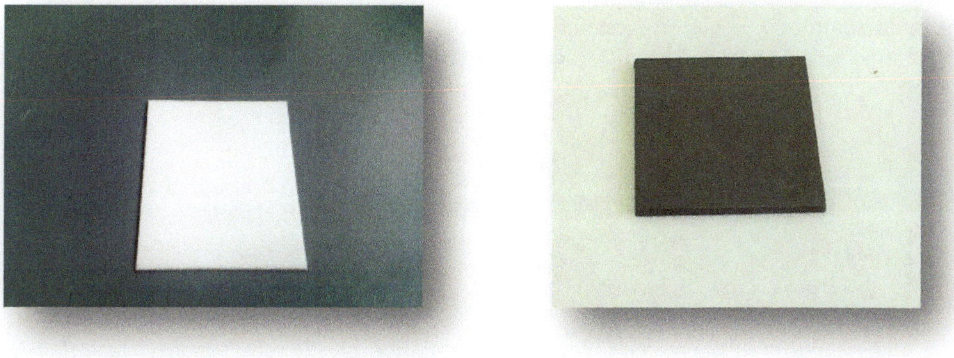

Figure 1. (a) Alumina tile; (b) Ni/YSZ coated sample.

can also reside in the pores thus barring intrusion medium to enter the pores during analysis.

Same procedure is followed for alumina tile which is acting as a substrate material. Here it should be noted that alumina is used instead of YSZ as substrate material for Ni/YSZ coating, because of the low price of Alumina and similarities of properties as both being dense ceramics.

3. Experimental Density Calculation

In this method first weight of composite is measured in air, then weight in water is measured by wrapping the sample in cling film and hanging the sample through a thin piece of metal wire in a beaker filled with distilled water. Weights are measured in Oertling NA 164 balance with values up to four decimal places. These weights are then used to calculate the density using the Archimedean formula.

$$\rho_{comp} = M_{\text{air}} / M_{\text{air}} - M_{\text{water}} \quad (1)$$

where

$$\rho = \text{Density}$$

And

$$M = \text{mass}$$

Similar process is repeated for alumina tile to calculate its density. Density of alumina is repeated five times and means value is used. Once density of alumina tile and sample are calculated then rule of mixture is applied to calculate the density of the coating.

$$\rho_{comp} = \rho_a V_a + p_{cot} V_{cot} \quad (2)$$

where

$$V_a = v_a / v_{com}$$

$$V_{cot} = v_{cot} / v_{com}$$

where "V" represents volume fraction and "v" represents only volume. By putting all these values in Equation (2) we will get the experimental density of the coating.

Theoretical Density Calculation

After calculating experimental density, theoretical density is calculated. As the coating is made up of Ni and YSZ so theoretical density is calculated by using the rule of mixture

$$\rho_{cot} = \rho_{\text{Ni}} V_{\text{Ni}} + p_{ysz} V_{ysz} \quad (3)$$

Density of Ni is 8.9 gm/cm^3 while that of YSZ is 6.10 gm/cm^3. These values are taken from literature. Now for volume fraction we must know the volume of each constituent. For that the samples are analyzed under the SEM and elemental analysis is carried out, which gave the percentage weight of each constituent. That weight percentage is then converted in to volume percentage and then further in to volume fraction. Weight percentage Ni content in all samples is given in Appendix 6(a)-(c). By putting all these values in Equation (3) theoretical density of Coating is calculated. The difference in theoretical density value and experimental density value is because of porosity which is calculated by formula given below;

$$\Phi = \left(\rho_{th} - \rho_{exp} / \rho_{th}\right) \times 100 \tag{4}$$

$$\Phi = \text{percent porosity}$$

$$\rho_{th} = \text{Theoratical density}$$

$$\rho_{exp} = \text{Experimental density}$$

For better understanding of the method, complete calculation of one representative sample is given here.

Example

Sample

Weight of sample in air = 0.4335 gm

Weight in water = 0.3177 gm

Put these values in Equation (1)

$$\rho_{comp} = 0.4335/0.4335 - 0.3177$$

$$\rho_{comp} = 3.74 \text{ gm/cm}^3$$

Similarly density of alumina comes out 3.74 gm/cm^3. This is a mean value of five alumina samples given in Appendix 5(b).

$$\text{Dimensions of the sample} = (1.105 \times 1.053 \times 0.098)\ \text{cm}^3$$

$$\text{Volume of samle} = 0.1140\ \text{cm}^3$$

$$\text{Volume of coating} = 0.000226\ \text{cm}^3$$

$$\text{Volme fraction of coating} = 0.000226/0.1140 = 0.00198$$

$$\text{Volume fraction of Alumina} = 1 - 0.00198 = 0.99802$$

Put all these values in Equation (2)

$$3.74 = 3.74 \times 0.99802 + 0.00198 \times \rho_{cot}$$

$$\rho_{cot} = 3.74 \text{ gm/cm}^3$$

To calculate the theoretical density elemental percentage of all the constituent element is required which is calculated by SEM (**Figure 2** and **Table 3**).

$$Wt\ \%\ \text{age of Ni} = 81.92\%$$

$$Wt\ \%\ \text{age of YSZ} = 18.08\%$$

These weight percentages are converted in volume by dividing them with density.

$$\text{Volume of Nickel} = 81.92/8.9 = 9.2045\ \text{cm}^3$$

$$\text{Volume of YSZ} = 18.08/6.10 = 2.9639\ \text{cm}^3$$

$$\text{Total volume of the coating} = 9.2045 + 2.9639 = 12.1684\ \text{cm}^3$$

$$\text{Volume fraction of Ni} = 9.2045/12.1684 = 0.7564$$

$$\text{Volume fraction of YSZ} = 1 - 0.7564 = 0.2463$$

Put these values in Equation (3)

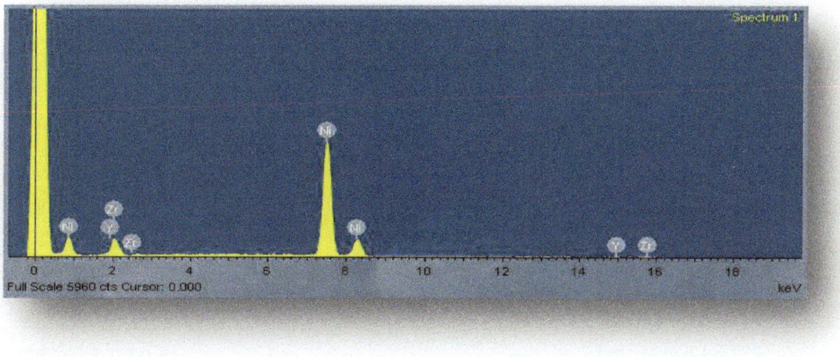

Figure 2. EDXA spectrum of Ni/YSZ anode.

Table 3. Elemental analysis of Ni/YSZ anode from SEM.

Element	Weight %	Atomic %
Ni K	81.92	90.46
Y L	3.43	1.85
Zr L	14.65	7.69
Totals	100.00	100.00

$$\rho_{cot} = 8.9 \times 0.7564 + 6.1 \times 0.2463$$

$$\rho_{cot} = 8.218\ \text{gm}/\text{cm}^3$$

Now we know the theoretical density and experimental density, putting these values in Equation (4) we get the percent porosity of the coating

$$\Phi = (8.218 - 3.74/8.218) \times 100$$

$$\Phi = 54.49\%$$

This method is tried on ten different samples; results are discussed in chapter 4.

4. Results and Discussion

These results are beyond rational limits, giving negative and above 100 percent porosity simultaneously. The reasons for these unacceptable results are either water ingress into the sample or the sample became too buoyant due to air in the cling film. Therefore a different approach has to be adopted

4.1. Relation between Thickness and Porosity

Results obtained from this approach are quite encouraging and revealed some very important information. For replicates carried out on Archimedean Porosity a general trend is observed between coating thickness and porosity as shown in **Figure 3** and **Figure 4**. The trend is that in water Archimedean porosimetry higher the porosity lower will be coating thickness and vice versa Perhaps this trend could be because at high porosity there are less growth sites available for the coating to grow on, while for thicker coating more active growth sites are available which encourages the coating to grow and in that growth it sometimes covers the pores in their path thus closing the open pores. A technique used here only measures the open porosity so these pores are not detected and gave lower percentage porosity.

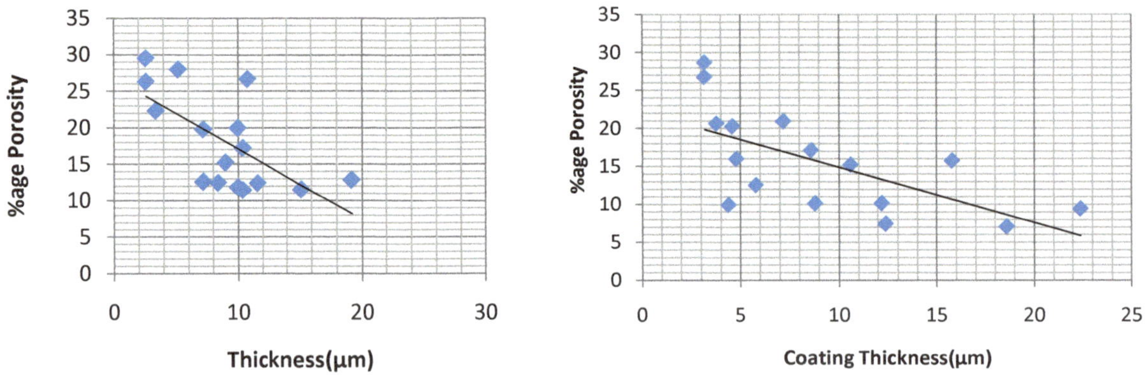

Figure 3. Graph thickness vs. percentage porosity.

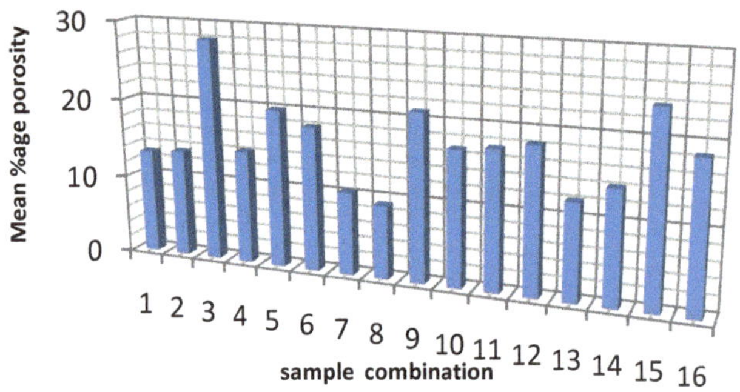

Figure 4. Comparison of mean porosities of two replicates measured.

4.2. Optimum Set of Variable

To analyse the optimum set of variables a bar graph is plotted by taking the mean porosities of all three replicates through water Archimedean porosimetry. Among all the three replicate it is observed that combination 3 and combination 15 always gave the highest porosities, as predictable the thickness of these combinations are among lowest. On observing the thickness it is revealed that combination 15 of replicate 1 has thickness a lot higher than for the same combination of other two replicates, which resulted in its low porosity.

$$Combination3 = A1B2C1D1$$

$$Combination15 = A2B2C1D2$$

It can be seen that both values of A and D gave high porosity, so they are not the ones which actually contribute in increasing the porosity.B2 and C1 these are the variables which are involve in both combination. so it can be inferred that whenever B2 and C1 are used with any combination of A and D we will get high porosity. In literal meaning when bath pH is low and stirrer used as agitator with any combination of particle size and surface treatment we will get the high porosity, hence these are e the optimum set of variables for porosity of SOFC anode made by electro less deposition.

5. Conclusion

Porosities of SOFC anodes made by electro less deposition are successfully calculated by water Archimedean porosimetry. In electro less deposition anodes made with bath pH 4.9 and stirrer used as agitator gives the maximum porosity the particle size and surface treatment may have any value. A general rule is established which gives you an idea about porosity qualitatively. Porosity is inversely proportional to coating thickness. Results from water Archimedean porosimetry should only be used qualitatively due to errors associated with the results. AP is not the ideal technique for measuring the porosity of coated samples.

References

[1] Davidson, A.M. and Waugh, W. (2008) Method of Manufacturing of an Electrode for Fuel Cell. New International (PCT) Patent Application Claiming Priority from U.K. Application No. 0719260.2.

[2] Duncan, R.C. (2005-2006) The Social Contract, the Olduvai Theory Energy, Population, and Industrial Civilization.

[3] Sayigh, A. (2009) Worldwide Progress in Renewable Energy.

[4] Kelly. M.J. (2010) Energy Efficiency, Resilience to Future Climates and Long-Term Sustainability: The Role of the Built Environment. *Philosophical Transactions of the Royal Society A*, **368**, 1083-1089.

[5] EIA U.S. Energy Administration Independent Statistics and Analysis. http://tonto.eia.doe.gov/energyexplained/index.cfm?page=about_energy_units

[6] Bradley, J. (2010) The Scotsman, North Sea Oil Discovery.

[7] Boudghene, A. and Traversa, S.E. (2002) Solid Oxide Fuel Cells (SOFCs): A Review of an Environmentally Clean and Efficient Source of Energy. Renewable and Sustainable Energy Reviews, **6**, 433-455

[8] Cleveland, C.J. Encyclopaedia of Energy. Vol. 1, Elsevier Academic Press, 151, 158-160.

[9] O' Hayre, R., Cha, S.-W., Colella, W. and Prinz, F.B. (2006) Fuel Cell Fundamentals. John Willie and Sons.

[10] www.lifeaftertheoilcrash.net

[11] King, B. (2007) Bakhtiari's Event of the Century. Whiskey & Gunpowder, Pittsburgh.

[12] Davidson, A. Lecture Notes.

[13] READE Speciality Chemical Resource. www.reade.com/Particle_Briefings/mohs_hardness_abrasive_grit.html

20

Electrical Performance of Porcelain Surge Arrester in 22 kV Distribution System under Contaminated Conditions

Wichet Thipprasert, Prakasit Sritakaew

Department of Electrical, Faculty of Engineering, Rajamangala University of Technology Lanna Chiang Rai, Chiang Rai, Thailand
Email: wichet_thip@rmutl.ac.th, Prakasit.sri@gmail.com

Abstract

Stability and reliability of distribution system are able to be improved by many protective devices including insulators, relay and surge arrester (SA). SA is used for protection of lightning and switching surge in Provincial Electricity Au-thority (PEA). This paper presents an experimental investigation of 22 kV distribution SA electrical performances under pollution conditions according to IEC 60507. The experimental results of leakage currents more than 6 mA at Equiva-lent Salt Deposit Density (ESDD) 0.6 mg/cm^2 at nominal voltage. These data will be useful to be guideline for solving problems and reducing power loss from leakage current on the surface of SAs as a result of surface dirt and pollution. Moreover, it will be useful to select or design suitable SAs for using in places with salt conditions, high rainfall, high wind speeds and high humidity.

Keywords

Polluted Condition, Equivalent Salt Deposit Density, Lightning Surge, Switching Surge

1. Introduction

Electric power is an important ingredient for the industry and all round development of a country. With the increasing of electrical demand, the leads to a growing interconnection of electrical distribution system. And to satisfy of electrical demand, SAs of distribution system had been improved efficiency or withstand ability of distribution system. The efficiency of the system is based mainly on the continuity of the service, avoiding faults that supposed economic losses for distribution system and users.

Over voltages in power system may occur due to lightning, fault or switching operation. These over voltages could reach dangerous amplitudes for power system apparatus. To protect the system electrical equipment and to guarantee an economic and reliable operation, SAs are applied in almost all types of electrical power network. Gapless zinc oxide (ZnO) surge arresters or Metal Oxide Surge Arresters (MOSAs) are widely used. SAs are usually connected between the phase and ground terminals. They limit the voltage level in equipments such as transformers below the withstand voltage level.

MOSAs are exposed to heavy thermal and electrical stresses if the porcelain housings of the arrester units are polluted. Together with other environmental stresses such as temporary overvoltages or high temperatures, this can lead to thermal runaway of MOSA. In the past in pollution research of MOSAs, a lot of attention was paid to the coupling in of currents at the flanges of a multi-unit arrester with the porcelain housings non-uniformly polluted. This effect has also been taken into account in the testing standard [1] [2] for pollution tests of MOSAs. The effect of coupling in currents at the flange due to in homogeneously polluted housings can be avoided if the total varistor column is included in a single housing [2]-[4]. Another problem caused by pollution of MOAs is the partial discharge phenomena inside the MOSAs [3]. These discharges have been responsible for the destruction of a complete arrester [5] [6].

Problem of dry band formation and elongation on the polluted surface of housing of the MOSA may not only cause the temperature rise but may distort the voltage distribution and leakage current flow externally and internally. This may sometimes become serious in extreme pollution conditions. Formation of dry band on the housing may result in to [3]-[5]:

- Increase in internal leakage current, causing local heating and thermal instability of varistors.
- Puncture or crack on the surface.
- Flashover on the housing of a unit or the complete MOSA.
- Chemical reaction inside MOSA and formation of corona.

Adiabatic heating due to leakage current and dry band arcing causing flashover on the housing and imbalance in heat energy distribution with absorption in varistors, leads to Stress degradation and finally thermal runaway. Internal ionization resulting into the decomposition of the medium surrounding the varistor reduces the oxygen concentration and as a result varistor characteristics might get changed. Dry band formation is a dynamic process and depends on contaminant conductivity, wetting conditions, voltage levels, ambient conditions etc. In artificial pollution tests, there are three polluting methods frequently used, including quantitative Brushing Method, Dipping Method and Spraying Method. Base on measurements of ac flashover performance of four types of porcelain and composite insulators in the artificial climate chamber. In this paper to study of leakage current and temperature absorption of porcelain surge arrester under polluted condition using spraying method [2]-[5]. The equivalent salt deposit density (ESDD) is the most commonly used method to characterize pollution quality and quantity on the surface of porcelain surge arresters. The method determines the salt deposition density by washing down the SAs surface with a known amount of water and then measuring the conductivity of the water. A major disadvantage of the ESDD method is that the SAs must be removed from the distribution line for an exact measurement [1] [2] [4]. Separately this method is lab or intensive and is subject to the experience level and attention to detail of the field personnel. Determination of ESDD on insulators installed on distribution line towers is possible but it requires skilled personnel [4]-[6].

Therefore to guideline for solving problems above, in this work to study Evaluation of Leakage Current Measurement of Porcelain Surge Arrester for Site Pollution Severity Assessment testing by setup the artificial pollution chamber based on IEC 60507.

2. Properties of MOASs in Polluted Conditions

MOASs operating in polluted areas may be subject to enhanced environmental stress, causing and increased risk of external flashover, internal partial discharges, and increased varistor temperature. The influence of these phenomena on the design and operating performance of MOSAs is discussed in the following. Regarding the risk of external flashovers, the MOSAs should be treated in the same way as other high voltage insulators. Hence, the requirements on the external insulation, e.g. creepage distance and shed shape, shall be based on the actual pollution severity, in order to keep the risk of flashover at a sufficiently low level. The second point of interest is related to the Occurrence of partial discharges (PD) inside the porcelain housing. Although the arrester may be completely free from partial discharges when dry, internal PD may appear during periods of pollution [5] [6]. The PD is caused by the difference between the internal voltage distribution (as determined by the varistor column), and the external voltage distribution (as determined by the wet pollution layer). However, the ZnO varistors can be efficiently protected against the additional chemical stress resulting from the internal PD. Furthermore, proper selections of insulating materials for other internal parts cans are the complete surge arrester resistant to PD. The third basic phenomenon related to MOSAs operating in polluted conditions is the temperature rise that may occur due to the uneven voltage distribution created by surface leakage currents. Varistor temperature, energy absorption capability and temporary overvoltage (TOW withstand, are closely related for MOSAs; there is a certain temperature limit above which the arrester is not thermally stable when energized at the continuous operating voltage. Hence, a considerable temperature rise of the varistors due to severe pollution will reduce the TOV and/or energy absorption capability. Among the three basic properties of MOSAs described above, the temperature rise phenomenon is of the greatest interest today. The amount of field data on temperature rise in polluted conditions is still limited, and the most appropriate laboratory test methods are not yet identified. The studies presented in this paper deals, therefore, entirely with the temperature rise of MOSAs during artificial and natural pollutions.

3. Measurement of Contamination Severity [7]-[9]

Contamination severity on the surface of MOSA can be given in terms of ESDD as explained earlier. The measurement of ESDD in case of porcelain and glass has been standardized in the International Electrotechnical Commission document 60507 ESDD is measured by dissolving the contaminants on the surface of MOSAs, in deionized water and then measuring the conductivity of the water. The ESDD is then calculated using the following formula

$$\sigma_{20} = \sigma_{\theta}\left(1 - b(\theta - 20)\right) \tag{1}$$

σ_{θ} = the measured conductivity in S/m.
b = 0.01998 (constant).

$$ESDD = \left(5.7 \times \sigma_{20}\right)^{1.03} \times \frac{V_0}{A_R} \tag{2}$$

V_0 = Volume of dissolvent (distilled water) in m^3.
A_R = Area of MOSA surface in m^2.
This method is observed to be good for measuring ESDD for porcelain surge arresters, as these are completely wettable [8].

4. Test Specimens, Setups, and Procedures

4.1. Test Specimens

The specimens are the Porcelain Surge Arrester in 22 kV distribution systems as shown in **Figure 1**. **Table 1** shows the dimensional details of the samples test.

Figure 1. Porcelain surge arrester.

4.2. Test Setup

The test circuit of ZnO SA is provided schematically in **Figure 2**. The voltage elevation transformer is supplied by a Variac, thus resulting in a controlled gradual voltage increase. The transformer secondary coil is connected in series resistance R0 = 2 MΩ and parallel with a capacitive voltage divider comprised of capacitors C1 and C2 respectively. In parallel with the voltage divider is the series connected of MOSA. The voltage across C2 is measured by a scope meter (190B Scope Meter Series: Fluke) and connected to a computer where it is recorded and visually displayed with the use of appropriate software. The technical characteristics of the capacitors C1 and C2 were 224 pF, 100 kV 50 Hz and 2 μF, 1000 V/50 Hz, respectively.

4.3. Test Procedures

The research was carried out on Porcelain Surge Arrester of 22 kV distribution system of PEA. The experimental set-up is schematically shown in **Figure 2**. Referring to the test standard [7]-[9], the test procedures in the paper were as follows. After the connected circuit in **Figure 2** completely, ac voltage was applied to the specimens and increased at a constant rate till the voltage is 12.8 kVrms. Then ac voltage supplied to specimens is a long time 3 hours. The thermal image camera (Fluke Ti32) records the temperature profile and the leakage current shows in **Figure 3** and **Figure 4**.

5. Test Results of Porcelain Surge Arrester

The results form experiments are show in **Figure 5** and **Figure 6**. At an A.C. test voltage of 12.8 kVrms no partial arcs were burning across the dry zone and at the pollution layer. The wetting of the dried pollution layer was accomplished by relative air humidity in the test room of more than 75%.

The leakage current (LC) of MOSA increased to 20 mA at ESDD = 0.8 mg/cm^2 at nominal voltage. Because of the capacitive coupling between the varistor column and the pollution layer the current in the arrester discs is influenced by the voltage distribution along the external pollution layer. This leads to locally higher stresses of

Table 1. Dimensional details of porcelain surge arrester.

Arrester Rating (kVrms)	MCOV (Kv rms)	Front of Wave Protective Level (kV crest)	Maximum Discharge Voltage (kV Crest) 8/20 μs Current Wave					
			1.5 kA	3 kA	5 kA	10 kA	20 kA	40 kA
21	17.0	71.3	60	64	68	76	89	111

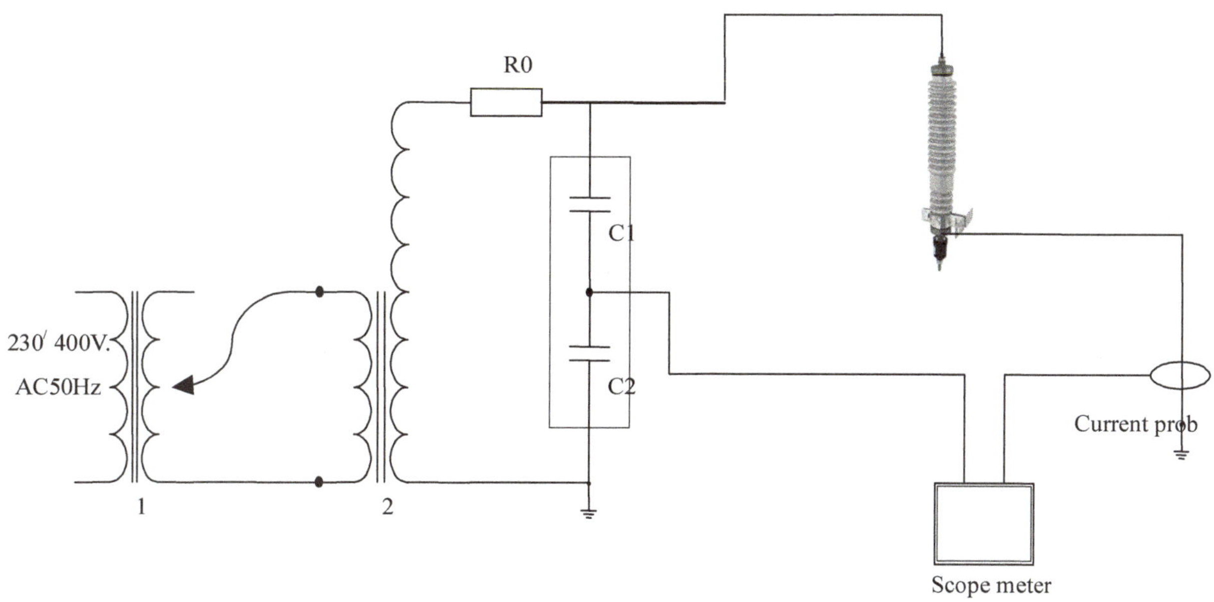

Figure 2. Schematic of testing.

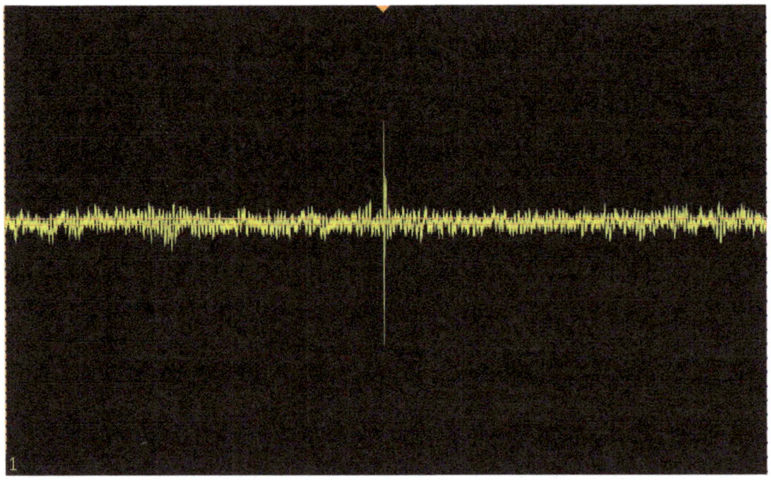

Figure 3. Leakage Current of testing MOSA at 12.8 kVrms and ESDD at 0.415 mg/cm^2.

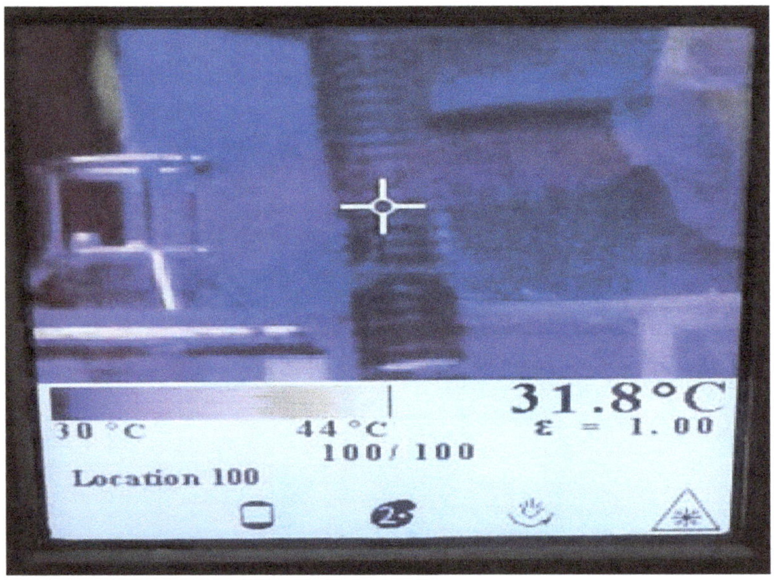

Figure 4. Temperature of arrester with 12.8 kVrms ac voltages applied and ESDD at 0.285 mg/cm^2.

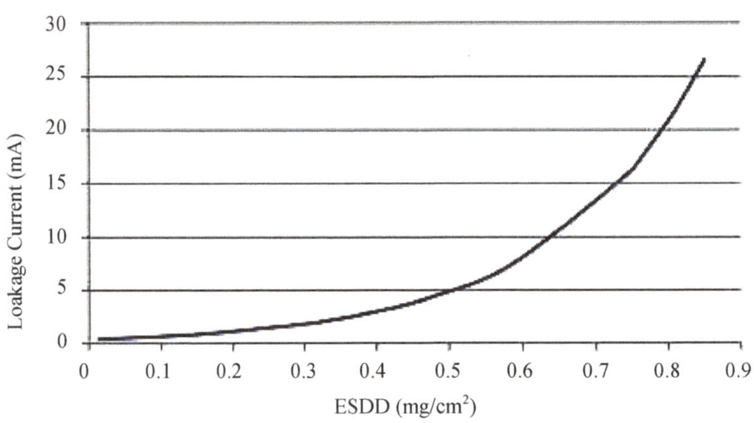

Figure 5. Leakage Current versus ESDD at 12.8 kVrms ac voltages applied.

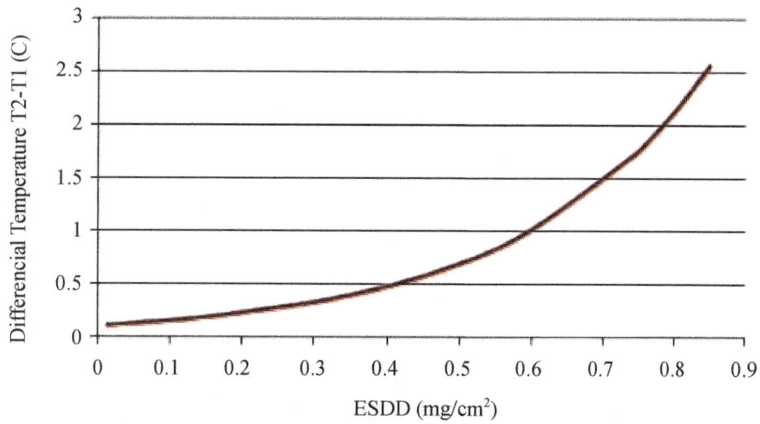

Figure 6. Temperature versus ESDD at 12.8 kVrms ac voltages applied.

varistor elements near the dry zone. And **Figure 6** shows the evaluation of temperature of the central ZnO element versus ESDD. When a voltage is applied to the electrodes, the heat dissipated by the leakage current flowing in the contamination will raise the surface temperature above ambient and moisture will be lost from the surface. This loss will become significant when it causes a measurable increase in the surface resistance; this is the first step in dry-band formation. This heat is lost from the surface by convection, by conduction through the contamination and by conduction through the body of the insulator. In the present work, heat transfer by conduction was found to be small or negligible by comparison with the heat transfer by convection.

6. Conclusions

In this paper, experimental investigation of 22 kV distribution SA electrical performances under pollution conditions according to IEC 60507 with porcelain surge arrester. The following conclusions are obtained.

- The leakage current (LC) of MOSA increased to 20 mA at ESDD = 0.8 mg/cm^2 at nominal voltage. Because of the capacitive coupling between the varistor column and the pollution layer the current in the arrester discs is influenced by the voltage distribution along the external pollution layer. This leads to locally higher stresses of varistor elements near the dry zone.
- The evaluation of temperature of the central ZnO element versus ESDD. The heat dissipated by the leakage current flowing in the contamination will raise the surface temperature above ambient and moisture will be lost from the surface.

These data will be useful to be guideline in case to take porcelain surge arrester of distribution system to reduce the investment cost. Moreover, it will be useful to select or design the suitable insulators for using in pollution area in Thailand as well.

References

[1] Haddad, A. and Warne, D. (2004) ZnO Surge Arresters. *Advance in High Voltage Engineering, LET Power & Energy Series*, **40**, 191-215.

[2] Feser, K., Kohler, W., Qiu, D. and Chrzan, K. (1991) Behaviour of Zinc Oxide Surge Arresters under Pollution. *IEEE Transactions on Power Delivery*, **6**, 688-695. http://dx.doi.org/10.1109/61.131128

[3] Bargigia, A., de Nigris, M., Pigini, A. and Sironi, A. (1993) Comparison of Different Test Methods to Assess the Thermal Stresses of Metal Oxide Surge Arresters under Pollution Conditions. *IEEE Transactions on Power Delivery*, **8**, 146-155. http://dx.doi.org/10.1109/61.180330

[4] Kado, H., Izumi, K., Shirakawa, S., Komatsu, K., Watanabe, H., Yamaguchi, M., Nakajima, M., Xobayashi, M. and Nishimura, S. (2000) Artificial Pollution Tests on Porcelain-housed Metal-Oxide Surge Arresters for 275kV Power Systems. *IEEE Power Engineering Society Winter Meeting*, **3**, 2087-2092.

[5] Lee, S.B., Lee, S.L. and Lee, B.K. (2009) Analysis of Thermal and Electrical Properties of ZnO Arrester Block. *Current Applied Physics*, **10**, 176-180.

[6] Nishiwaki, S., Kimura, H., Satoh, T., Mizoguchi, H. and Yanabu, S. (1984) Study of Thermal Runaway/Equivalent Prorated Model of a ZnO Surge Arrester. *IEEE Transaction on Power Apparatus and Systems*, **PAS-103**, 413-421.

http://dx.doi.org/10.1109/TPAS.1984.318260

[7] (2004) Selection and Dimensioning of High-Voltage Insulators for Polluted Conditions-Part I: Definitions, Information and General Principles, IEC Standard 60815.

[8] International Electrotechnical Commission (1991) Artificial Pollution Tests on High-Voltage Insulators to Be Used on AC Systems. International Standard IEC 60507.

[9] International Standard of Surge Arrester IEC60099-4 Edition 2.1:2004 Consolidated with Amendment 1: 2006 Section 6.9 Operating Duty, Section 10 Test Requirement on Polymer-Housed Arresters and Annex B Test to Verify Thermal Equivalency between Complete Arrester and Arrester Section, IEC60099-4.

21

Improved Design of a 25 MW Gas Turbine Plant Using Combined Cycle Application

Barinaadaa Thaddeus Lebele-Alawa, Anthony Kpegele Le-ol

Department of Mechanical Engineering, Rivers State University of Science and Technology, Port Harcourt, Nigeria

Email: lebele-alawa.thaddeus@ust.edu.ng, le-ol.anthony@ust.edu.ng

Abstract

This paper presents the improved design of a 25 MW gas turbine power plant at Omoku in the Niger Delta area of Nigeria, using combined cycle application. It entails retrofitting a steam bottoming plant to the existing 25 MW gas turbine plant by incorporating a heat recovery steam generator. The focus is to improve performance as well as reduction in total emission to the environment. Direct data collection was performed from the HMI monitoring screen, log books and manufacturer's manual. Employing the application of MATLAB, the thermodynamics equations were modeled and appropriate parameters of the various components of the steam turbine power plant were determined. The results show that the combined cycle system had a total power output of 37.9 MW, made up of 25.0 MW from the gas turbine power plant and 12.9 MW (an increase of about 51%) from the steam turbine plant, having an HRSG, condenser and feed pump capacities of 42.46 MW, 29.61 MW and 1.76 MW respectively. The condenser cooling water parameters include a mass flow of 1180.42 kg/s, inlet and outlet temperatures of 29.8°C and 35.8°C respectively. The cycle efficiency of the dry mode gas turbine was 26.6% whereas, after modification, the combined cycle power plant overall efficiency is 48.8% (about 84% increases). Hence, SIEMENS steam turbine product of MODEL: SST-150 was recommended as the steam bottoming plant. Also the work reveals that a heat flow of about 42.46 MW which was otherwise being wasted in the exhaust gas of the 25 MW gas turbine power plant could be converted to 12.9 MW of electric power, thus reducing the total emission to the environment.

Keywords

Gas Turbine, Steam Turbine, HRSG, Combined Cycle, Power Output, Overall Efficiency, Condenser Cooling Water

1. Introduction

Energy is a basic requirement for human existence, and thus a driving force of civilization as almost all of our everyday activities and productive processes involve energy in one form or another [1]. The economic development and living standard of any society is a function of the availability and accessibility of electrical power to her. Therefore, the acknowledgement of the importance of increasing access to commercial electricity is fundamental for the future and sustainable development of any society [2].

The increased world population and the quest to improve living conditions as well as global economic growth have resulted in a continuous increase in energy demand [3]. Accommodating the projected increase in energy demand must be approached with caution as it is now generally accepted that human activities such as burning of fossil fuels involved in energy systems are the main sources of acid rain and greenhouse gases which are the major contributor to global climatic change [4]. Yadav [5] in his work, observed that one of the greatest challenges facing humanity during the twenty-first century is that of giving everyone on the planet access to safe, clean and sustainable energy supplies.

Anheden [6] did a study on the analysis of gas turbine systems for sustainable energy conversion and suggested that one way to provide the energy service demand by the growing populations, and yet reduce the CO_2 emissions, is to increase the energy conversion efficiency of existing and future energy conversion processes by various technological advancements. Nkoi *et al.* [7] in their work on investigations encompassing comparative assessment of gas turbine (GT) cycle optional also observed that a modified gas turbine cycle configurations incorporating unconventional components such as engine cycle with low pressure compressor (LPC) zero-staged, recuperated engine cycle, and intercooled/recuperated (ICR) engine cycle exhibited better performances in terms of thermal efficiency and specific fuel consumption than the traditional simple cycle engine. A study on power increase and economical results of different efficiency improvement methods has revealed that fogging and evaporative method are the most effective methods of efficiency improvement in Khoramshahr power plant [8].

Severally, there have often been suggestions from literatures of the combination of two or more thermal cycles within a single power plant with the intention of increasing efficiency over that of single cycles. In a study on the performance of combined gas turbine-steam cycle for power generation, it was noted that thermal processes can be combined in this way whether they operate with the same or with differing working media. The study concluded that a combination of cycles with different working media as in combined cycle is more interesting because their advantage can complement one another [9]. Korobitsyn [10] in his analysis of cogeneration, combined and integrated cycles observed that improved efficiency can be achieved by modifications to the original simple cycle to recover heat from the turbine exhaust for generation of steam in a steam turbine bottoming cycle as found in combined cycle.

A combined cycle power plant (CCPP) is a power plant where electricity production is done from the combination of both gas and steam turbines. Previous work has been done on rotor-blades' profile influence on a gas compressor effectiveness. The study stated that in the gas turbine, atmospheric air is drawn in through an intake duct and compressed to a high pressure in the compressor consisting of a cascade of several stages of blades located on a single axle in radial form [11] [12]. The high pressure air is delivered to the combustor where fuel is sprayed and the high pressure and temperature product of combustion impinges on the turbine blades which rotates the shaft that drives an electrical generator to produce electricity and the exhaust gas energy from the gas turbine is used to generate steam in a heat recovery steam generator (HRSG) which then produces more electricity from a steam turbine. There is a great potential for increased power output and efficiency, as well as reduction of emitted pollutants to the environment, in energy converting devices by recovering waste heat to a greater extent. It was observed that the development of combined cycle power plant is the most efficient and effective effort in this direction. This is gaining increasing acceptance as an alternative to conventional gas or steam cycle due to high thermal efficiency of about 60% utilizing natural gas as fuel [5].

Ragland *et al.* [13] did a study on combined cycle heat recovery optimization and observed that combustion turbine performance has the primary impact on combined cycle plant efficiency and the next most important piece of equipment that impacts efficiency is the heat recovery steam generator (HRSG). The study continued that in the design of CCPP, the HRSG parameters to optimize include steam pressure, temperature, flows, pinch points, approach temperatures and HRSG exit gas temperature. The HRSG is a heat exchanger comprising three components namely the economizer, the evaporator, and the superheater and can be designed with typical configurations of single-pressure or multiple-pressure levels, with or without supplementary firing. In a study on the optimization of Maputo power plant, it was observed that the most important design parameter of a steam turbine is the steam data which include the mass flow and the steam turbine input pressure and temperature, as well as the condenser pressure [14].

To effectively harness heat from the HRSG so as to have a higher combined cycle thermal efficiency, the stack temperature should be reduced as much as possible but not below the acid dew point of the flue gas to avoid levels of condensation. Acid dew point of exhaust gas is within the range of 104˚C - 160˚C [15]. In a study on combined cycle gas and steam turbine power plants, the typical steam turbine outlet pressure in condensing mode ranges from 0.03 to 0.25 bar [16] while Aref [17] in his work on development of framework for thermo-economic optimization of simple and combined gas turbine cycles indicated that pinch point is usually between 8˚C and 15˚C and approach temperature is in the range 8˚C - 12˚C.

This paper considers an improved design of the Omoku gas turbine power plant using combined cycle application, with the aim of increasing its performance (power output and efficiency) as well as reducing the total emission to the environment. It analyzes a suitable HRSG capable of producing steam at an effective pressure which will utilize the exhaust gas from the gas turbine to power the steam turbine.

The Omoku gas turbine is located in the Niger Delta area of Nigeria. It has six (6) units, each consisting of a gas turbine for electricity generation of 25 MW and exhaust gas temperature of 487˚C. The manufacturer of the gas turbine system is General Electric (GE) and it is an MS 5001 (GE Frame 5), single shaft model having a double bus single breaker; 3 × 80 MVA transformer and a 33 kV to 132 kV line switchyard system which is capable of evacuating 200 MW of power [18].

It is observed that the hot exhaust gas from the Omoku gas plant is vented to the atmosphere, as common to all conventional open cycle gas turbines; hence, considerable amount of heat energy goes as waste with the exhaust of the gas turbine, thus, contributing to environmental pollution. Rather than flaring the hot exhaust gas from this Brayton cycle plant, this heat can be utilized for other useful purposes using heat recovery steam generator (HRSG). This reason prompted the need for the improved design of the Omoku gas turbine power plant to a combined-cycle system such that the otherwise wasted heat from the hot exhaust gas is captured and channeled into the HRSG and used to generate steam to drive another generator (steam generator) to produce more electricity, thus improving the performance (increased total output and efficiency) with less environmental pollution compared to the single cycle turbine plant. Hence, a combined cycle power plant (CCPP) usually consists of a Gas turbine plant, a HRSG and a Steam turbine plant. The gas turbine plant operating on Brayton cycle and the steam turbine plant operating on rankine cycle are often called topping and bottoming cycles respectively [19]. As common to all conventional CCPP, this modification requires a source of cooling water to be utilized in the condensing unit of the proposed steam turbine system. However, it is interesting to know that there is a flowing river close to the Omoku gas plant which enhances the availability of water source for this purpose, justifying the feasibility of the study.

2. Materials and Methods

The research methodologies adopted are as follows:

1) Data collection of the Omoku Gas turbine power plant (GE FRAME 5—MS 5001 PA—Single Shaft) from direct observation from the monitoring screen of the human machine interface (HMI), log books and manufacturer's manuals.

2) Assessment of the current plant condition.

3) MATLAB software was used to model the standard thermodynamic relations and equations for the analysis of the turbo-machinery components of the gas turbine to determine the heat energy rejected in the exhaust gas of the existing plant so as to ascertain the corresponding power that can match with the exhaust gas temperature (EGT).

4) Determination of a suitable capacity (sizing) of a heat recovery steam generator (HRSG).

5) From the relevant design calculations, the capacities of other components of the proposed steam turbine system were determined. They include the turbine, condenser, boiler and feed pump. Hence the steam turbine components that will match with the existing gas turbine were selected.

6) The power output and efficiency of the steam turbine, hence, power output and efficiency of the combined cycle gas turbine plant were determined.

Table 1 displays the necessary data required for the performance analysis of the power plant which equally serve as the input data for the design/conversion to a combined cycle. Also, the ambient conditions of the physical environment of the plant vicinity were noted $\eta_{th_{gT}}$.

2.1. Analytical Model of the Combined Gas and Steam

2.1.1. Performance of a Theoretical Heat Balance on HRSG

This will give us the relationship between the tube side and shell side processes. It is also important that tube side components which will make up the HRSG unit be decided. Even though these components may include other heat exchange services, at this time we will only consider the three primary coil types namely; Economizer, Evaporator and Super heater as shown in **Figure 1**.

From **Figure 1**, the steam temperature increases form point 4 to 1. Process 4 - 5 occur in the economizer; 5 - 6 occur in the evaporator, and 6 - 1 is superheating.

The exhaust gas is cooled from point T_d of the gas turbine to a temperature of T_a well above its acid dew point temperature. The pinch points are x-5 and d-1. A pinch and approach temperature of 10°C is used in order to

Table 1. Average operating data for Omoku gas turbine plant.

Component	Parameter	Symbol	Unit	Value
Turbo compressor	Inlet temperature	T_1	°C	30.4
	Outlet temperature	T_2	°C	367
	Inlet pressure	P_1	bar	1.013
	Outlet pressure	P_2	bar	10
	Mass flow rate (Air)	$\dot{m}_a$	kg/s	122.9
Combustion chamber	Fuel consumption (flow rate)	$\dot{m}_f$	kg/s	1.2
	Inlet temperature	T_3	°C	959
Gas turbine	Outlet temperature	T_4	°C	487
	Exhaust gas flow (flow rate of gas)	$\dot{m}_{eg}$	kg/s	124.1
Other data	GT Thermal efficiency	$\eta_{th_{gT}}$	%	26.6
	GT Power output	P_{gT}	MW	25

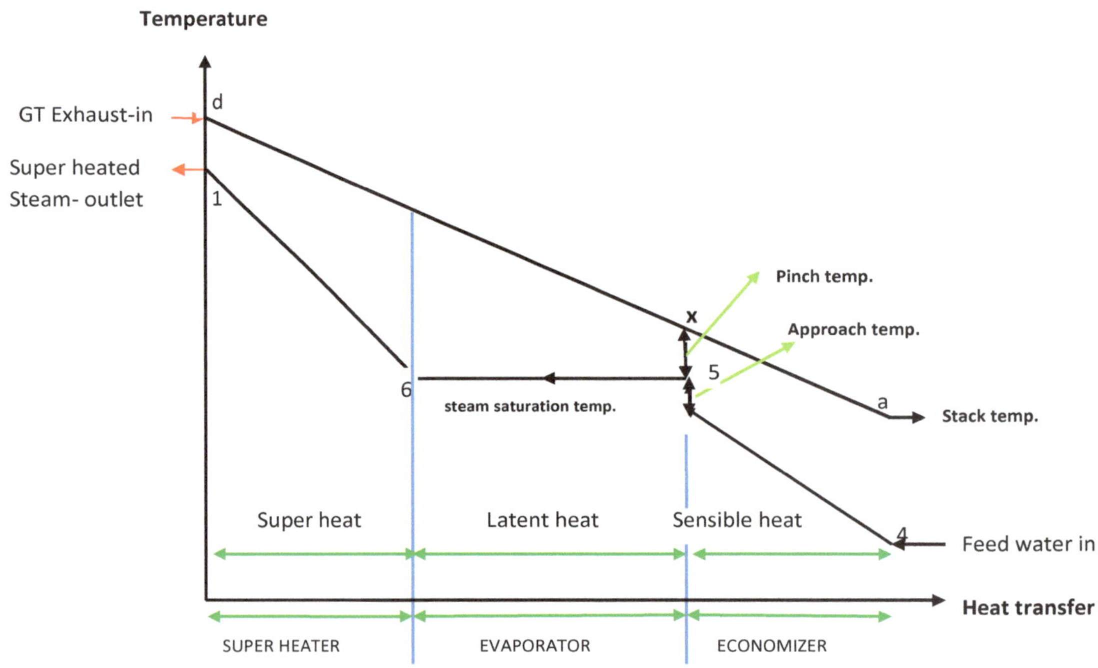

Figure 1. Variation of heat flow with temperature between exhaust gas and HRSG.

minimize heat loss between exhaust gas and steam and at the same time ensure heat transfer for maximum power output.

Applying the heat balance equations between the gas turbine exhaust and HRSG as given below:

$$\dot{m}_s(h_1 - h_4) = \dot{m}_{eg} c_{peg} (T_d - T_a) \quad (1)$$

$$\dot{m}_s(h_5 - h_4) = \dot{m}_{eg} c_{peg} (T_x - T_a) \quad (2)$$

Heat rejected by exhaust gas is given by

$$\dot{Q}_{eg} = \dot{m}_{eg} c_{peg} (T_d - T_a) \quad (3)$$

where;

$\dot{m}_s$ = Mass flow rate of steam in kg/s

$\dot{m}_{eg}$ = Mass flow rate of exhaust gas in kg/s

T_a = Cooling temperature of exhaust gas (HRSG stack temperature)

T_x = Temperature of exhaust gas at pinch point

T_d = Temperature of exhaust gas at gas turbine outlet

m_{eg} = Specific heat of exhaust gas

The capacities of the various components of the HRSG are thus estimated as follows:

Economizer:

$$Q_{ec} = Q_{eg_{ec}} \times \dot{m}_{eg} = (h_5 - h_4)\dot{m}_s \quad (4)$$

Evaporator:

$$Q_{ev} = Q_{eg_{ev}} \times \dot{m}_{eg} = (h_6 - h_5)\dot{m}_s \quad (5)$$

Superheater:

$$Q_{sh} = Q_{eg_{sh}} \times \dot{m}_{eg} = (h_1 - h_6)\dot{m}_s \quad (6)$$

HRSG capacity:

$$Q_{HRSG} = Q_{ec} + Q_{ev} + Q_{sh} \quad (7)$$

Heat Loss

$$H_L = \dot{Q}_{eg} - Q_{HRSG} \quad (8)$$

Thermal Efficiency of HRSG,

$$\eta_{HRSG} = \frac{Q_{HRSG}}{\dot{Q}_{eg}} \quad (9)$$

2.1.2. Determination of Net Power Output at Generator Terminal of Steam Turbine

The actual power developed by the steam turbine P_{ST} can be calculated from the relation:

Isentropic efficiency

$$\eta_{ST_{is}} = \frac{P_{ST}(\text{actual})}{P_{ST}(\text{isentropic})} = \frac{\dot{m}_s(h_1 - h_{2^1})}{\dot{m}_s(h_1 - h_2)} \quad (10)$$

Steam turbine net power output

$$P_{NET} = P_{ST}(\text{actual}) - P_p \quad (11)$$

where

$$P_p = \text{pump power input (actual)} = \dot{m}_s(h_{4^1} - h_3) \quad (12)$$

The steam turbine cycle efficiency

$$\eta_{ST} = \frac{P_{NET}}{Q_{HRSG}} \tag{13}$$

The capacity of a steam plant is often expressed in terms of steam rate or ***specific steam consumption (S.S.C).*** Nag defined it as the rate of steam flow kg/s required to produce unit shaft output (1 kW) [19]

Specific Steam Consumption (SSC)

$$\text{S.S.C} = 1/W_{net}\ (\text{kg/kW}\cdot\text{h}) \tag{14}$$

Where $W_{net} =$ specific network output

Specific steam turbine work

$$W_{ST} = \frac{P_{ST}(\text{actual})}{\dot{m}_s} \tag{15}$$

Specific Pump Work

$$W_p = \frac{P_p}{\dot{m}_s} \tag{16}$$

$\therefore$ ***Net specific work***

$$W_{NET} = W_{ST} - W_P \tag{17}$$

The condenser cooling water system heat transfer (heat rejected to the condenser) gives

$$Q_2 = \dot{m}_s (h_2 - h_3) \tag{18}$$

Applying the energy balance equation between the condenser and the cooling water;

$$\dot{m}_w = c_{p_w}(t_{\text{out}} - t_{\text{in}}) = \dot{m}_s(h_2 - h_3) \tag{19}$$

The condenser Efficiency η, $\eta_{\text{cond}} = \dfrac{\text{Actual temp.rise of cooling water}}{\text{maximum temp.rise of cooling water}}$

$$\eta_{\text{cond}} = \frac{t_{\text{out}} - t_{\text{in}}}{T_2 - t_{\text{in}}} \tag{20}$$

Where; $\dot{m}_s$

$\dot{m}_s =$ Mass flow rate of steam in kg/s

$\dot{m}_w =$ Mass flow rate of cooling water

$c_{p_w} =$ Specific heat of water

$t_{\text{in}} =$ Temperature of cooling water into condenser

$t_{\text{out}} =$ Temperature of cooling water out of the condenser

$T_2 =$ Condenser saturated temperature

2.1.3. Combined Gas and Steam

The combined-cycle unit combines the Rankine (steam turbine) and Brayton (gas turbine) thermodynamic cycles using HRSG. The HRSG comprising the economizer, the evaporator, and the superheater, trapped the energy in the exhaust gases of the turbine to generate steam for the steam turbine to generate more power. The schematic diagrams are shown in **Figure 2** and **Figure 3**.

From **Figure 2**:

Omoku Gas turbine Cycle ≡ a → b → c → d → a

Steam Turbine Cycle ≡ 1 → 2 → 3 → 4 → 1

Process 4 → 1 is the heat transfer process from exhaust gas of gas turbine to the Heat Recovery Steam Generator of the Steam turbine part.

The combined-cycle overall efficiency and power output are given by:

$$\eta_{CC} = \eta_{ST} + \eta_{gT} - \eta_{ST}\eta_{gT} \tag{21}$$

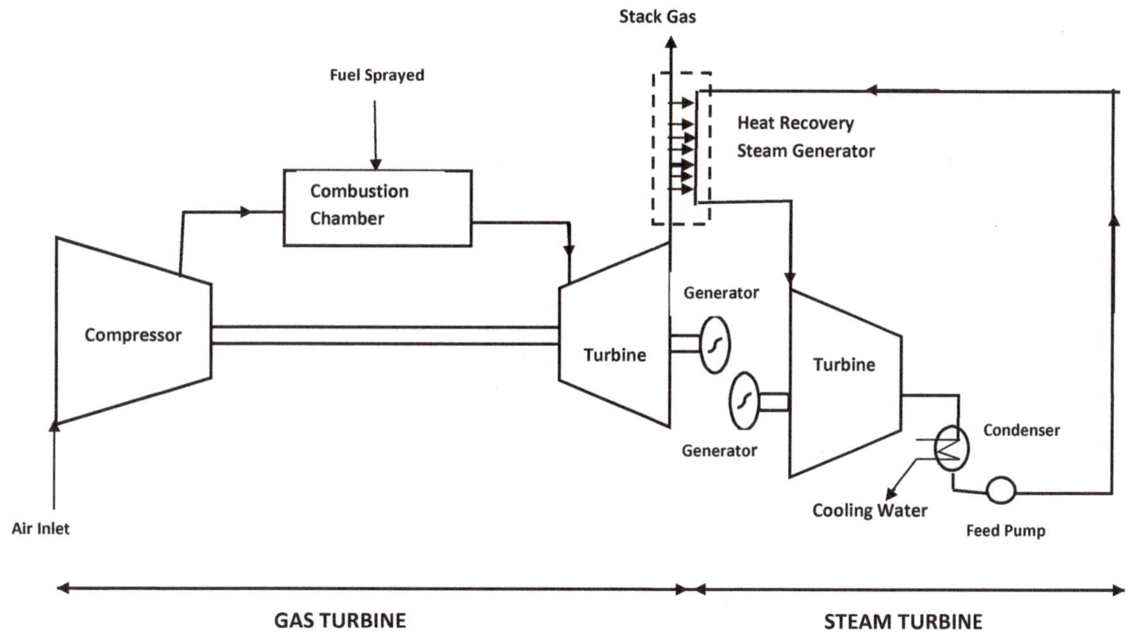

Figure 2. Schematic diagram of combined gas and steam cycle.

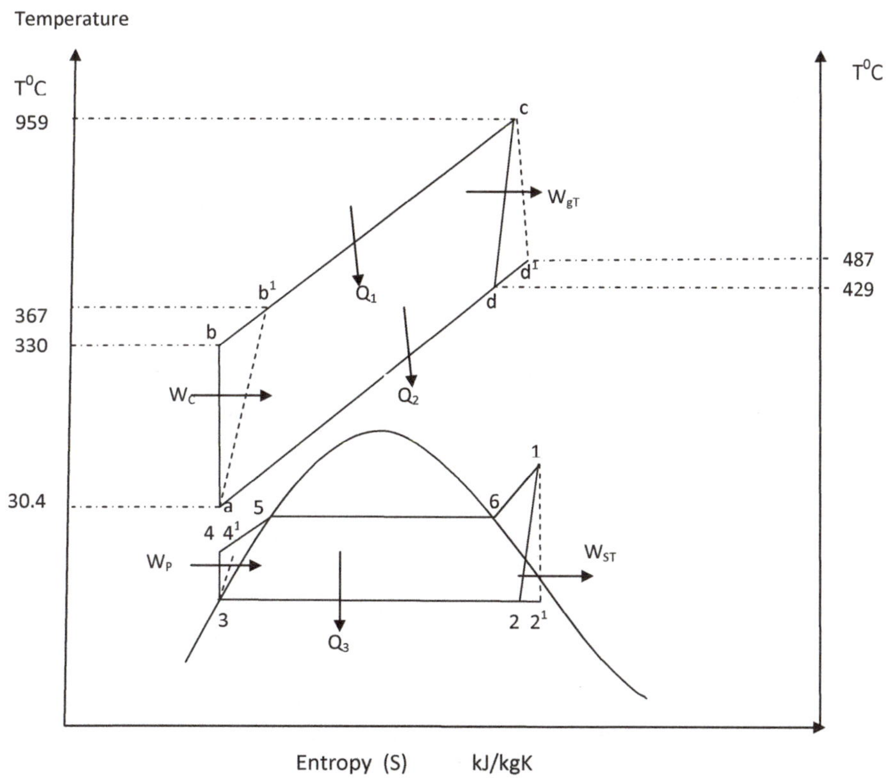

Figure 3. Combined gas and steam cycle power plant on T-S diagram.

$$P_{CC} = P_{ST} + P_{gT} \tag{22}$$

where η_{CC} and P_{CC} are combined cycle overall efficiency and power output respectively.

P_{ST} = Steam turbine power output, P_{gT} = gas turbine power output

η_{ST} = Steam turbine thermal efficiency

η_{gT} = Gas turbine thermal efficiency

2.2. HRSG and Steam Turbine Design Analysis

Ragland *et al.* [13] in his study on combined cycle heat recovery optimization indicated that as a general guideline in the design of combined cycle power plant (CCPP), the gas turbine will represent 66% of the plant's electrical output assuming that the HRSG does not employ a duct burner. The remaining output will be supplied by the steam turbine.

This implies that for the Omoku gas turbine power plant having a capacity of 25 MW, the required output of the proposed combined cycle power plant will be 37.9 MW (whose 66% gives the 25 MW of the existing Omokugas turbine) and the remaining output will be supplied by the steam turbine which is 34% of the CCPP. Hence, the power output of the steam turbine that will match the existing Omoku gas turbine is 12.9 MW.

A single pressure HRSG without supplementary firing is employed in this design. A pinch point and approach temperature of 10˚C, and a condenser pressure of 0.1 bar were used in this work while the exhaust gas cooling temperature (HRSG stack temperature) was chosen above 160˚C to avoid condensation of the exhaust gas at its acid dew point. For the purpose of this design, the inlet temperature difference between the condenser and the cooling water is taken as 16˚C and the pinch point of 10˚C is taken as the exit temperature difference resulting to a cooling water temperature rise of 6˚C is chosen.

To arrive at the required steam turbine capacity of 12.9 MW resulting to a more efficient combined cycle system with increased output, MATLAB software was used to model the thermodynamics equations, and appropriate parameters of the various components of the steam turbine power plant were determined considering Nine (9) different conditions of iterating the steam saturation pressure from 20 bar to 50 bar and HRSG stack temperature within the range of 170˚C to 190˚C, in applying the pinch technology for the HRSG and steam turbine analysis and the result is presented in **Table 2**.

3. Results and Discussion

3.1. Results

In order to determine the combined cycle components, all parameters were determined using the MATLAB model. The summary of the results obtained are shown in **Table 2**.

3.2. Discussions

Table 2 gives the result of the analysis after nine iterations. Iterations 1, 3 & 7 show that as the saturation pressure increases from 20 bar - 40 bar and HRSG stack temperature increases from 170˚C - 190˚C, the steam turbine net power output increases by 13.2% and the combined cycle overall efficiency increases by 6.2%. This indicates that a simultaneous increase in steam saturation pressure and stack temperature improve the combined cycle power output and efficiency.

Iterations 2, 3 & 6 show that as the stack temperature increases from 170˚C to 190˚C at constant saturation pressure of 30bar, the steam turbine net power output decreases from 13.5007 MW to 10.6399 MW and the combined cycle efficiency decreases from 48.4805% to 44.9933%. This indicates that for every 20˚C increase in the stack temperature, the turbine net power output and efficiency decreases by 21% and 7.2% respectively.

The 3rd to 5th iterations show that as the saturation steam pressure increases from 30 bar to 40 bar at constant HRSG stack temperature of 180˚C, the steam turbine net power output increases from 12.0690 MW to 13.6630 MW and the combined cycle efficiency increase from 46.7923% to 49.4494%. This show that for every 10 bar increase in the saturation steam pressure, there exist about 13.2% increase in the steam turbine net power output (which in turn increases the combined cycle power output) and 5.7% increase in the combined cycle efficiency.

Iterations 6 - 9 show that at constant HRSG stack temperature of 190˚C, P_{CC} increases as the saturation pressure increases. It indicates that for every 20˚C increase in the saturation pressure, the P_{CC} increases by 8%. This behaviour is represented in **Figure 4**.

Figure 5 represents the variation of combined cycle efficiency with saturation pressure at different HRSG stack temperature. It shows a progressive increase in the combined cycle efficiency with increase in steam pressure and HRSG stack temperature to about 45 bar, 190˚C. Further increase in the steam pressure at the stack temperature shows a decline in the overall cycle efficiency. It therefore indicate that for the modified design

Table 2. Summary of combined cycle analysis.

					HRSG CAPACITY		TOTAL											
Iteration	P_1 (bar)	T_a (°C)	$\dot{m}_s$ (MW)	Q_{ec} (MW)	Q_{ev} (MW)	Q_{sh} (MW)	Q_{HRSG} (MW)	$\dot{Q}_{eg}$ (MW)	H_L (MW)	η_{HRSG} (%)	P_p (MW)	P_{ST} (actual) (MW)	$P_{NET_{ST}}$ (MW)	$P_{NET_{gT}}$ (MW)	P_{CC} (MW)	η_{gT} (%)	η_{ST} (%)	η_{CC} (%)
1	20.0000	170.0000	14.9550	7.4911	28.2650	9.5623	45.3184	45.3193	0.0009	99.9980	3.2320	14.5840	11.3520	25.0000	36.3520	26.6000	25.0549	44.9867
2	30.0000	170.0000	14.3750	10.5512	25.8031	8.9645	45.3188	45.3193	0.0005	99.9988	1.1784	14.6791	13.5007	25.0000	38.5007	26.6000	29.8070	48.4805
3	30.0000	180.0000	14.3750	9.1181	25.7956	8.9619	43.8760	43.8900	0.0140	99.9680	2.6092	14.6782	12.0690	25.0000	37.0690	26.6000	27.5070	46.7923
4	35.0000	180.0000	14.1680	10.3000	24.8300	8.7600	43.8900	43.9000	0.0100	99.9800	1.8553	14.6852	12.8299	25.0000	37.8299	26.6000	29.2300	48.0548
5	40.0000	180.0000	13.9212	11.4789	23.8589	8.5483	43.8860	43.8897	0.0040	99.9919	0.9796	14.6436	13.6630	25.0000	38.6630	26.6000	31.1260	49.4494
6	30.0000	190.0000	14.3760	7.6915	25.8045	8.9651	42.4600	42.4601	0.0001	99.9999	4.0392	14.6791	10.6399	25.0000	35.6399	26.6000	25.0590	44.9933
7	40.0000	190.0000	13.9210	10.0499	23.8590	8.5483	42.4575	42.4601	0.0026	99.9940	2.4088	14.6436	12.2348	25.0000	37.2348	26.6000	28.8170	47.7754
8	45.0000	190.0000	13.7470	11.0219	23.0324	8.4041	42.4600	42.4601	0.0001	99.9999	1.7625	14.6085	12.8460	25.0000	37.8460	26.6000	30.2544	48.8108
9	50.0000	190.0000	13.5550	12.0000	37.8882	8.2523	58.1400	42.4601	−15.6800	136.9000	1.0590	14.5548	13.4958	25.0000	38.4958	26.6000	23.2126	43.6381

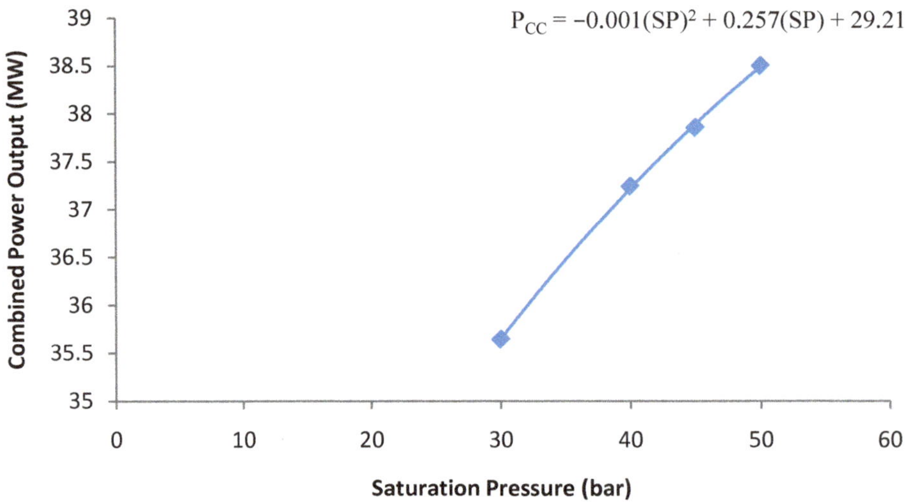

Figure 4. Variation of combined cycle power output with saturation pressure at constant HRSG stack temperature of 190°C.

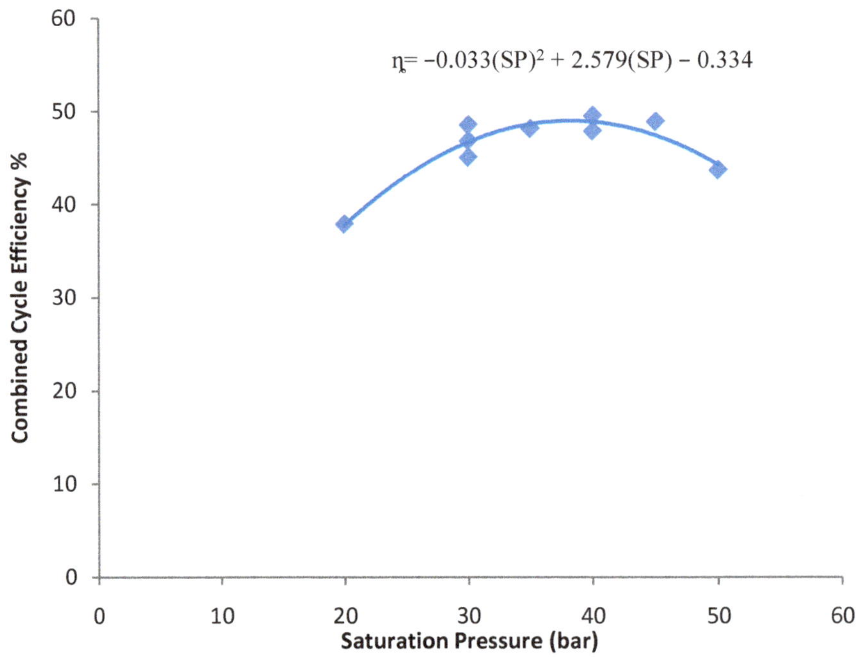

Figure 5. Variation of combined cycle efficiency with saturation pressure at different HRSG stack temperature.

process, a steam saturation pressure of 45 bar and HRSG stack temperature of 190°C is used as the maximum limiting condition.

Figure 6 shows the variation of steam turbine power output with saturation pressure at different HRSG stack temperature. It shows that the maximum steam power is generated at the saturation pressure of 35 bar and HRSG stack temperature of 180°C.

Again, observation from iterations 2, 3 & 6 show that at constant saturation steam pressure and increased HRSG stack temperature from 170°C to 190°C, the HRSG capacity decreases from 45.3188 MW to 42.4575 MW as presented on **Figure 7**. This shows that for every 1°C increase in the stack temperature at a constant saturation pressure of 30 bar, the HRSG capacity decreases by 0.3%. It is an indication that for higher HRSG capacity, the stack temperature should be within a considerable average value. This is represented in **Figure 7**.

As previously stated, a steam turbine capacity rating 12.9 MW (34% of the required CCPP) is required to

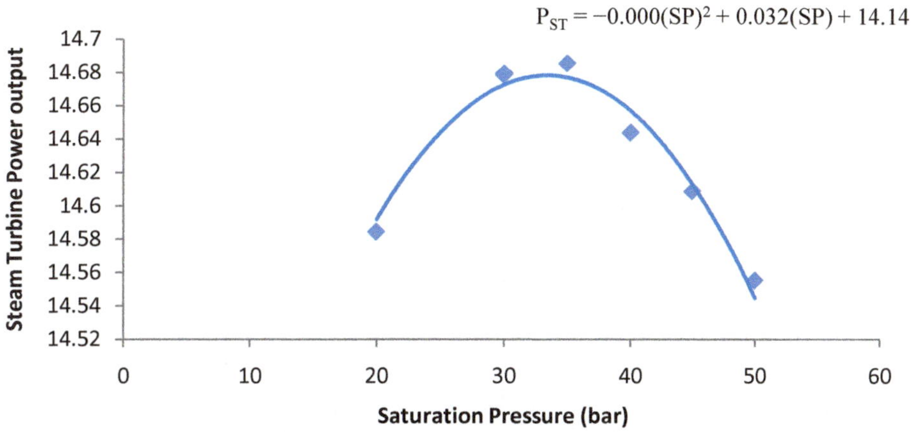

Figure 6. Variation of steam turbine power output with saturation pressure at different HRSG stack temperature.

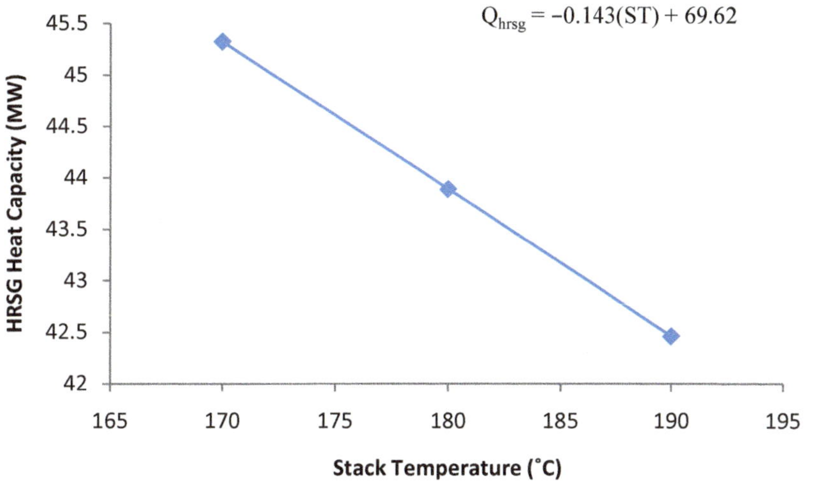

Figure 7. Variation of HRSG capacity with stack temperature.

Table 3. Steam cycle specification for combine cycle.

HRSG

- Stack temperature, $T_a = 190$ ˚C
- Saturation temperature in Evaporator $T_5 = 257.1$ ˚C
- Pinch point temperature difference $\Delta_{t_{pp}} = 10$ ˚C
- HRSG capacity $Q_{HRSG} = 42.46$ ˚C

Steam Turbine Generator

- Poweroutput $P_{ST_{gen}} = P_{NET} = 12.85\ \text{MW} \approx 12.9\ \text{MW}$
- Thermal efficiency $\eta_{gen} = 30\%$

Condenser Cooling Water Parameters

- Inlet temperature $t_{in} = 29.8$ ˚C
- Outlet temperature $t_{out} = 35.8$ ˚C
- Mass flow rate $\dot{m}_w = 1180.42$ kg/s
- Condenser efficiency $\eta_{cond} = 37.50\%$

STEAM TURBINE

- Inlet temperature $T_1 = 467$ ˚C
- Steam saturation pressure $P_1 = 45$ bar
- Isentropic efficiency $\eta_{is} = 89.06\%$
- Thermal efficiency $\eta_{th} = 30.25\%$
- Poweroutput $P_{ST}(\text{actual}) = 14.61$ MW
- Mass flow rate $\dot{m}_s = 13.747$ kg/s
- Specific steam consumption $\text{S.S.C} = 3.85\ \text{kg/kW}\cdot\text{h}$
- Steam outlet pressure (condenser pressure) $P_2 = 0.1$ bar
- Condenser capacity $\dot{Q}_3 = 29.61$ MW
- Feed pump capacity $P_P = 1.76$ MW

Combined Cycle Analysis

- Overall power output $P_{CC} = 37.9$ MW
- Overall efficiency $\eta_{CC} = 48.81\%$

match the existing Omoku gas turbine of 25 MW (66% of the required CCPP), making it a combined cycle power plant (CCPP) of 37.9 MW capacity.

Table 3 shows the detailed specification of the required 12.9 MW capacity steam turbine (attained from the 8th iteration of **Table 2**) for the combined cycle power plant. The steam turbine thermal efficiency is determined to be 30.25%, with HRSG capacity of 42.46 MW which is equivalent to the heat rejected by the exhaust gas except for a loss of 0.0001 MW. It also shows the condenser and feed pump capacities to be 29.61 MW and 1.76 MW respectively, and specific steam consumption (S.S.C) of 3.85 kg/kWh, indicating that a steam rate of 0.064 kg/s is required for every 1 kW of shaft output. Again, results presented in **Table 3** show that for the steam turbine capacity of 12.9 MW, to enhance condensation process, the cooling water enters the condenser at 29.8˚C and leaves at 35.8˚C, indicating an actual cooling water temperature rise of 6˚C and a maximum cooling water temperature rise of 16˚C resulting to a condenser efficiency of 37.5%. The cooling water flowing at 1180.42 kg/s with the steam mass flow of 13.75 kg/s indicates that for every kg of steam condensed, 85.55 kg of cooling water is required. Thus, to meet the huge demand of water, the power plant should be located where there is surplus water supply.

Deductions from all the analysis and discussions result to a combined cycle power output of 37.9 MW and an overall efficiency of 48.8%. This higher efficiency of the combined cycle achieved compared with that of the dry mode Omoku gas turbine indicates that the amount of emissions discharged into the atmosphere per unit mass of fuel burnt is less.

4. Conclusions

The Omoku gas turbine power plant design has been improved from a simple cycle operating on a dry mode to a combined cycle operation by retrofitting a steam bottoming plant through the incorporation of a HRSG.

This produces additional electric power of 12.9 MW (an increase of about 51%) from steam power plant to the existing 25 MW of the present gas turbine power plant, making a total power output of 37.9 MW for the combined cycle power plant; the steam turbine plant having a HRSG, condenser and feed pump capacities of 42.46 MW, 29.61 MW and 1.76 MW respectively. The condenser cooling water parameters include a mass flow of 1180.42 kg/s, inlet and outlet temperatures of 29.8˚C and 35.8˚C respectively with a condenser efficiency of 37.5%. Moreover, the conversion increases the efficiency of the Omoku gas turbine power plant from 26.60% to 48.81% of the combined cycle power plant.

Based on the above parameters, SIEMENS steam turbine product of MODEL: SST-150 was recommended as the steam bottoming plant.

It is thus discovered that a heat flow of about 42.46 MW which was otherwise being wasted in the EGT of the dry mode engine of the Omoku gas turbine could be converted to 12.9 MW of electric power thus, reducing the total emission to the environment. In effect, the plant performance is enhanced.

References

[1] Rao, S. and Parulekar, B.B. (2007) Energy Technology: Non-Conventional, Renewable and Conventional. Khanna Publishers, Naisarak, Delhi.

[2] Mohamed, K.M. (2005) Parametric Analysis of Advanced Combined Power Generation Systems. M.Sc. Thesis, University of New Brunswick.

[3] German Advisory Council on Global Change (WBGU) (2011) World in Transition-Towards Sustainable Energy Systems. http://www.wbgu.de

[4] Goswami, D.Y. and Kreith, F. (2008) Energy Conversion. CRC Press, Taylor & Francis Group, Boca Ranton, London, New York.

[5] Yadav, R. (2009) Steam and Gas Turbine and Power Plant Engineering. Central Publishing House, Allahabad.

[6] Anheden, M. (2000) Analysis of Gas Turbine System for Sustainable Energy Conversion. Ph.D. Thesis, Royal Institute of Technology, Stockholm.

[7] Nkoi, B., Pericles, P. and Theorklis, N. (2013) Performance Assessment of Simple and Modified Cycle Turboshaft Gas Turbine. *Propulsion and Power Research*, **2**, 96-106. http://dx.doi.org/10.1016/j.jppr.2013.04.009

[8] Espanni, R., Ebrahimi, S.H. and Ziaeimoghadam, H.R. (2013) Efficiency Improvement Methods of Gas Turbine. *Energy and Environmental Engineering*, **1**, 36-54.

[9] Ahmed, S.Y. (2013) Performance of the Combined Gas Turbine-Steam Cycle for Power Generation. *Mathematical*

Theory and Modeling, **3**, 12.

[10] Korobitsyn, M.A. (1998) New and Advanced Energy Conversion Technologies. Analysis of Cogeneration, Combined and Integrated Cycles. Febodruk BV.

[11] Lebele-Alawa, B.T., Hart, H.I., Ogagi, S.O.T. and Probert, S.D. (2008) Rotor-Blades' Profile Influence on a Gas Turbine Compressor Effectiveness. *Applied Energy*, **85**, 494-505. http://dx.doi.org/10.1016/j.apenergy.2007.12.001

[12] Lebele-Alawa, B.T. and Jo-Appah, V. (2015) Thermodynamic Performance Analysis of a Gas Turbine in an Equatorial Rain Forest Environment. *Journal of Power and Energy Engineering*, **3**, 11-23. http://dx.doi.org/10.4236/jpee.2015.31002

[13] Ragland, A. and Stenzel, W. (2000) Combined Cycle Heat Recovery Optimization. *Proceedings of* 2000 *International Joint Power Generation Conference*, Miami Beach, 23-26 July 2000, 1781-1787.

[14] Armando, A. (2013) Optimization of Maputo Power Plant. Master of Science Thesis KTH School of Industrial Engineering and Management Energy Technology, STOCKHOLM.

[15] Jehar and Associates. Introduction to HRSG design. www.hrsgdesign.com

[16] Kehlhoffer, R. (1997) Combined Cycle Gas and Steam Turbine Power Plants. Penn Well Publishing Company, Oklahoma.

[17] Aref, P. (2012) Development of Framework for Thermo-Economic Optimization of Simple and Combined Gas Turbine Cycles. Ph.D. Thesis, School of Engineering, Cranfield University, Bedfordshire.

[18] Service Manual (Gas Turbine MS 5001) of 25 MW Unit of Omoku Power Generation Station. GEPS Oil & Gas, Nuovo Pignone, Volume 1; G.T Description, Instruction & Operation.

[19] Nag, P.K. (2011) Power Plant Engineering. 3rd Edition, Tata McGraw-Hill Education Private Limited, New Delhi.

Nomenclature

c_{peg}	Specific heat of exhaust gas at constant pressure, kJ/kg·K
c_{p_w}	Specific heat of water, kJ/kg·K
h_1	Enthalpy at steam turbine inlet, kJ/kg
h_2	Enthalpy at steam turbine outlet, kJ/kg
h_{2^1}	Enthalpy at steam turbine outlet (actual), kJ/kg
h_3	Enthalpy at condenser outlet, kJ/kg
h_4	Feed water enthalpy, kJ/kg
h_{4^1}	Feed water enthalpy (actual), kJ/kg
h_5	Enthalpy at economizer exit, kJ/kg
h_6	Enthalpy at evaporator exit, kJ/kg
H_L	Heat loss, MW
$\dot{m}_{eg}$	Mass flaw rate of exhaust gas in, kg/s
$\dot{m}_w$	Mass flow rate of cooling water, kg/s
$\dot{m}_s$	Mass flow rate of steam in, kg/s
P_1	Steam saturation pressure, bar
P_2	Condenser pressure, bar
P_C	Compressor Power, MW
P_T	Turbine Power, MW
P_{CC}	Combined cycle power output, MW
P_{gT}	Gas turbine power output, MW
P_{NET}	Steam turbine net power output, MW
P_P	Pump power input (actual), MW
P_{ST}	Steam turbine power output, MW
$\dot{Q}_1$	Heat Supplied, MW
$\dot{Q}_2$	Heat Rejected, MW
$\dot{Q}_{ec}$	Heat recovered by economizer (economizer capacity), MW

$\dot{Q}_{eg}$	Heat rejected by exhaust gas, MW
$Q_{eg_{ec}}$	Exhaust gas energy at the economizer, kJ/kg
$Q_{eg_{ev}}$	Exhaust gas energy at the evaporator, kJ/kg
$Q_{eg_{sh}}$	Exhaust gas energy at the superheater, kJ/kg
Q_{HRSG}	Heat recovered by HRSG (HRSG capacity), MW
Q_{sh}	Heat recovered by superheater (superheater capacity), MW
Q_{ev}	Heat recovered by evaporator (evaporator capacity) ,MW
T_a	HRSG stack Temperature, °C
T_d	Gas turbine exhaust Temperature, °C
T_2	Condenser saturated temperature, °C
T_x	Temperature of exhaust gas at pinch point, °C
t_{in}	Temperature of cooling water into condenser, °C
t_{out}	Temperature of cooling water out of the condenser, °C
W_{net}	Net work output, kJ/kg

Greek Symbols

η	Efficiency, %
η_{th}	Thermal efficiency %
η_c	Compressor isentropic efficiency, %
η_{ST}	Steam turbine thermal efficiency %
η_{gT}	Gas turbine thermal efficiency %
η_{CC}	Combined cycle overall efficiency, %
η_{cond}	Condenser efficiency, %
$\Delta_{t_{pp}}$	Pinch point temperature difference, °C

Subscripts

G	Gas
P	At constant pressure
th	Thermal
is	Isentropic
in	Inlet
out	Outlet

Abbreviations

CCPP	Combined cycle power plant
CO_2	Carbondioxide
EGT	Exhaust gas temperature
GE	General electric
GT	Gasturbine
HMI	Human machine interface
HRSG	Heat recovery steam generator
ICR	Intercooled/recuperated
kg	Kilo gram
kW	Kilo watts
LPC	Low pressure compressor
MW	Mega watts
OPR	Overall pressure ratio
SST	Siemens steam turbine
ST	Steam turbine
WBGU	German advisory council on global change

22

A Novel Wind Energy Conversion System with Storage for Spillage Recovery

J. Cheng*, F. Choobineh

Department of Electrical and Computer Engineering, University of Nebraska-Lincoln, Lincoln, USA
Email: *chengjiecc@huskers.unl.edu

Abstract

This paper proposes a new system configuration for integrating a compressed air energy storage system with a conventional wind turbine. The proposed system recycles the mechanical spillage of blades and stores it for later electricity generation with assistance from a rotary vane machine. The configuration and operational policy is explained, and a comparative case study shows that the proposed system recovers investment costs through savings on electricity procurement and revenue through power export.

Keywords

Wind Energy Conversion System, Compressed Air Energy Storage, Rotary Vane Machine

1. Introduction

The use of renewable energy technologies is receiving additional attention by power system planners and regulators. Renewable energy sources, such as wind and solar, are considered to be promising and sustainable solutions for the future energy needs. Among the renewable sources, wind energy has been the fastest-growing source of electricity in the world [1]. Its ubiquitous availability, environmental friendliness, and strong output make it a competitive energy resource compared to traditional resources.

A variety of wind turbines (WTs) have been designed and installed in wind farms for converting wind power to electricity. Most turbine designs rely on the traditional wind power profile, as shown in **Figure 1**, mapping various wind speeds to the output power. As can be seen, the domain is divided into four regions. In Region I, the power output pegs at zero because of low wind speed. In Region II, output power increases with a cube of the wind speed until generation capacity of turbine is reached. In Region III, as wind speed reaches and surpasses the rated speed, the WT keeps output power constant, while excessive energy in the wind is trimmed away by blade pitching. In Region IV, the output goes down to zero because wind speed is too high for the turbine to survive while operating.

However, a careful scrutiny of the power curve reveals that, in Region II, a WT is unable to utilize its full electrical capacity because of low wind speed, while in Region III, a great amount of wind energy is trimmed away because of the power limitation of the turbine generator, causing mechanical spillage. A mechanical spil-

*Corresponding author.

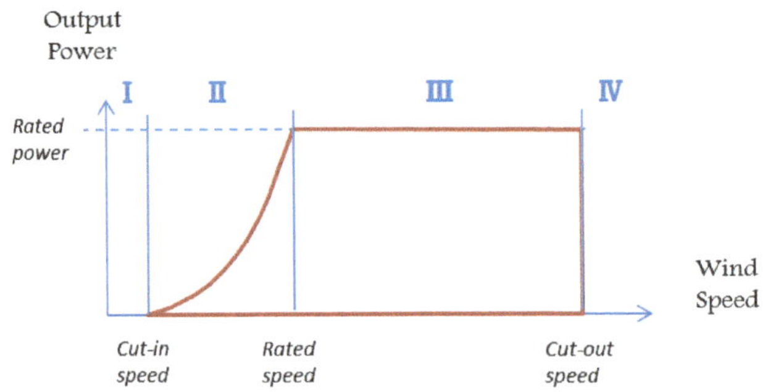

Figure 1. Traditional wind power profile.

lage is defined as the difference between potential power that blades can extract from the wind and the rated power of the generator. It should be noted that in the case of a grid-connected infinite bus system, although no electric spillage is present, mechanical spillage still exists.

Since we have identified the capacity vacancy in Region II and mechanical spillage in Region III, our idea is to recycle the mechanical spillage into storing and to regenerate electricity from the storage, therefore increasing overall wind power generation.

This paper is organized as follows. Section 2 describes the system configuration, operational procedures and evaluation metrics. Section 3 studies the performance of the system numerically. Finally, Section 4 draws conclusions.

2. Configuration and Procedures

A traditional WT generation system usually consists of blades, a gearbox, a generator, and some electronics. In the proposed design, the traditional speed-up gearbox is replaced by a differential planetary gearbox, shown in **Figure 2**. A differential planetary gearbox has one primary shaft to the left, and two coaxial secondary shafts to the right. The secondary shafts consist of a solid shaft to the generator and a hollow shaft to the Rotary Vane Machine (RVM). ARVM is coupled to the hollow shaft so as to regulate the power delivery between primary shaft and secondary shafts. A compressed air tank is linked to the RVM via a switching valve. The wind turbine is connected to the grid through a point of common coupling (PCC), forming a community-scale microgrid. The fundamentals of the structure and usage of each component will be discussed in the following subsections.

2.1. Blade

Blades are mechanical parts that are optimally designed to capture energy from the wind. According to the classic formula for a wind power curve, wind energy is the product of wind energy density, swept area of the blade, and the C_p coefficient. The C_p coefficient is the function of tip speed ratio and pitching angle [2]. For a real wind turbine system, the limit on the deliverable blade power is extrapolated by removing generator capacity constraints from the classic power curve and subtracting a mechanical loss.

2.2. Differential Gearbox

A differential gearbox is a gear system that enables uneven delivery of power from the primary shaft to two different secondary shafts. A planetary gearbox is one type of differential gearbox consisting of ring gear in the out layer, planetary gears in the middle, and sun gear in the center.

In the proposed application, shaft on the left-hand side is connected to the blade shaft; and the solid shaft on the right-hand side is connected to the generator. The hollow shaft on the right which links to the planetary gear set is connected to the RVM, providing positive or negative power to compensate for the difference between input and output, as desired. In steady state analysis, the contributive/input power and consumptive/output power have to match at all times.

2.3. Rotary Vane Machine

An RVM has an off-center shaft and spring-engaged vanes that jointly create a series of uneven chambers in its

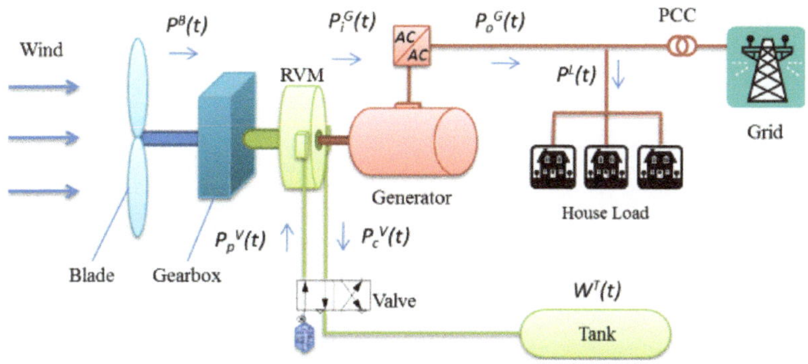

Figure 2. Proposed wind conversion system.

housing [3]. An RVM is incorporated into the system through a planetary gearbox to recapture power while wind power is in surplus, and to generate power while wind power is in deficit. The details of its two working modes are described below.

While the switching valve stays at the left-side position, as depicted in **Figure 3(a)**, the inlet of the RVM is connected to atmosphere while the outlet is connected to the compressed air tank. The RVM is working in the compressor mode in which the vanes are driven by a shaft, compressing the air to its outlet. On the contrary, while the switching valve switches to the right-hand side position, as in **Figure 3(b)**, the tank outlet air flow is reversed. The RVM works as an expander, in which the vanes are driven by compressed air, thereby exporting power.

2.4. Air Tank Storage

An air tank is used as a storage component, working jointly with the RVM. Each type of tank has its threshold of safety pressure. Thus, for a given volume and safety pressure, and a specified constant temperature, the maximum energy in a tank is determined. In this sense, the characteristics of compressed air storage are analogous to battery storage, where capacity can be expanded by simply parallelizing component units to constitute bulk storage.

2.5. Control Policy

An energy management policy is developed to maximize the benefit of energy production from the proposed system. The control logic is summarized in **Figure 4**.

The blade power is calculated based on the given wind speed. The difference between the blade power and the generator capacity becomes an indicator of the working status of the RVM. If the deviation is positive, the RVM is switched to compressor mode, in which the aforementioned mechanical spillage is recycled and stored in the form of compressed air. If the deviation is negative, RVM switches to expander state to supplement the power generation through the gearbox.

Through the operation procedure, it is assumed that the power output of both the generator and RVM cannot go beyond their capacity.

2.6. Cost Model

The total energy cost (EC) consists of two parts: 1) the facility costs, consisting of the initial investment cost and annual operation and maintenance (O & M) cost and 2) the procurement costs, the net cost of buying electricity from the grid, minus selling electricity to the grid. Those costs have to be converted into the same planning horizon H through discount factors. Net metering policy is applied with a flat electricity rate within H.

3. Performance Study of the System

In order to illustrate the performance of the proposed system, a comparative case study is presented in this section. The following three cases will be studied:

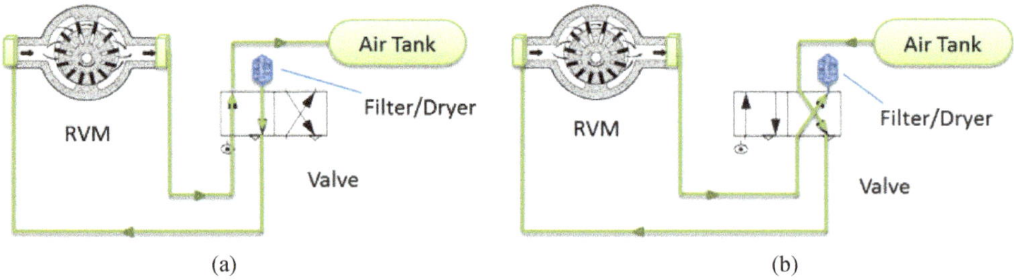

Figure 3. (a) RVM as a compressor and (b) RVM as an expander.

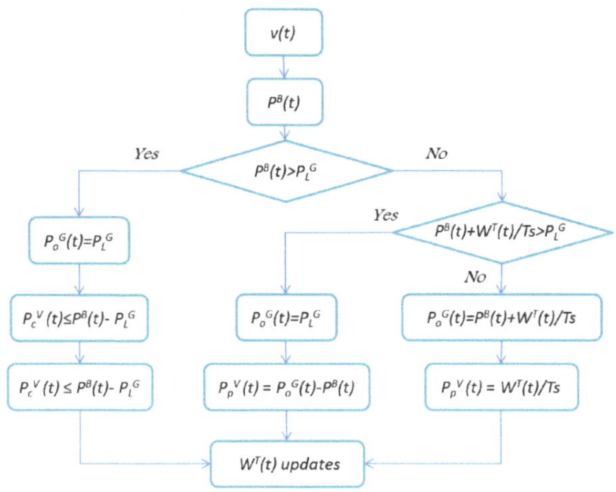

(a)

Symbol	Description	Symbol	Description
$v(t)$	Wind speed at time t	$P_c^V(t)$	Compression power at time t
$P^B(t)$	Blade power at time t	$P_p^V(t)$	Expansion power at time t
$P_i^G(t)$	Input power of generator at time t	P_L^V	Rated output power of RVM
$P_o^G(t)$	Output power of generator at time t	$W^T(t)$	Energy in the tank at time t
P_L^G	Rated power of wind turbine	W_L^T	Rated capacity of tank storage

(b)

Figure 4. (a) Energy management flow chart and (b) Term discription.

- Case 1: A community without wind energy
- Case 2: A community with a single WT; and
- Case 3: A community with the proposed system

3.1. Conditions and Parameters

The planning horizon H is set at 7 days, consisting of 168 one-hour time step. The hourly wind speed is retrieved from the historical data at Springview, Nebraska. And the hourly load demand was obtained from the NE-ISO power pool [4] that has been scaled down to the 0 - 400 kW range to simulate the consumption pattern of a small community.

The system component parameters are listed in **Table 1**. The annual interest rate is assumed to be 6%. The electricity rate is constant at 3.6 cents/kWh. The economic life is 20 years for the WT, RVM, and tank, respectively.

3.2. Results

As shown in **Figure 5(a)**, the total energy costs of the three cases are $1.813, $1.807, and $1.798, respectively. This indicates a decreasing cost trend as one moves from Case 1 to Case 3. Moreover, the average power flow through the PCC is also reduced from the original 300 kW to 78 kW and 52 kW, representing a 74% and 83% decrease, respectively. The aforementioned trend proves that the proposed system not only reduces the kWh cost but also reduces the power flow through the PCC. This indicates microgridself-sufficiency improves as one moves from Case 2 to Case 3.

The proposed system is designed to increase the capacity factor. **Figure 5(b)** shows that the capacity factor has been improved by 8.4%, from 86.8% to 95.2%, providing additional electricity generation of 3.829 kWh during period H.

The detail of the hourly generation is presented in **Figure 6**. As it can be seen, the proposed system gives full

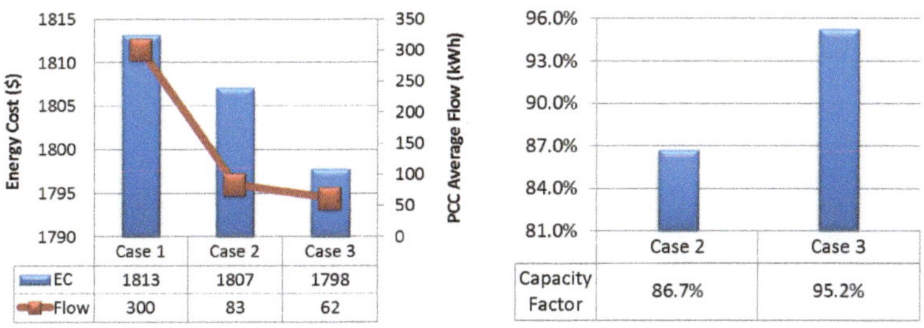

(a) (b)

Figure 5. (a) Comparison of energy cost and PCC flow and (b) Comparison of capacity factor.

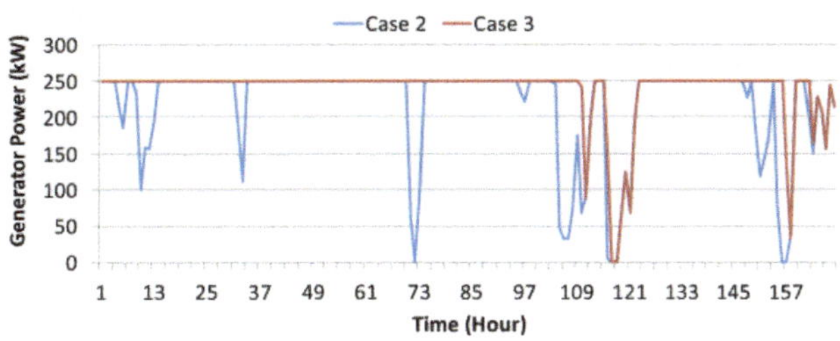

Figure 6. Comparison of generation.

Table 1. Component parameters.

Symbol	Description	Value	Symbol	Description	Value
P_L^G	Rated power of WT	250 kW	P_L^V	Rated power of RVM	375 kW
v^{in}	Cut-in wind speed	2.5 m/s	W_L^T	Rated capacity of tank	2000 kWh
v^{rated}	Rated wind speed	6.0 m/s	η^G	Generator efficiency	95%
C_p	Coefficient of power	0.39	η^{Vc}	Compression efficiency	60%
D	Blade diameter	81.5 m	η^{Vp}	Expansion efficiency	70%
f^{loss}	Friction loss	19.5 kW	η^T	Air tank efficiency	95%

capacity generation within the first 109 hours, while Case 2 has several negative gaps (spikes) in its power supply. However, after 109th hour, Case 3 also experiences strong fluctuation on its output, due to the lack of wind resource and depletion of tank storage.

The hourly operation of the RVM is shown in **Figure 7**. Its performance follows the pattern defined in the control policy. While wind is in surplus, the RVM is constantly running as a compressor to store energy as well as to compensate for tank leakage. While capacity vacancy appears, the RVM runs to support generation of electricity as long as enough pressure in the tank exists. The RVM stops running while wind is insufficient and storage is empty, creating the aforementioned negative output spikes in **Figure 6**.

The comparison, under the described circumstances, represents a tentative conclusion that the proposed system is able to recover the investment cost through the savings of electricity procurement as well as the sale of electricity exports.

4. Conclusion

This paper has proposed a novel WECS configuration that integrates a compressed air assisted storage system into a traditional WT system, recycling the mechanical spillage previously trimmed by the traditional wind power curve and regenerating electricity from storage as needed. Compared to a single WT or conventional CAES, the proposed system increases the capacity factor and, furthermore, the overall electricity generation. Therefore the proposed system renders an economic solution to fulfill energy demand of the community.

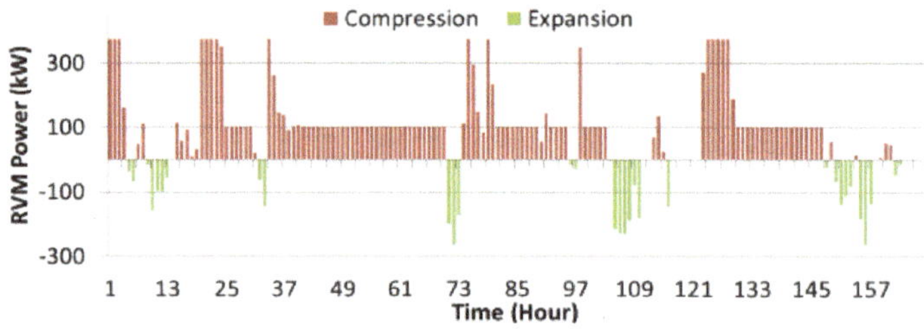

Figure 7. RVM performance.

Acknowledgements

The authors gratefully acknowledge the financial support by American Public Power Association (APPA) under Scholarship of Demonstration of Energy & Efficiency Development (DEED): Compressed Air Assisted Wind Power Generation and Regulation (Student Research Grant number: CS-1087).

References

[1] Catalao, J.P.S., Pousinho, H.M.I. and Mendes, V.M.F. (2012) Optimal Offering Strategies for Wind Power Producers Considering Uncertainty and Risk. *IEEE Systems Journal*, **6**, 270-277. http://dx.doi.org/10.1109/JSYST.2011.2163009

[2] Miller, N.W., Sanchez-Gasca, J.J., Price, W.W. and Delmerico, R.W. (2003) Dynamic Modeling of GE 1.5 and 3.6 MW Wind Turbine-Generators for Stability Simulations. 2003 *IEEE Power Engineering Society General Meeting* (IEEE Cat. No. 03CH37491).

[3] DOE Fundamentals Handbook: Mechanical Science. Volume 2. United States. http://energy.gov/sites/prod/files/2013/06/f2/h1018v2.pdf

[4] ISO New England, Energy, Load, and Demand Reports. http://www.iso-ne.com/isoexpress/

23

Development of a Fast Fluid-Structure Coupling Technique for Wind Turbine Computations

Matias Sessarego, Néstor Ramos-García, Wen Zhong Shen

Department of Wind Energy, Technical University of Denmark, Lyngby, Denmark
Email: matse@dtu.dk

Abstract

Fluid-structure interaction simulations are routinely used in the wind energy industry to evaluate the aerodynamic and structural dynamic performance of wind turbines. Most aero-elastic codes in modern times implement a blade element momentum technique to model the rotor aerodynamics and a modal, multi-body, or finite-element approach to model the turbine structural dynamics. The present paper describes a novel fluid-structure coupling technique which combines a three-dimensional viscous-inviscid solver for horizontal-axis wind-turbine aerodynamics, called MIRAS, and the structural dynamics model used in the aero-elastic code FLEX5. The new code, MIRAS-FLEX, in general shows good agreement with the standard aero-elastic codes FLEX5 and FAST for various test cases. The structural model in MIRAS-FLEX acts to reduce the aerodynamic load computed by MIRAS, particularly near the tip and at high wind speeds.

Keywords

Fluid-Structure-Interaction, Wind-Turbine, Aero-Elasticity

1. Introduction

Fluid-structure interaction (FSI) is the mutual action and reaction between a moveable or deformable structure with an internal or surrounding fluid flow [1]. FSI simulations are routinely performed for the analysis of wind turbine operation in the wind energy industry using what are known as aero-elastic codes. Examples for horizontal-axis wind turbines (HAWTs) are FLEX5 [2], FAST [3], and HAWC2 [4]. Most aero-elastic codes in modern times implement a blade element momentum (BEM) technique to model the aerodynamics, see e.g. Hansen [5], while mode shapes [2], multibody dynamics or finite element techniques [4] are used for the structural dynamic modelling. This paper presents a novel aero-elastic code that uses a three-dimensional (3D) viscous-inviscid solver for the aerodynamic analysis, called MIRAS [6], and a structural dynamic model based on mode shapes from FLEX5. The new code is called MIRAS-FLEX.

This paper begins with a brief explanation of MIRAS and FLEX5 in Section 2. Sections 3 and 4 describe the coupling methodology and simulation results, respectively. Finally, conclusions are given in Section 5.

2. Description of MIRAS and FLEX5

2.1. MIRAS

Method for Interactive Rotor Aerodynamic Simulations, MIRAS, is a 3D viscous-inviscid solver for HAWT rotor computations. The solver predicts the aerodynamic behavior of wind turbine wakes and blades for steady and unsteady conditions. The MIRAS code consists of inviscid and viscous parts. The inviscid part is a 3D panel method using a surface distribution of quadrilateral sources and doublets. The inviscid part is coupled to the viscous part through a viscous boundary layer solver, Q^3UIC [7]. A free-wake model simulates the wake behind the wind-turbine rotor using vortex filaments that carry the vorticity shed by the blades trailing edges. These features give MIRAS a more detailed aerodynamic description than the BEM technique and at a much lower computational cost than Navier-Stokes solvers. The reader is referred to [6] for a comprehensive description of MIRAS.

2.2. FLEX5

FLEX5 is an aero-elastic code that runs in the time-domain producing time-series of loads and deflections. The program relies on the BEM technique to model rotor aerodynamics, while the structural behavior of the wind turbine is modelled using carefully selected degrees of freedom (DOFs). The deflections of the blades and tower are given by shape functions, while the nacelle, rotor shaft, and hub are modelled as stiff bodies connected by flexible hinges. FLEX4 was originally developed with 17-20 DOFs and later upgraded to 28 DOFs in what is known as FLEX5. **Figure 1** depicts the parameters and coordinate systems in MIRAS (left) and FLEX5 (right).

3. Coupling Methodology

3.1. Fluid-Structure-Interaction Terminology

FSI problems can be solved by the monolithic or partitioned approaches. In the monolithic approach, the equations governing fluid flow and structural displacement are solved simultaneously with a single solver. The monolithic approach is advantageous in terms of stability because the mutual influence from the fluid and structural parts are taken into account [8]. However, a code needs to be developed incorporating both physical domains to allow a single solver to be used.

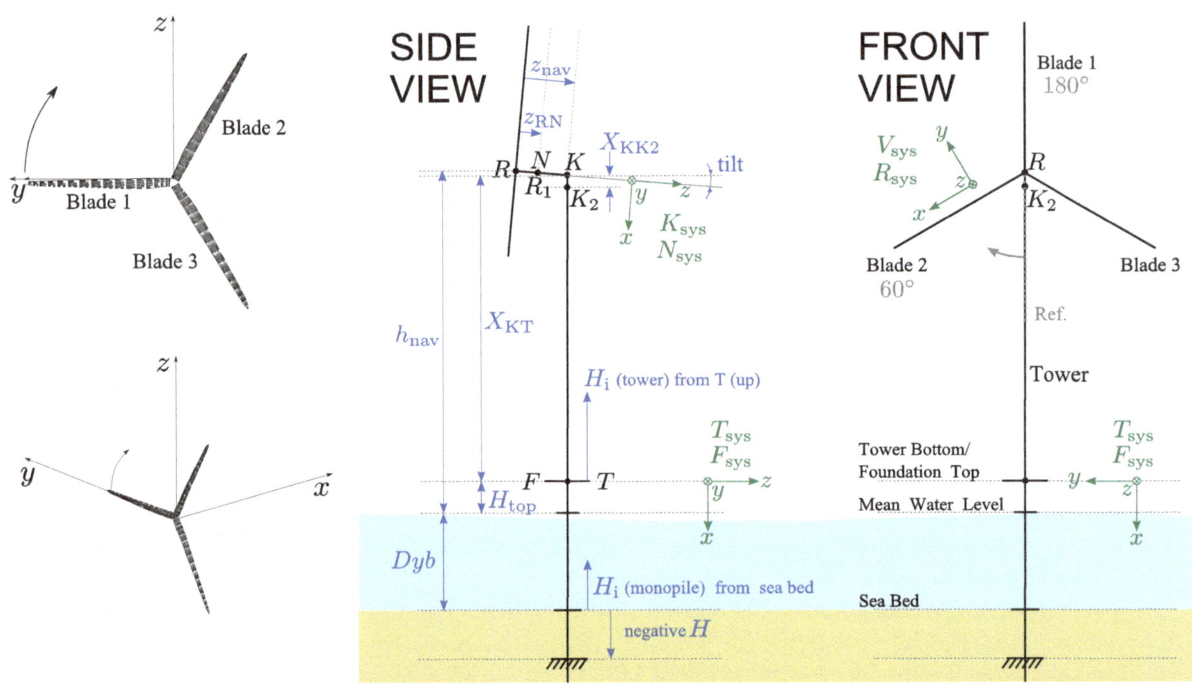

Figure 1. MIRAS (left) and FLEX5 (right) parameters and coordinate systems.

Conversely, the equations governing fluid flow and structural displacement in the partitioned approach are solved separately using two distinct solvers. Solving the equations from each domain separately means that the flow solution does not change while the solution of the structural part is being updated and vice versa. The partitioned approach allows existing computational codes to be used for each part of the FSI problem. Due to the sequentially staggered approach, however, partitioned approaches are susceptible to stability and accuracy issues. Partitioned approaches require a carefully designed loosely- or strongly-coupled scheme for proper function.

Loosely coupled schemes require only one evaluation per time step, but are more likely to experience numerical instabilities than strongly-coupled schemes. Strongly-coupled schemes may have the same staggered solution algorithm as loosely-coupled algorithms, but differ in that they perform sub-iterations at each time step. The additional sub-iterations improves the stability, but at an additional computational cost. See [1] and [9] [10] for details on FSI.

3.2. Coupling Scheme

MIRAS and the dynamic model in FLEX5 are used for the fluid and structural parts, respectively, in a partitioned and loosely-coupled approach. Loose-coupling was chosen because the approach was successfully applied to aero-elastic problems in [9] [10] and to make MIRAS-FLEX as computationally inexpensive as possible. **Figure 2** depicts the coupling scheme in MIRAS-FLEX, which is similar to the ones used in [9] [10]. Only the blade DOFs are considered in the coupling, *i.e.* 1st and 2nd flap-wise and 1st edge-wise blade modes. Changes in the angle of attack in MIRAS due to blade motion are taken into account in the coupling.

3.3. Aerodynamic Load Transfer

Load transfer from MIRAS to FLEX5 is performed through interpolation between the 3D mesh of MIRAS (*i.e.* x_1, x_2 and x_3) and the one-dimensional beam of FLEX5. **Figure 2(right)** illustrates the concept, where the loads defined at the center-points of each panel with coordinates c_1, c_2 and c_3 from the MIRAS mesh are converted into equivalent loads in FLEX5 (diamonds). Both the beam in FLEX5 and the x_1-axis in MIRAS coincide with the blade pitch-axis.

3.4. Deflection Transfer

FLEX5 uses Euler-Bernoulli beam theory (no torsion), see e.g. Bauchau and Craig [11], to predict the elastic deformation of the turbine blades. Thus, the displacement field of the blade cross-sections in MIRAS are based solely on rigid body translations ($\overline{u}_1$, $\overline{u}_2$ and $\overline{u}_3$) and rotations (ϕ_2 and ϕ_3):

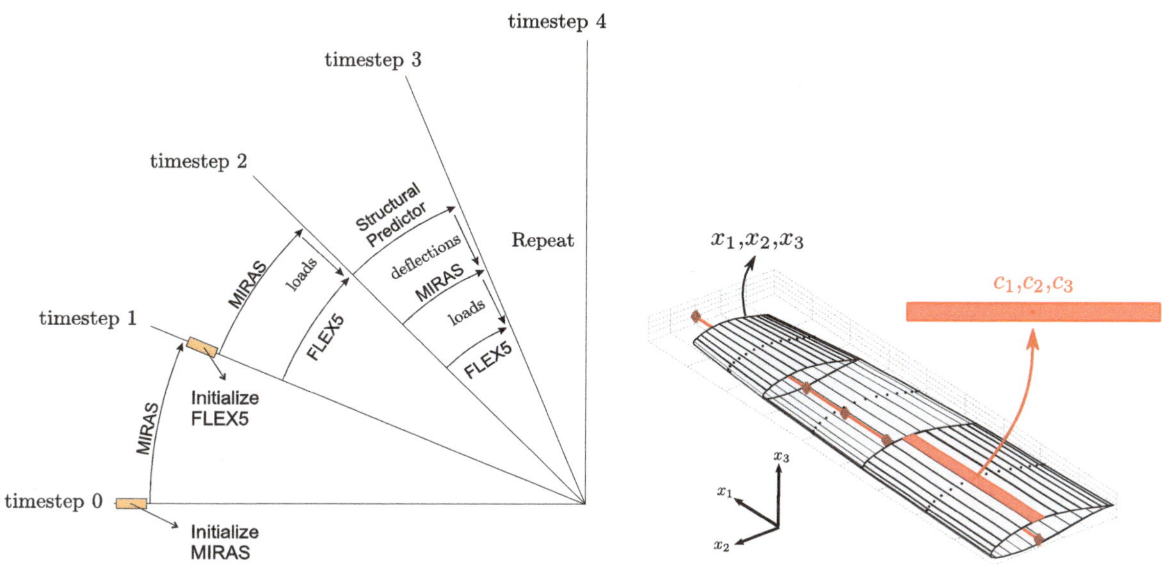

Figure 2. Loose-coupling methodology in MIRAS-FLEX (left) and aerodynamic load transfer from the 3D mesh of MIRAS and the 1D beam of FLEX5 for a blade section (right).

$$u_1(x_1,x_2,x_3) = \bar{u}_1(x_1) + x_3\phi_2(x_1) - x_2\phi_3(x_1),$$
$$u_2(x_1,x_2,x_3) = \bar{u}_2(x_1), \quad (1)$$
$$u_3(x_1,x_2,x_3) = \bar{u}_3(x_1)$$

where u_1, u_2 and u_3 are the total displacements and x_1, x_2 and x_3 are the coordinates of the blade mesh, see **Figure 2(right)**.

4. Simulations

4.1. Description of Simulation Tests

The NREL 5MW reference wind turbine [12] with a 126 m rotor diameter was chosen to perform simulation tests. Only steady wind conditions ranging from 5 to 14 m/s in an onshore environment are considered here. The tip-speed ratio is 8, pitch angle is 0, and the maximum rotational speed is 12.1 rpm. The codes used for the study are FLEX5, FAST, MIRAS-FLEX and MIRAS alone. For the FLEX5 simulations, the aerodynamics is evaluated using the built-in BEM technique. FAST is included as a third-party code to validate MIRAS-FLEX and FLEX5. In Section 4.2, the aerodynamic loads normal (F_N) and tangential (F_T) to the rotor plane are compared. All quantities are plotted versus the blade radius, *r*. **Figure 3** displays the rotor wake of the MIRAS (left) and MIRAS-FLEX (right) simulations, where a total of 3000 mesh points (20 span-wise × 150 chord-wise) were used for each blade.

4.2. Results and Discussion

All results are presented in **Figure 4**, which shows that MIRAS and MIRAS-FLEX are generally in good agreement with the standard aero-elastic codes FLEX5 and FAST. All results from FLEX5 and FAST are identical because both implement the BEM technique for rotor aerodynamics and a modal approach for the structural dynamics. The tangential aerodynamic loads computed by MIRAS and MIRAS-FLEX are slightly higher than FLEX5 and FAST, especially at low wind speeds, because of the very distinct aerodynamic models used. As a consequence, the rotor torque in MIRAS and MIRAS-FLEX are also slightly higher than FLEX5 and FAST. In terms of the FSI coupling, the structural model in MIRAS-FLEX acts to reduce the aerodynamic load computed by MIRAS, particularly near the tip and at high wind speeds.

5. Conclusion

MIRAS has been coupled with FLEX5 using a partitioned and loosely-coupled methodology to predict the aero-elastic response of wind turbines. The novel code is called MIRAS-FLEX. In general, MIRAS-FLEX results are in good agreement with the standard aero-elastic codes FLEX5 and FAST. MIRAS and MIRAS-FLEX give higher values of tangential load in comparison with FLEX5 and FAST due to the distinct aerodynamic models used. MIRAS and MIRAS-FLEX implement a 3D viscous-inviscid solver, while FLEX5 and FAST relies on the blade element momentum technique. MIRAS-FLEX gives a lower prediction of loads in comparison with MIRAS, especially near the tip where deflections are the highest, due to the inclusion of structural dynamic modelling.

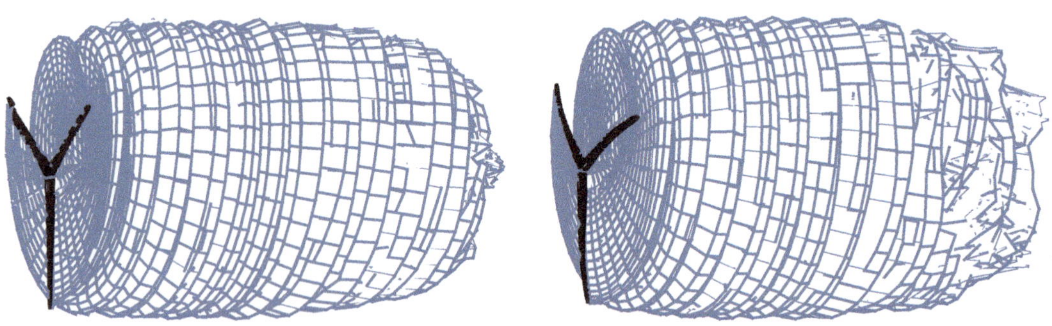

Figure 3. Simulations of the NREL 5MW wind turbine rotor using MIRAS (left) and MIRAS-FLEX (right).

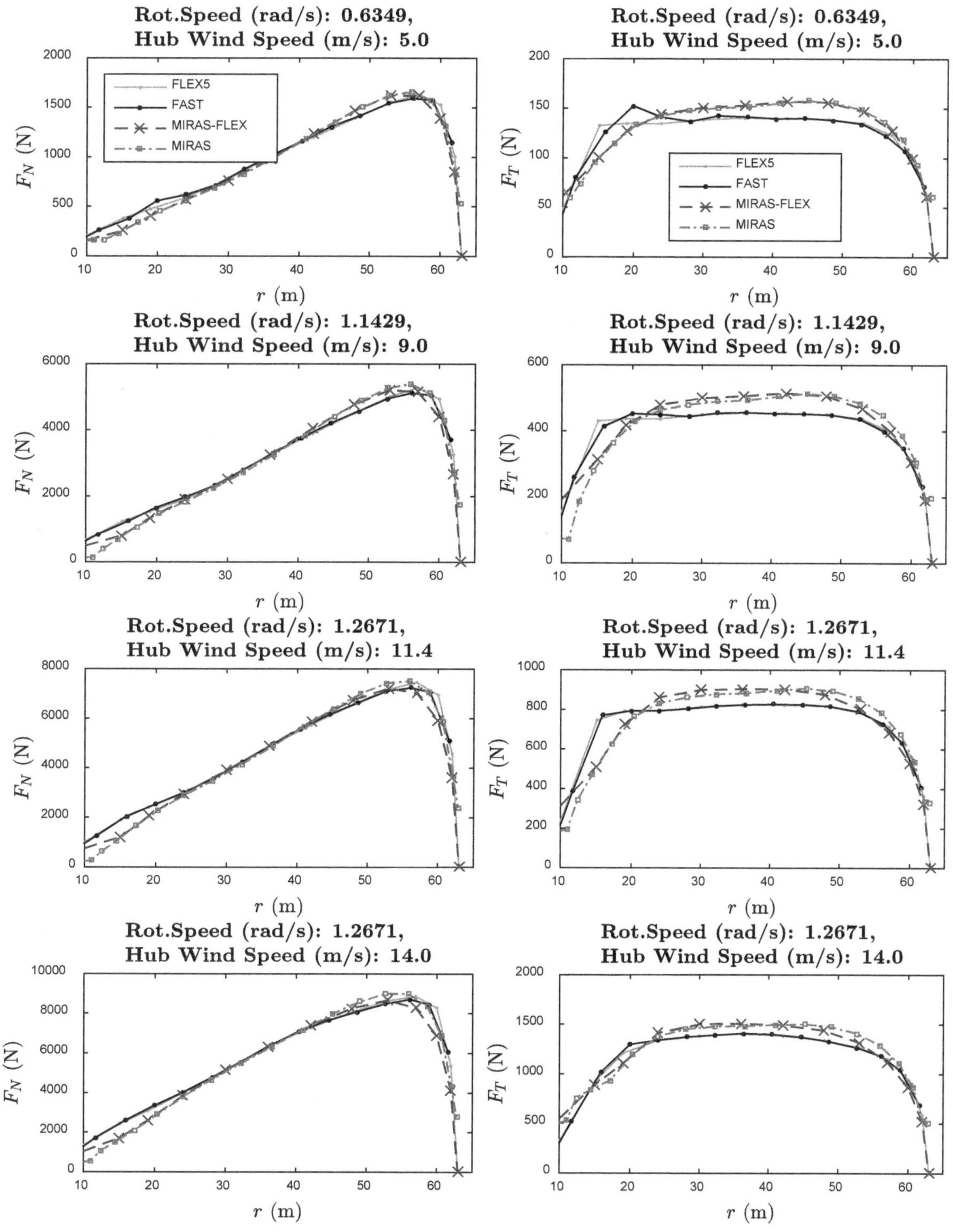

Figure 4. Comparison of the aerodynamic loads normal (left) and tangential (right) to the rotor plane computed by FLEX5, FAST, MIRAS-FLEX and MIRAS.

Acknowledgements

The present work was funded by the Danish Council for Strategic Research under the project name Off Wind China (Sagsnr. 0603-00506B). The support is gratefully acknowledged.

References

[1] Bungartz, H.J., Mehl, M. and Schäfer, M. (2010) Fluid Structure Interaction II: Modelling, Simulation, Optimization, 1439-7358. *Springer Berlin Heidelberg*, **53**. http://dx.doi.org/10.1007/978-3-642-14206-2

[2] Øye, S. (1996) FLEX4 Simulation of Wind Turbine Dynamics. *Proceedings of 28th IEA Meeting of Experts Concerning State of the Art of Aeroelastic Codes for Wind Turbine Calculations*, Lyngby, 1996.

[3] NWTC Information Portal (FAST v8). https://nwtc.nrel.gov/FAST8

[4] Kim, T., Anders, A.M., Hansen, M. and Branner, K. (2013) Development of an Anisotropic Beam Finite Element for Composite Wind Turbine Blades in Multibody System. *Renewable Energy*, **59**, 172-183. http://dx.doi.org/10.1016/j.renene.2013.03.033

[5] Hansen, M.O.L. (2008) Aerodynamics of Wind Turbines. 2nd Edition, Earthscan, London.

[6] Ramos-García, N., Sørensen, J.N. and Shen, W.Z. (2014) Three-Dimensional Viscous-Inviscid Coupling Method for Wind Turbine Computations. *Wind Energy*. http://dx.doi.org/10.1002/we.1821

[7] Ramos-García, N., Sørensen, J.N. and Shen, W.Z. (2014) A Strong Viscous-Inviscid Interaction Model for Rotating Airfoils. *Wind Energy*, **17**, 1957-1984. http://dx.doi.org/10.1002/we.1677

[8] Degroote, J., Bathe, K.J. and Vierendeels, J. (2009) Performance of a New Partitioned Procedure versus a Monolithic Procedure in Fluid-Structure Interaction. *Computers & Structures*, **87**, 793-801. http://dx.doi.org/10.1016/j.compstruc.2008.11.013

[9] Farhat, C., van der Zee, K.G. and Geuzaine, P. (2006) Provably Second-Order time-Accurate Loosely-Coupled Solution Algorithms for Transient Nonlinear Computational Aeroelasticity. *Computer Methods in Applied Mechanics and Engineering*, **195**, 1973-2001. http://dx.doi.org/10.1016/j.cma.2004.11.031

[10] Heinz, J. (2013) Partitioned Fluid-Structure Interaction for Full Rotor Computations Using CFD. Ph.D. Thesis, Technical University of Denmark, Denmark.

[11] Bauchau, O.A. and Craig, J.I., Eds. (2009) Structural Analysis: With Applications to Aerospace Structures, Solid Mechanics and Its Applications. *Springer Netherlands*. http://dx.doi.org/10.1007/978-90-481-2516-6

[12] Jonkman, J., Butterfield, S., Musial, W. and Scott, S. Definition of a 5-MW Reference Wind Turbine for Offshore System Development. Technical Report NREL/TP-500-38060, National Renewable Energy Laboratory, Golden.

24

Power Quality Study of an Isolated Northwest Grid of Saudi Arabia with PV and Storage

Hani Albalawi, Elham Makram

Department of Electrical and Computer Engineering, Clemson University, Clemson, USA
Email: halbala@clemson.edu, makram@clemson.edu

Abstract

One main concern of power quality is harmonics because the distorted waveforms of current and voltage have a huge effect on electrical equipment. Due to the continuous increase of grid connected photovoltaic (PV) and nonlinear loads, as a result of the fast development and growth of power electronics application, power quality becomes more important since it introduces harmonics to the power system. This paper presents a power quality study to the isolated northwest grid of Saudi Arabia in presence of PV system and battery storage. Moreover, the study includes nonlinear loads for more analysis regarding harmonics penetration and the design procedure for passive filters to eliminate the harmonics.

Keywords

PV, Battery, Nonlinear Load, Harmonics, Passive Filter, Tabuk Grid, Saudi Arabia

1. Introduction

Supplying energy at constant frequency and magnitude of voltage is always the objective for any electrical utility. This can only happen in ideal power system but in practice this condition is hard to achieve. Many studies are required for planning and operation of power system in order to assist system performance, reliability and safety. Power quality is one important aspect of power system. It is the concept for many types of power disturbances such as voltage variation, and transient and harmonic distortion. The most unlikely condition for power quality is the harmonic distortion as it is a steady state condition. With high growth of installing grid connected photovoltaic (PV), power quality becomes more important as the DC-AC inverter can affect the quality of the power delivered to the grid due to the switching mechanism responsible for additional harmonics. Moreover, the widespread nonlinear loads due to power electronics applications result in distorted current and voltage in transmission and distribution networks. A solution for this power quality issue can be achieved by using harmonic filters which are designed to reduce the harmonic penetration to the network [1].

All appliances that use diode rectifiers to convert AC to DC have a nonlinear characteristic that can lead to introduce harmonics. Examples of nonlinear loads include television, computers, fluorescent lamp and adjustable speed drives (ASD) for air condition and heat pump. Harmonic distortion is unwanted phenomena as it can cause additional line losses and interference in communication system. Therefore, analysis of harmonics and nonlinear loads is vital. A common way to analyze and study harmonics is to represent a load by linear model, then adding nonlinearity by using voltage or current harmonic sources [2].

This paper presents a complete study of power quality for the isolated northwest grid of Saudi Arabia. A detailed PV model with battery storage is modeled and a combination of nonlinear loads is introduced. Also, the paper includes filter design to reduce harmonics. The outline of this paper is as follows. Section 2 shows PV and battery model. Section 3 explains the filter design. Section 4 describes the nonlinear load modeling. Section 5 provides a brief description of the test system. Section 6 shows the simulation results.

2. PV and Battery Model

2.1. PV Model

Figure 1(a) shows a diagram of a three phase grid connected PV system with battery storage. It consists of a PV array, Inverter and 3 phase reactor which is connected in series with the ac terminal of the inverter. The PV array is connected in parallel to the dc link capacitor and the dc terminal of the inverter. The inverter is controlled by voltage source converter (VSC) modulation strategy based on sinusoidal pulse-width modulation (S-PWM). The main element of the PV array is the solar cell, which has a very small power. In order to increase and have sufficient power to supply a load, single cells are connected in series to form a module. Therefore, PV array is a combination of parallel *Np* module and series Ns module. By increasing *Np* PV current increases and the PV voltage rises by increasing Ns. the Capacitor C_f is used to reduce the harmonic produced by the PV system. **Figure 1(a)** illustrates that, synchronizing the VSC and control to the grid voltage is achieved by a phase-locked loop (PLL). Also, the error between the V_{dc} reference given by MPPT and the dc-link voltage is passed through PI compensator whose output provides the d-axis reference current $\left(i_{dref}\right)$. Then, the dq-frame current loop control is fed by i_{dref} to force d-axis current to track the i_{dref} to control the PV output power, for unity Power factor the q-axis reference current i_{qref} is set to zero [3].

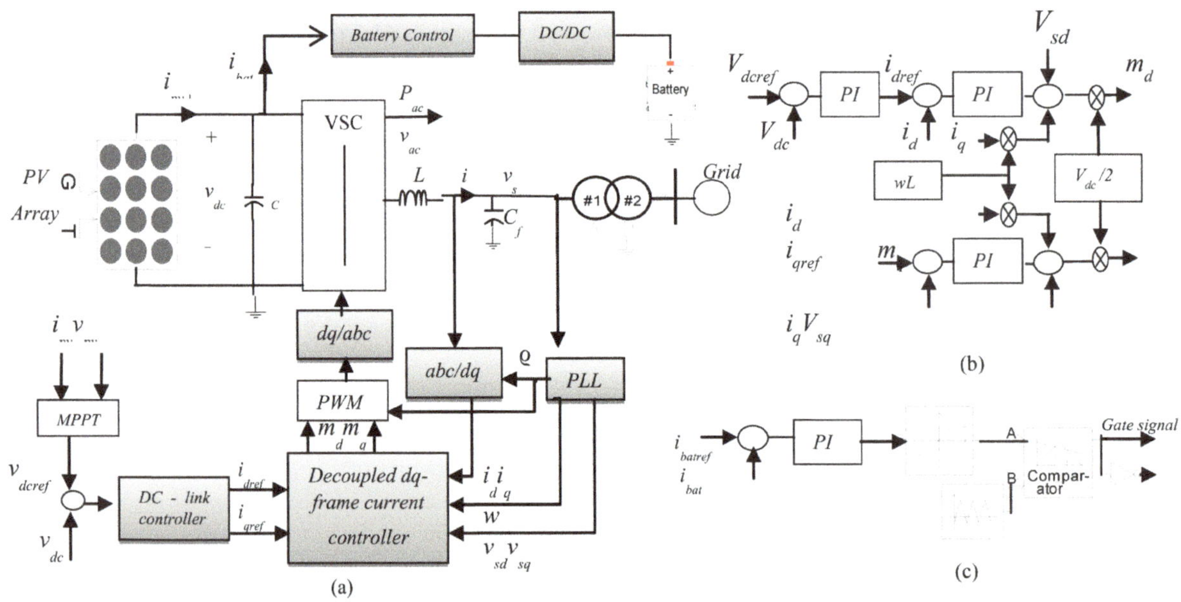

Figure 1. (a) Single line diagram of PV system; (b) PV control; (c) Battery control.

2.1.1. Phase Locked Loop

The PLL is used to synchronize VSC's output voltage with electric-grid angular frequency in order to make dq-frame transformed variables constant and by regulating v_{sq} to zero the PV system active and reactive power are controlled as follows:

$$P = \frac{3}{2} v_{sd} i_d \tag{1}$$

$$Q = -\frac{3}{2} v_{sd} i_q \tag{2}$$

From (1) and (2), controlling the active power is archived by regulating i_d and the reactive power is controlled by regulating i_q.

2.1.2. DC Voltage Control

The control of the dc voltage link is achieved by simply comparing the dc voltage with the reference value. This reference is calculated from MPPT strategy to ensure that the PV works at its maximum output power. The error between these two signals is passed through a PI compensator whose output provide i_{dref}, which will be the input for the current controller as seen in **Figure 1(b)**. i_{qref} is set to zero to achieve unity power factor.

2.1.3. Current Controller

The current controller is comparing the d*q*-reference currents with the actual ones to obtain the modulation signals (m_d, m_q) the control scheme is achieved by

$$L\frac{di_d}{dt} = -Ri_d + Lwi_d + \frac{v_{dc}}{2} m_d - v_{sd} \tag{3}$$

$$L\frac{di_q}{dt} = -Ri_q - Lwi_d + \frac{v_{dc}}{2} m_q - v_{sq} \tag{4}$$

The state variables of (3) and (4) are i_d and i_q while the input control signals are m_d and m_q. v_{sd} and v_{sq} are disturbance inputs. The d-q frame dynamics is coupled and nonlinear due to the Lw term. Hence, it must be linearized according to the following control principals [3]:

$$m_d = \frac{2}{v_{dc}}\left(u_d - Lwi_q + v_{sd}\right) \tag{5}$$

$$m_q = \frac{2}{v_{dc}}\left(u_q + Lwi_d + v_{sq}\right) \tag{6}$$

By substituting (5) and (6) into (3) and (4), two control inputs produced which can control i_d and i_q independently, respectively, as following:

$$L\frac{di_d}{dt} = -Ri_d + u_d \tag{7}$$

$$L\frac{di_q}{dt} = -Ri_q + u_q \tag{8}$$

However, the m_d and m_q signals is converted back to *abc* reference-frame to generate the reference signals for the VSC. These reference waveforms are compared with a triangular carriers to generate the pulsing signals to modulate the VSC [4].

2.2. Battery Model

This paper used the developed battery model in PSCAD [5]. Details of this battery model are described in [6]. It is modeled by a control voltage source with a series resistance. The charging and discharging process is controlled by constant current charging mode. Positive current reference is for charging process while negative reference is for discharging process. When the battery reaches a certain level of state of charge (SOC), the reference current is set to zero to keep the voltage at that level. From **Figure 1(c)**, the DC/DC converter is operating in current control mode; the error signal is passed through a limiter and then compared with a triangular wave carrier to generate the pulse signals for the converter [7].

3. Filter Design

One way to improve the power quality of power system is to resolve the harmonics issue. This can be done by many approaches described in [8]. One common way for harmonic filtering is to use passive filters, these kinds of filters are designed to provide low impedance path at the tuned frequency to prevent the harmonic current of flowing into the system. In this paper passive filter is used because it is one of the cheapest and economic way-for harmonic filtering [1]. Harmonic filters are used to reduce the harmonics penetration in power system. A well designed harmonic filter is important to keep the harmonic within acceptable level. One of the most types used of passive filter is the single tuned filter, which is a capacitor designed to eliminate a certain harmonic by adding a reactor so that $X_C = X_L$ at the tuned frequency (*fn*) [9]. The tuned frequency for harmonic filter designing usually should be between 2% to 10% below the desired harmonic frequency to avoid parallel resonance and near short circuit at tuned frequency as the filter provides low impedance. However, since the passive filters also provide a reactive compensation, they can also provide a power factor correction to a desired level in parallel to its filtering action.

For a tuned harmonic (*hn*) with a capacitor (Q_c), the capacitor reactor at *fn* is

$$X_c = \frac{KV^2}{Q_c} \tag{9}$$

And to eliminate the *hn* harmonic, the reactor size should be

$$X_L = \frac{X_c}{h_n^2} \tag{10}$$

The reactance resistor can be calculated as follows

$$R = \frac{h_n X_L}{Q} \tag{11}$$

where Q is the quality factor and its typical value is between 30 and 100 [10].

4. Nonlinear Load Model

The increase of nonlinear loads due power electronics applications has introduced more effect on power system quality. Therefore, analysis of nonlinear loads and its affects is important. In power system, one common way to model a nonlinear load is to model it as a load for the fundamental current and as a current source for the harmonics current [11]. **Figure 2** shows the configuration of the nonlinear load. In this paper three types of nonlinear loads are used, which are fluorescent lamp, computer and adjustable speed drive (ASD). ASD is used to represent air condition and heat pump. One important parameter is the individual harmonic distortion as it shows the contribution of each harmonic frequency to the distorted waveform. And by knowing the contribution of each harmonic frequency for a specified load, a harmonic current source can be used to inject the magnitude current for each harmonic frequency. Hence, the nonlinear behavior can be modeled by doing so. The contribution of each individual harmonic distortion for the three types of the nonlinear loads is shown in **Table 1**.

5. Study System

Saudi Arabia grid is divided into four operation areas, West area (WA), East area (EA), South area (SA) and Central Area (CA). Each one has its own control center and operation however they are now under interconnection projects. The WA consists of two parts, the west part and the northwest isolated area which has Tabuk region and its surrounding towns. The grid data was provided from Saudi Electricity Company, Tabuk sector. Line and Peak load data are shown in **Table 2**. It observed form **Figure 3** that, the grid has 12 buses with voltage level of 132 KV, 10 substations are distributed around the city and two substations are connecting two different towns, Timma and Albir. It has two generation stations at bus 1 and 5 with a capacity of 102 MW and 857.2

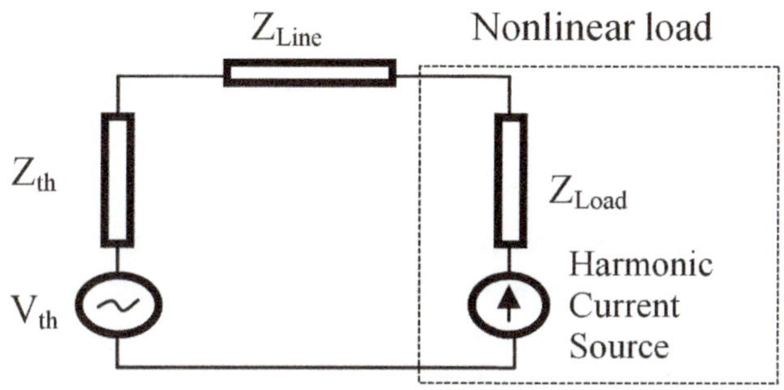

Figure 2. representation of nonlinear load.

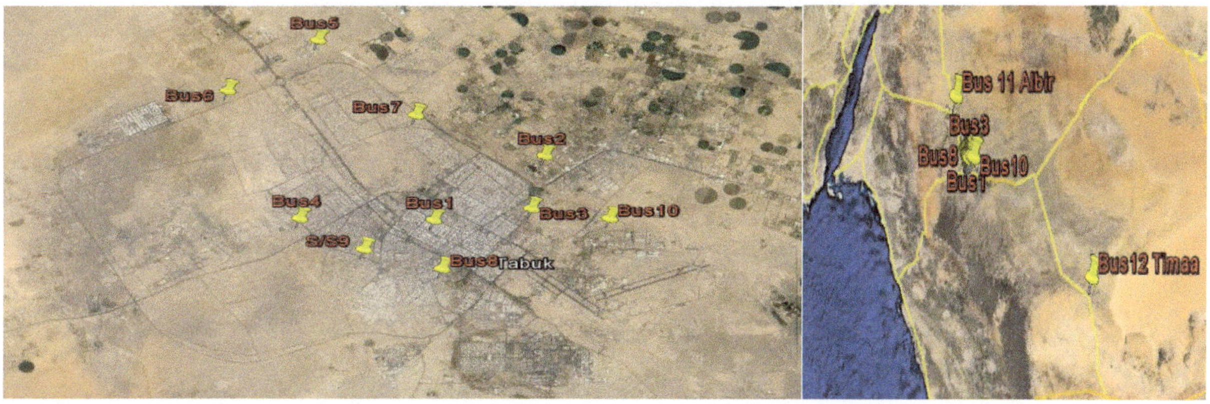

Figure 3. Tabuk Substations.

Table 1. Harmonic Number h (n) VS Individual Harmonic distortion [12] .

H (n)	IHD (%)		
	Fluorescent lamp	Computer	ASD
1	100	100	100
2	0.3	3.3	4.12
3	13.9	87.2	0.78
4	0.3	5.1	1.79
5	9	64.1	35.01
6	0.2	1.6	0.215
7	3.3	41.1	2.62
8	0	0	1
9	3.2	17.9	0.06
10	0.1	1.1	0.73
11		10.3	9.99
12	0.3	1.2	0.03
13	1.7	10.3	0.19
14	0.3	0	0.48
15	1.9	10.3	0.07
16	0.3	0	0.52
17	0.8	5.1	4.85
THD (%)	18	118.3	37.3

MW respectively. **Figure 4** provides an over view of the power consumed in the city for a summer and winter day.

6. Simulation Result

The PSCAD software is used to build and simulate the grid with a detailed PV model and Battery storage. A peak loading condition for a summer day, 07/16/2013, with a total load of 720 MW is selected for the simulation. Bus 5 has been selected as a slack bus due to its largest capacity. The PV Farm is connected to bus 6 because it has enough land that can accommodate two PV farms connected in parallel with a total capacity of 40 MW, 20 MW each. The battery is assumed to have a capacity of 50% of the total PV Power. A harmonic current source is used to represent the nonlinear loads; the harmonic injection based on **Table 1** is placed at the substation, at

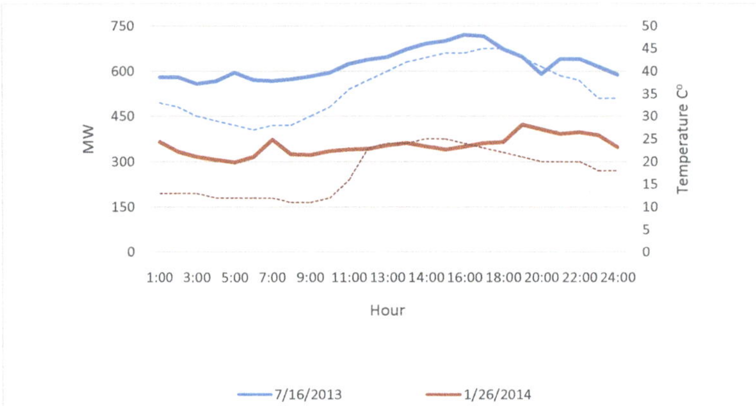

Figure 4. Load Profile for Tabuk. Solid lines is daily load, dashes lines is temprature.

Table 2. Line and Peak load data.

From Bus	To Bus	R (Ohm/Km)	X (Ohm/Km)	Bus	Peak Load (MW)
1	2	0.0493	0.2408	1	106
1	4	0.0493	0.2408	2	94
2	3	0.047	0.25	3	55
2	5	0.0773	0.4682	4	62
2	7	0.08606	0.5062	5	11
3	10	0.0493	0.2408	6	37
4	5	0.0773	0.4682	7	60
4	6	0.093	0.452	8	67
4	8	0.0493	0.2408	9	134
4	9	0.047	0.25	10	12
5	6	0.094	0.458	11	33
5	7	0.0794	0.4671	12	43
5	12	0.08	0.47		
6	9	0.0493	0.2408		
6	11	0.149	0.497		
10	8	0.0493	0.2408		

the low voltage side 13.8 KV. At each bus the load is divide into 80% linear load and 20% nonlinear load. The nonlinear loads that used are ASD, to represent air condition and heat pump, fluorescent lamp and a personal computer with percentage of 18%, 1% and 1% respectively. All loads at each bus has the same procedure therefore the highest load at bus 9 is selected to show the result of including the nonlinear loads, **Figure 5(a)** illustrates the distorted load current after adding the nonlinear loads. Also, **Figure 6(a)** shows that, the total harmonic distortion (THD) of the total load current is above the limits therefore a single tuned filter has been designed to eliminate the highest harmonic component which is the 5th component as shown in **Figure 6(b)**. **Figure 5(b)** shows the load current after installing the passive filter. Hence, **Figure 7** provides the THD for each bus during the simulation and it observed that as seen in **Table 3** the distortion is within the limits based on IEEE standard [13]. The initial irradiation applied for the PV is 1000 W/m^2 with initial SOC of 70% for the battery. **Figure 8** shows the different stages of PV penetration with battery storage. At 3 second, the charging process starts and stops when the SOC reaches 80% at 16 second. The SOC kept constant until the discharging of the battery begins at 18 second. During the discharge process the irradiation drops by 250 W/m^2 at 25 second and 30 second. The THD of the PV current injected to the grid for all penetration stages is within the acceptable limits. It is oscillating around 2.5% as observed from **Figure 9**; the two high values at the second 16 and 18 are instantaneous due to the switching of the reference current of the battery to keep the SOC at 80% and to start the discharging mode.

One advantage of the Battery model in PSCAD is that, it allows the user to define the characteristic charge and discharge curves. And since the PV farm output power is high and so the dc voltage link,it is hard to find a battery with such voltage level therefore a characteristic charge and discharge curve of lead acid battery has

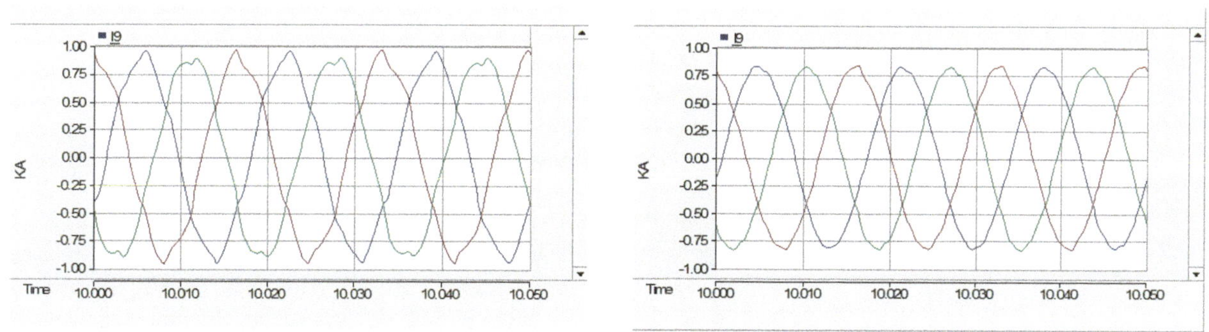

(a) (b)

Figure 5. (a) Load current without filter of bus 9; (b) Load current with filter of bus 9.

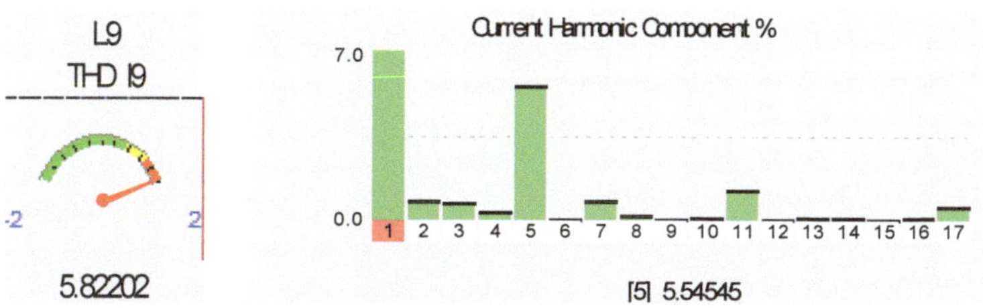

(a) (b)

Figure 6. (a) THD at bus 9; (b) Harmonic component.

Table 3. Load current THD compared with the standard Limits.

Bus No.	1	2	3	4	5	6	7	8	9	10	11	12
THD% for load current	2.11	1.96	2.03	2.18	5.87	2.33	1.90	1.97	1.98	5.63	1.83	2.49
$\frac{I_{sc}}{I_L}$	<20	<20	<20	<20	50 < 100	20 < 50	<20	<20	<20	50 < 100	<20	<20
IEEE limit % for 132 KV level	2.5%	2.5%	2.5%	2.5%	6%	4%	2.5%	2.5%	2.5%	6%	2.5%	2.5%

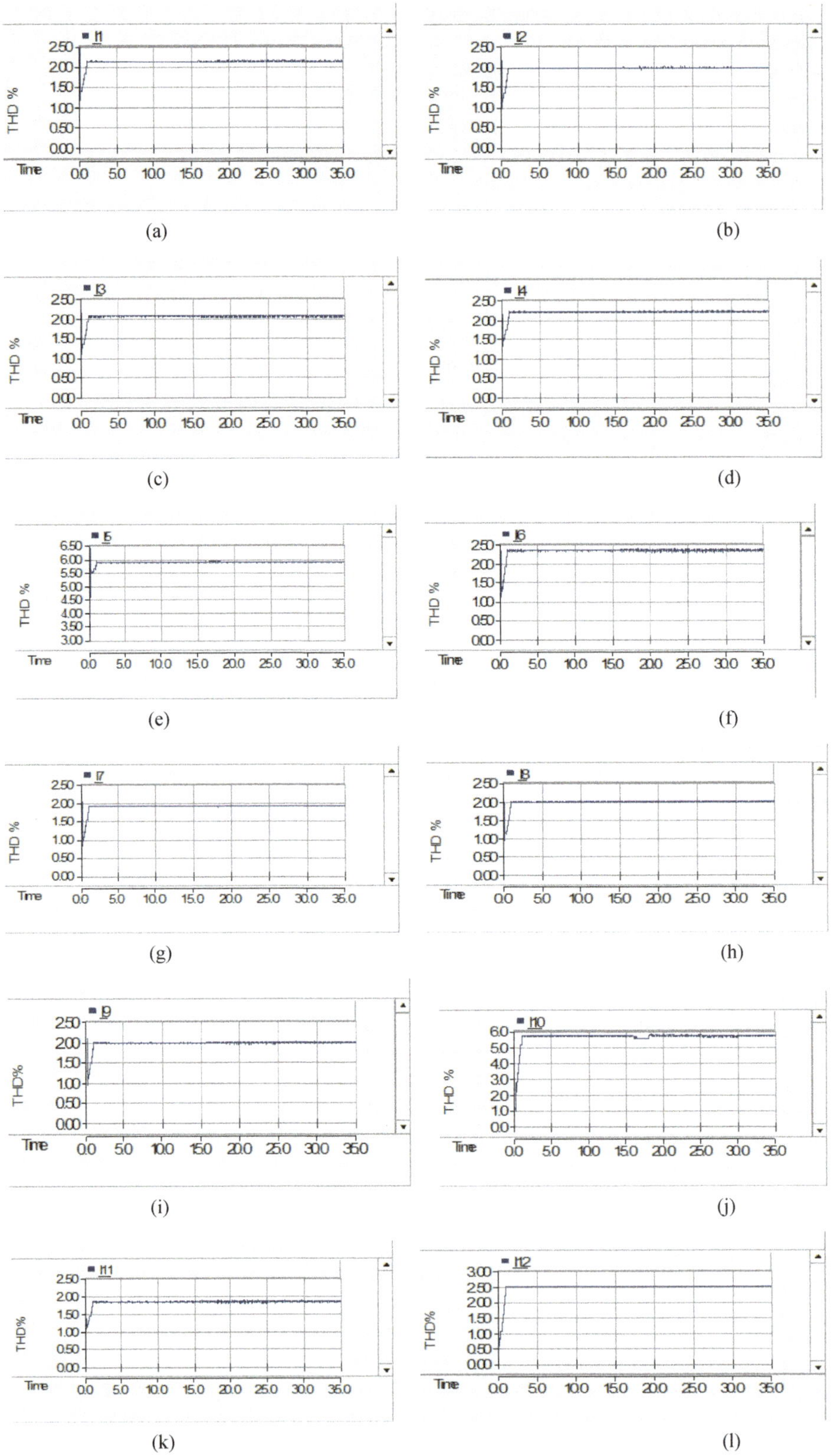

Figure 7. THD for all loads currents with filter. (a) THD of load current at bus 1; (b) THD of load current at bus 2; (c) THD of load current at bus 3; (d) THD of load current at bus 4; (e) THD of load current at bus 5; (f) THD of load current at bus 6; (g) THD of load current at bus 7; (h) THD of load current at bus 8; (i) THD of load current at bus 9; (j) THD of load current at bus 10; (k) THD of load current at bus 11; (l) THD of load current at bus 12.

been scaled up to match the DC link voltage and defined in the battery model. For the control scheme, the reference current for the battery can be obtained based on the DC link voltage and the amount of power to be stored. Since the battery assumed to have 50% of the PV power, the battery reference current is calculated as follows

$$I_{batref} = \frac{PV_{power} \times 50\%}{V_{dc}} \tag{12}$$

From **Figure 10** it observed that, the battery current follows the reference for both charging and discharging modes, for the discharge process a negative reference is applied. Also, when SOC of the battery reaches 80% the reference current is set to zero to keep the SOC of constant as seen in **Figure 11**. The battery power is shown in **Figure 12**.

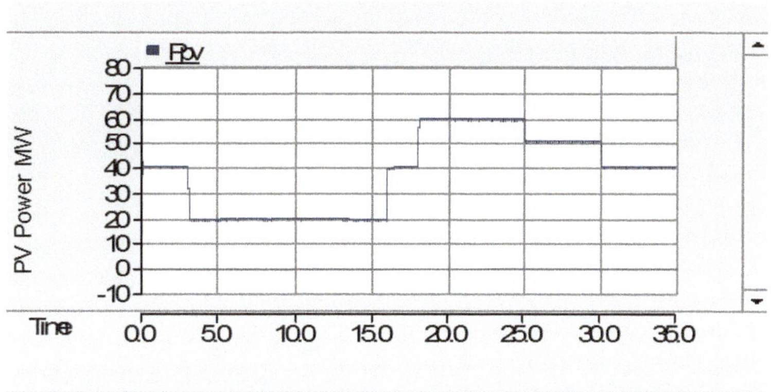

Figure 8. PV power at PCC.

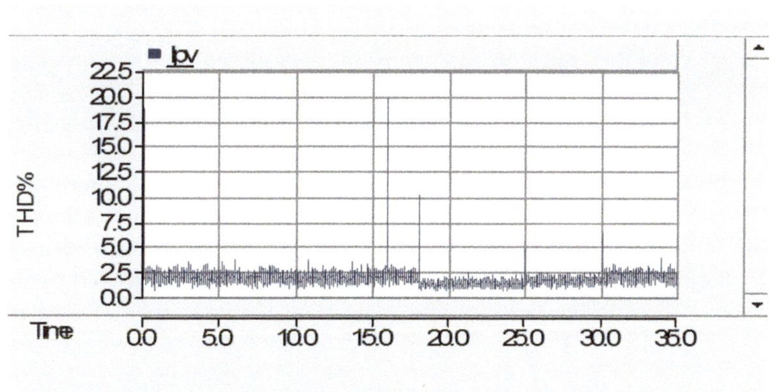

Figure 9. THD of PV Current at PCC.

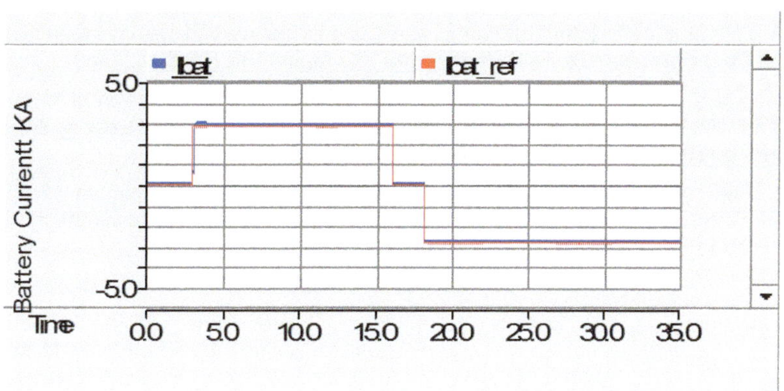

Figure 10. Battery current for one farm.

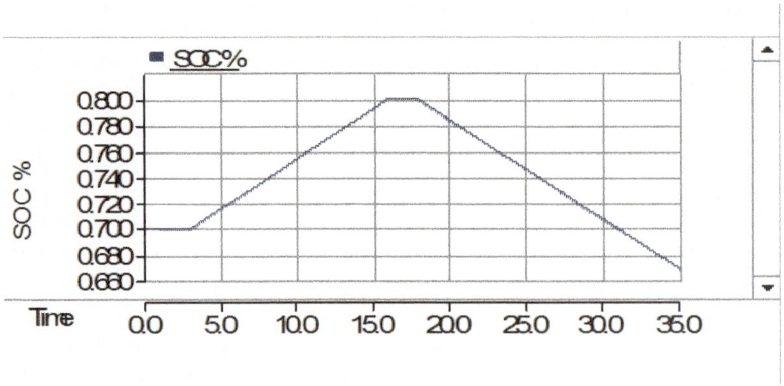

Figure 11. Battery SOC.

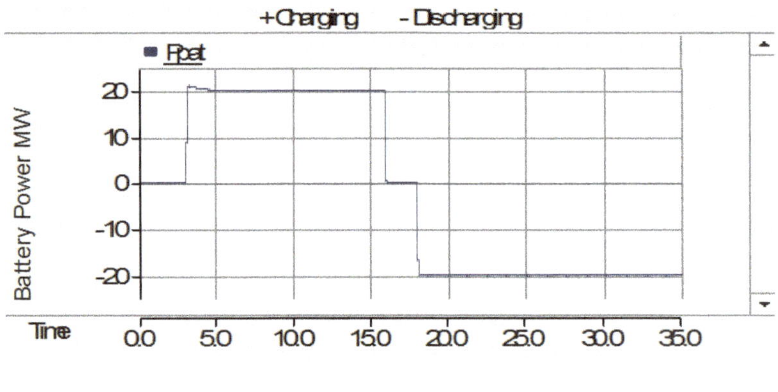

Figure 12. Battery Power.

7. Conclusion

This paper presents a power quality study for the northwest grid of Saudi Arabia in presence of PV and battery storage. Different PV power penetrations with battery storage are introduced and the THD of PV current is within the acceptable limit for all penetration stages. Also, nonlinear load has been included for more analysis and it shows that the harmonic impact of introducing such loads has been solved by designing passive filters. It also discusses the passive filter design producer.

References

[1] Rao, R.N. (2013) Harmonic Analysis of Small Scale Industrial Loads and Harmonic Mitigation Techniques in Industrial Distribution System. *International Journal of Engineering Research and Applications*, **3**, 1511-1540.

[2] Fauri, M. (1997) Harmonic Modeling of Non-Linear Load by means of Crossed Frequency Admittance Matrix. *IEEE Transactions on Power Systems*, **12**, 1632-1638. http://dx.doi.org/10.1109/59.627869

[3] Yazdani, A. and Dash, P.P. (2009) A Control Methodology and Characterization of Dynamics for Photovoltaic (PV) System Interfered with a Distribution Network. *IEEE Transactions on Power Delivery*, **24**, 1538-1551. http://dx.doi.org/10.1109/TPWRD.2009.2016632

[4] Mahmood, F., Yanfretti, L. and Hooshyar, H. (2014) Modeling of a Detailed Photovoltaic Generation System for EMT-Type Simulation. *IEEE International Energy Conference*, Cavtat, 13-16 May 2014, 916-921. http://dx.doi.org/10.1109/energycon.2014.6850535

[5] Jiang, S. (2012) Battery Component in PSCAD/EMTDC. Manitoba HVDC Research Centre, Winnipeg.

[6] Tremblay, O., Dessaint, L.A. and Dekkiche, A.I. (2007) A Generic Battery Model for the Dynamic Simulation of Hybrid Electric Vehicles. *IEEE Vehicle Power and Propulsion Conference*, Arlington, 9-12 September 2007, 284-289. http://dx.doi.org/10.1109/vppc.2007.4544139

[7] Clarke, A., Bihani, H., Makram, E. and Corzine, K. (2013) Fault Analysis on an Unbalanced Distribution System in the

Presence of Plug-In Hybrid Electric Vehicles. *Clemson Power System Conference*, March 2013.

[8] Balaraju, U.P., Kethineni, B., Shewale, R. and Grourishetti, S. (2012) Harmonic Effects and Its Mitigation Techniques for a Nonlinear Load. *International Journal of Advanced Technology & Engineering Research*, **2**, 122-127.

[9] Badrzadeh, B., Smith, K. and Wilson, R.C. (2011) Designing Passive Harmonic Filters for an Aluminum Smelting Plant. *IEEE Transactions on Industry Application*, **47**, 973-983. http://dx.doi.org/10.1109/TIA.2010.2103544

[10] Wakileh, G.J. (2001) Power System Harmonics: Fundamental, Analysis and Filter Design. Springer, New York. http://dx.doi.org/10.1007/978-3-662-04343-1

[11] Gonen, T. (2008) Electrical Power Distributed System Engineering. 2nd Edition, CRC Press, New York.

[12] Sankaran, C. (2008) Power Quality. CRC Press, New York.

[13] McGranaghan, M. (1988) Overview of the Guide for Applying Harmonic Limits on Power Systems—IEEE P519A. *8th International Conference on Harmonics and Quality of Power Proceeding*, Athens, 14-18 October 1998, 462-469.

25

Simulation Studies of Diesel Engine Combustion Characteristics with Oxygen Enriched Air

Heng Wang, Weijun Liu

Automotive Engineering Institute, Shanghai University of Engineering Science, Shanghai, China
Email: shgcwangheng@163.com

Abstract

Based on a six-cylinder direct injection diesel engine, the engine operating condition was simulated by application AVL-FIRE software coupling the n-heptane reduced mechanism containing polycyclic aromatic hydrocarbon (PAH) formation. The simulation and its verification test were both carried out under the maximum torque point. Then, the oxygen enriched combustion was simulated on the model, and the simulated condition was oxygen volume fraction from 21% to 30%. The simulation results show that, the oxygen enrichment (from 21% to 30%) increases the peak cylinder pressure of 3.32%, advances the start of combustion of 1.6 deg and rises the peak of average temperature in cylinder and wall heat flux. Among them, at the condition of 24% O_2, the change of the results is the most significant. Benzene (A1) is one of the precursors of soot generated, the analysis of its impress-cuts of the mass distribution field in cylinder shows that, the increase of oxygen concentration can significantly inhibit the formation of benzene. But the oxygen enrichment makes the combustion more sufficient, cased a rise in the cylinder temperature, an extension in high temperature area, and an increment in the NO_x emission.

Keywords

Diesel Engine, Numerical Simulation, Oxygen Enriched Combustion, Combustion Characteristic

1. Introduction

A lot of researches show that oxygen enriched combustion technology can obviously improve the combustion process of the cylinder, and have a significant effect on the power performance, fuel economy and emission characteristics of diesel engine [1]-[3]. Oxygen enriched intake increased the content of oxygen in the air, cased an exacerbation in the combustion, an increment in the flame speed, a rise in the combustion temperature and a decrease in the fuel consumption. In addition, it can also significantly reduce the emissions of soot, but will lead to a substantial increment in NO_x emissions [4]-[6]. The studies by GequnShu [7] and Peter L. Perez [8] show that oxygen enrichment combustion combining with some technologies, such as the controlling of fuel injection timing, water emulsion fuel and EGR technology etc., can make NO_x and soot emissions simultaneously get effective control. At present, there is little research on the oxygen enriched combustion characteristic of diesel engines with the intake oxygen concentration higher than 24%. Although some experimental studies [9] [10] show that, when the intake oxygen concentration is higher than 26%, the effective thermal efficiency of diesel engine will slow decline with the increasing of intake oxygen concentration. However, further exploring the effect of higher intake oxygen concentration on the combustion characteristics of the diesel engine is still significant for improving the combustion process and controlling the pollutant discharge of the diesel engine.

In the present work, the model built by application of AVL-FIRE software and coupling the n-heptane reduced mechanism containing the PAH formation were used for calculating some parameters, such as cylinder pressure curve, rate of heat release and the concentration distribution of emissions, etc. Then, these parameters were used to explore the influence of different inlet oxygen concentration on diesel engine combustion characteristic, so as to accumulate more theory basis in diesel engine.

2. Establishment of Coupling Model

2.1. Selection of Chemical Reaction Mechanism

Because of the cetane number of n-heptane and diesel are the same, the simulation study of diesel engine usually choose n-heptane as a substitute for diesel combustion calculation. Most of the reaction kinetics models are supplied from the Lawrence Livermore national laboratory in n-heptane detailed oxidation combustion mechanism [11] [12]. However, as a result of the mechanism is too complicated for coupling calculation, in this study, the n-heptane-butanol-PAH reduced mechanism which proposed by Hu Wang [13] was selected. The mechanism is verified by the experimental data of the shock tube, the constant volume combustion bomb and the engine , and it can well predict the process of the fuel combustion and the generation of the emissions. As fuel for pure diesel, the new mechanism should get rid of the reduced mechanism of n-butanol in the mechanism part, finally obtained a mechanism including 60 species and 272 elementary reactions to simulate the diesel combustion process and emissions generated.

2.2. Computational Grid and Model

According to the geometric parameters and operating conditions of the diesel engine which was selected by Haozhong Huang [14] for experimental research, the model is established. Diesel engine basic parameters see **Table 1**.

Because of the full size model calculation needs to consume too long and the combustion chamber is in a central symmetry, the 1/7 combustion chamber model is selected according to the number of injector holes. The computational grid is shown in **Figure 1**.

Due to just the combustion characteristics of diesel engine are researched, the calculation will start from the valve closing time of 583 deg to the exhaust valve opening time 845 deg. In the boundary conditions of the model, there are three surface temperatures needing to be set, including the wall of the combustion chamber, the piston head and the inner wall of the cylinder. The temperature of the three surfaces is assumed to be constant, and the temperature value is calculated by reference to empirical formula. In the initial calculation, the temperature and pressure of the cylinder are uniform, the pressure value is set according to the test values, and the temperature value is estimated by the corresponding empirical formula [15]. In addition, according to the engine speed, intake flow and other parameters, the initial kinetic energy intensity is calculated and its direction is defined. In the selection of calculation model, according to the structure characteristics of diesel engine, gas turbulence model selected k-zeta-f, spray model selected Spray Turbulent Diffusion, oil mist evaporation model selected Dukowicz and fuel crushing model selected Wave.

2.3. Combustion Chamber Numerical Model Verification

The accuracy of the model is verified by using the cylinder pressure curve measured by Haozhong Huang [14].

The fuel used in the experiment was pure diesel, and the test conditions were in agreement with the simulation conditions (see **Table 2**).

Table 1. Parameters of the diesel.

Parameters	Unit	Value
Bore × Stroke	mm	105 × 125
Connecting rod length	mm	210
Compression ratio	-	16
Displacement	L	6.5
Maximum torque engine speed	rpm	1400
Number of nozzle holes	-	7
Bore of nozzle hole	mm	0.17
Spray angle	deg	155

Table 2. Conditions of the simulation.

Projects	Value
Fuel	$C_{12}H_{26}$
Fuel delivery per cycle(mg)	47.25
Fuel injection timing(deg)	6.2 deg bTDC
Intake pressure(bar)	1.6
Engine speed(rpm)	1400
Injection rate	HD-Engine Model

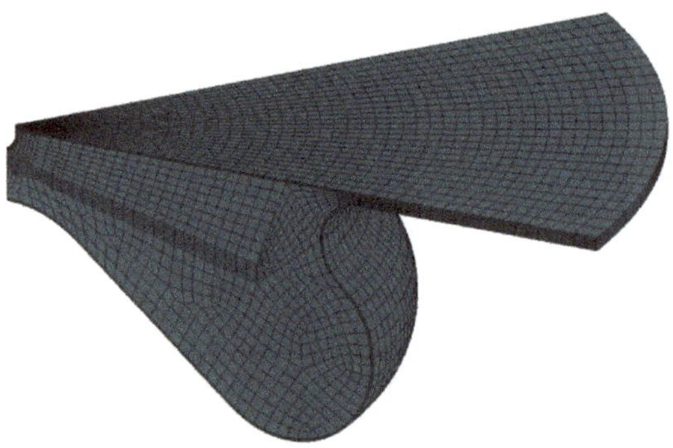

Figure 1. Simulation model of 1/7 combustion chambers.

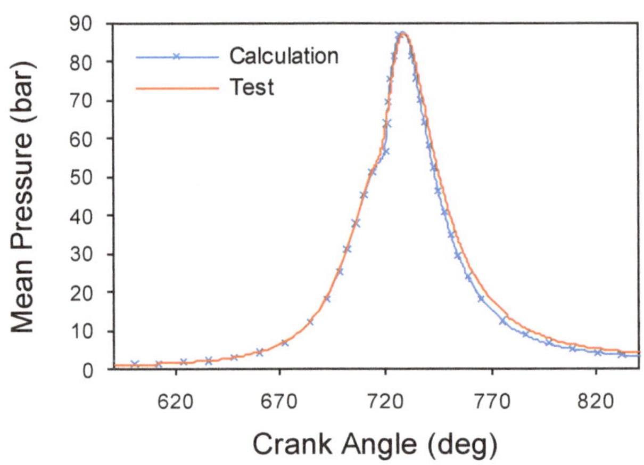

Figure 2. Comparison between the calculation and test in-cylinder pressure.

Figure 2 shows the comparison between the test data and the calculated values of the diesel engine at the intake air (21% O_2), It shows that the pressure in the cylinder is simulated value agrees well with the test results, especially the ignition timing prediction is accurate. It indicates that the boundary conditions are reasonable, calculation model and calculation result is believable. Therefore the model can be used to simulate the combustion process of diesel engine with different intake oxygen concentration.

3. Simulation Results and Discussion

By using the above model, the simulated working condition is kept unchanged, and according to the equal concentration interval, the oxygen mole fraction of the inlet air is respectively selected 21%, 24%, 27% and 30%. Then, some parameters of the diesel engine are simulated under 4 different inlet oxygen concentration conditions, such as cylinder pressure curve, rate of heat release , temperature field and the concentration distribution of emissions, etc., so as to explore the influence of different inlet oxygen concentration on diesel engine combustion characteristic.

3.1. Influence of Oxygen Enriched Combustion on Cylinder Pressure and Rate of Heat Release

Figure 3 is the calculation of cylinder pressure curve. It shows that with the increase of inlet oxygen concentration, there is a slight increment in cylinder peak pressure, a slight advancement in peak pressure point and ignition starting point accordingly. Because of the increase of the oxygen concentration, the fuel and oxygen mix better, the ignition period is shortened and fuel combustion more fully, resulting in the increase of peak cylinder pressure. In addition, in the process of inlet oxygen concentration increasing from 21% to 30%, with the oxygen concentration improving 3% each time, the peak cylinder pressure rise by 1.48%, 1.04% and 0.77% in turn. It can be seen that compared to other inlet oxygen concentration interval, the influence of inlet oxygen concentration on burning process is most significant when the oxygen concentration improve from 21% to 24%. This is because the effect of oxygen concentration decreases with the increase of inlet oxygen concentration, cased a declined in the influence of the oxygen concentration on the combustion process.

Figure 4 shows the rate of instantaneous heat release in the combustion chamber model of 1/7, which indicates that enhanced the oxygen concentration in the intake air makes the cylinder heat release slightly forward. With the promoting of the inlet oxygen concentration, the ignition time is advanced, and then, the heat release of the premixed combustion stage is reduced. As the combustion continued, it shows that the fuel injection continues while mixed diffusion combustion appears, so the heat release of the diffusion combustion stage is increased.

3.2. Influence of Oxygen Enriched Combustion on Mean Temperature Wall Heat Flux

Figure 5 and **Figure 6** respectively show the curve of the average temperature of the combustion chamber and the wall heat flux changes with the Crank Angle under the different inlet oxygen concentration. With the promoting of oxygen concentration, the number of high-energy collisions between molecules combustible mixture molecules and oxygen molecules increases, cased an exacerbation in the combustion and fuel combustion more fully. The average temperature of the cylinder in the ignition stage and the rapid burning stage show an increasing trend, which is favorable for the improvement of the combustion efficiency. At the same time, the temperature of the cylinder increases, cased an increment in the heat flux of the combustor wall. It indicate that the heat loss of the combustor wall is increased, which hinders the improvement of the effective thermal efficiency of the diesel engine.

3.3. Influence of Oxygen Enriched Combustion on Temperature Field and NO Mass Distribution

Figure 7 and **Figure 8** respectively show the combustion chamber temperature field and mole fraction distribution of the NO at the average temperature in cylinder peak point (13 deg a TDC). With the promoting of inlet oxygen concentration, the maximum temperature inside the cylinder is improved and the high temperature region is expanded. Thereinto, the change of oxygen concentration of 24% was the most significant. By comparing the temperature field and NO concentration distribution in cylinder, the changed trend of both is consistent

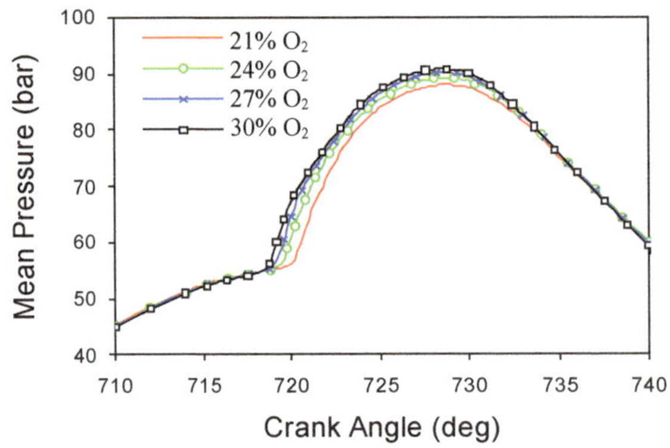

Figure 3. Influence of different O_2 concentration on cylinder pressure.

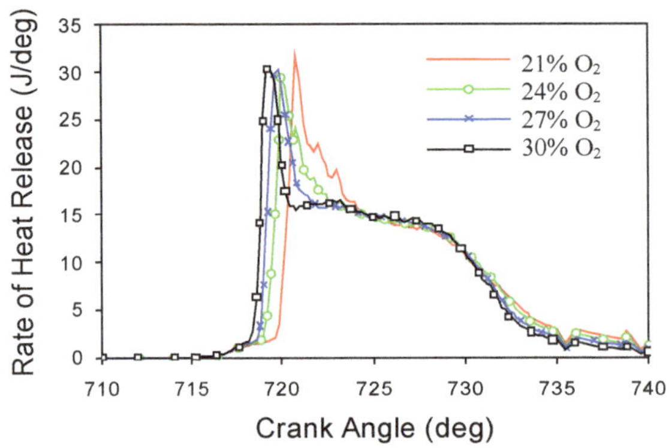

Figure 4. Influence of different O_2 concentration on the rate of heat release.

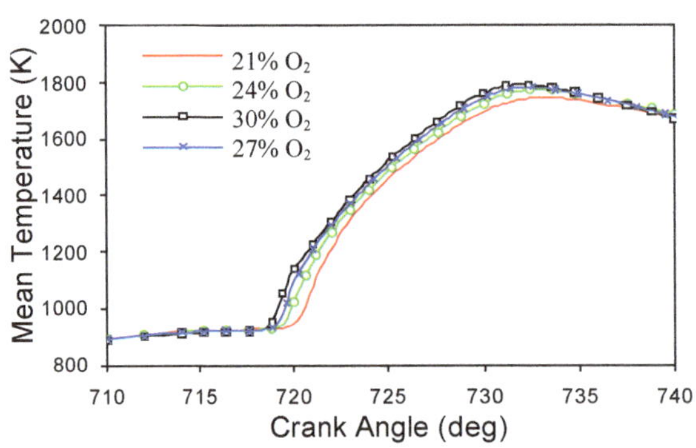

Figure 5. Influence of different O_2 concentration on mean temperature.

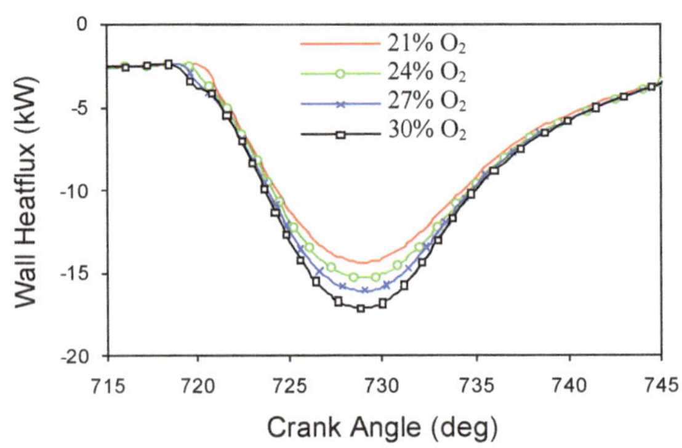

Figure 6. Influence of different O_2 concentration on wall heatflux.

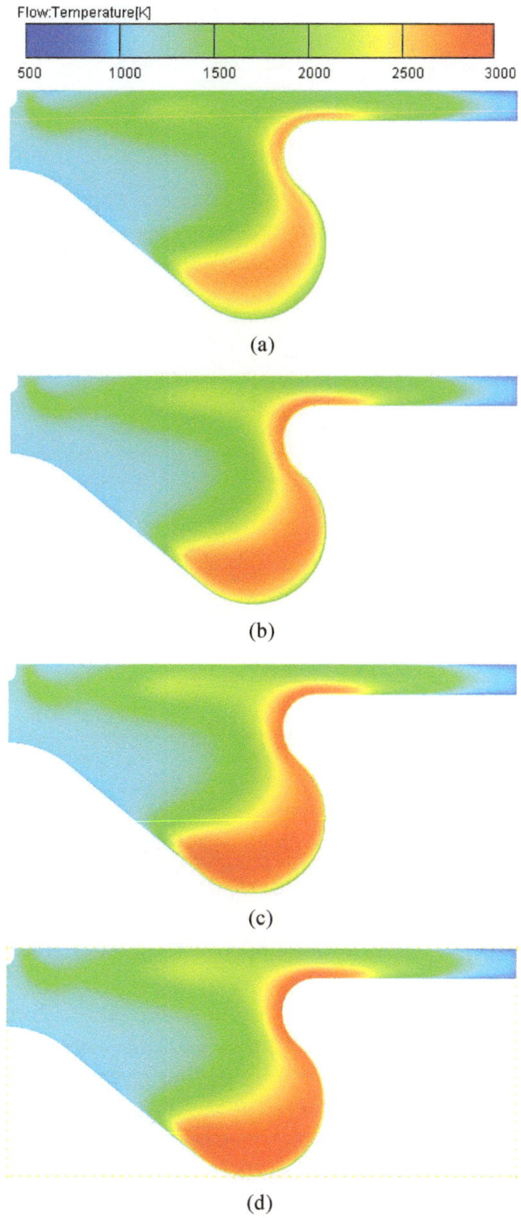

Figure 7. Temperature field in combustion chamber. (a) 13° CA ATDC (V_{O2} = 21%); (b) 13° CA ATDC (V_{O2} = 24%); (c) 13° CA ATDC (V_{O2} = 27%); (d) 13° CA ATDC (V_{O2} = 30%).

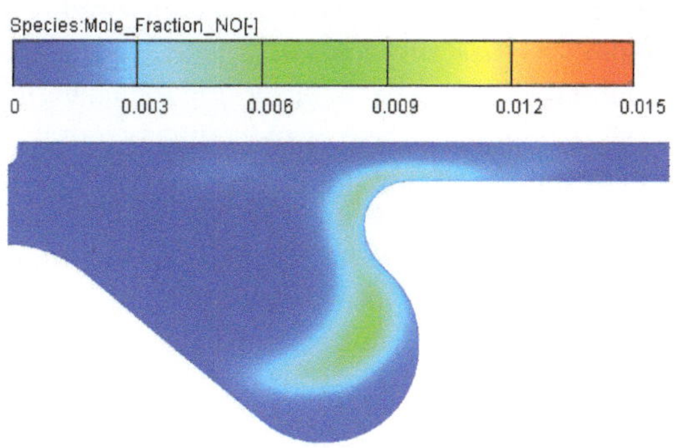

(a)

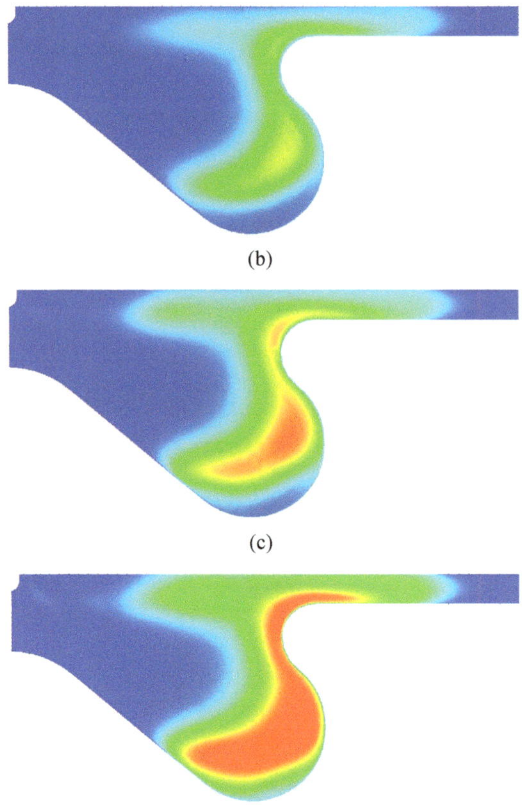

(b)

(c)

(d)

Figure 8. NO mass distribution in combustion chamber. (a) 13° CA ATDC (V_{O2} = 21%); (b) 13° CA ATDC (V_{O2} = 24%); (c) 13° CA ATDC (V_{O2} = 27%); (d) 13° CA ATDC (V_{O2} = 30%).

with the increasing of inlet oxygen concentration. Under oxygen enrichment, the formation rate of NO increases rapidly in the high temperature region.

Figure 9 shows the curve of the average mass fraction of the NO with the crank angle at different inlet oxygen concentrations. From the picture, it can be seen that with the increase of the oxygen concentration, the NO production volume increases sharply, and the curve is getting steeper. This is because the formation conditions of NO are high temperature, rich oxygen and the residence time in high temperature. See **Figure 7**, with the increase of the oxygen concentration, the range of N_2 in high temperature and oxygen enriched conditions becomes larger and the retention time is lengthened, so, the NO emissions increase greatly.

3.4. Influence of Oxygen Enriched Combustion on Soot Emissions

The molecular of the polycyclic aromatic hydrocarbon (PAH) generated by combustion is considered as the precursor of soot, so PAH oxidation reaction mechanism is used to predict soot emissions. However, the changing trend of A1 and soot generation is basically consistent [16], so the changes of soot formation in different oxygen concentrations can be qualitatively studied by observing the changing trend of A1.

As can be seen from **Figure 10**, with the increase of oxygen concentration, the formation of A1 is greatly reduced and the generation time in advance. Among them, when the oxygen concentration increased from 21% to 24%, oxygen enrichment had the most significant effect on the inhibition of A1 formation. This is because the increase of the intake oxygen enriched oxygen makes the fuel burning more fully and the lack of oxygen burning area significantly smaller, caused a decrement in the formation of soot.

4. Conclusions

Under the constant inlet total, with the increase of the inlet oxygen concentration, the peak of the pressure and the average temperature of the cylinder are increased, and the engine power of the diesel engine is enhanced. In

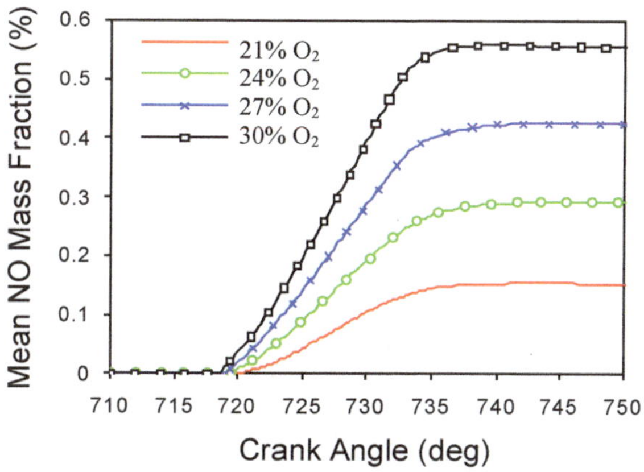

Figure 9. Influence of different O_2 concentration on mean NO mass fraction.

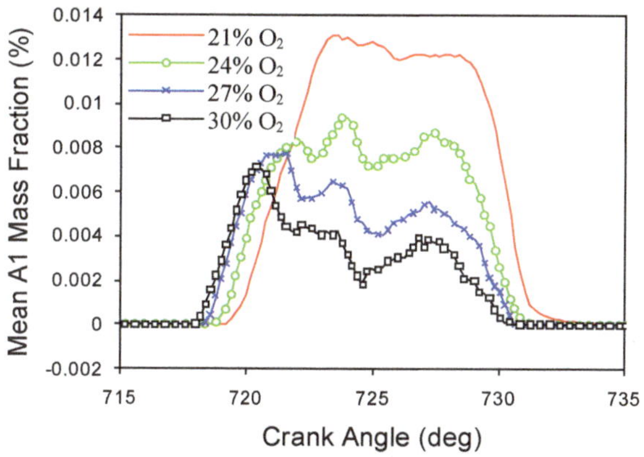

Figure 10. Influence of different O_2 concentration on mean A1 mass fraction.

addition, oxygen enriched combustion can improve the combustion efficiency, and the emissions of soot can decrease greatly. Thus, oxygen enriched combustion is an effective method to improve the diesel engine performance.

However, the performance of the diesel engine is not linear improvement with the increase of the intake oxygen concentration. When the oxygen concentration is over 24%, the effect of oxygen concentration on the combustion of the cylinder is gradually weakened. The cylinder wall heat loss increases greatly, which hinders the improvement of the effective thermal efficiency of the diesel engine. Therefore, the practical application of oxygen enriched combustion in diesel engine also needs to calibrate the best inlet oxygen concentration under different working conditions.

With the improvement of inlet oxygen concentration, the soot emissions are greatly reduced, however, the NO_x emissions increased sharply. Therefore, the future research emphasis should focus on the combination of oxygen enriched combustion and NO_x emissions control technology, in order to achieve the purpose of controlling the emissions of NO_x and soot of diesel engine simultaneously.

Acknowledgements

My deepest gratitude goes first and foremost to Professor Liu Weijun, my supervisor, for his considerable help by means of suggestion, comments and criticism. Without his painstaking efforts in revising and polishing my drafts, the completion of the present thesis would not have been possible. In addition, I would like to express my gratitude to all those who have helped me during the writing of this thesis.

References

[1] Ghojel, J., Hilliard, J.C. and Levendis, Y.A. (1983) Effect of Oxygen-Enrichment on the Performance of IDI Diesel Engines. SAE Paper 830245.

[2] Assanis, D.N., Baker, D., Sekar, R.R., Siambekos, C.T., Cole, R.L. and Marciniak, T.J. (1990) Simulation Studies of Diesel Engine Performance with Oxygen Enriched Air and Water Emulsified Fuels. CONF-900102-6, 1991.

[3] Desai, R.R., Gaynor, E., Watson, H.C. and Rigby, G.R. (1993) Giving Standard Diesel Fuels Premium Performance Using OxygenEnriched Air in Diesel Engines. SAE Technical Paper No. 932806.

[4] Lida, N., Suzuki, Y., Sato, G.T. and Sawada, T. (1986) Effects of Intake Oxygen Concentration on the Characteristics of Particulate Emissions from a D.I. Diesel Engine. Society of Automotive Engineers Paper 861233.

[5] Assanis, D.N., Poola, R.B., Sekar, R. and Cataldi, G.R. (2001) Study of Using Oxygen Enriched Combustion Air for Locomotive Diesel Engines. *Journal of Engineering for Gas Turbines and Power*, **123**, 157-166. http://dx.doi.org/10.1115/1.1290590

[6] Poola, R.B. and Sekar, R. (2003) Reduction of NO_x and Particulate Emissions by Using Oxygen-Enriched Combustion Air in a Locomotive Diesel Engine. *Journal of Engineering for Gas Turbines and Power*, **1252**, 524-533. http://dx.doi.org/10.1115/1.1563236

[7] Shu, G.Q., Zhao, W., Zhang, W. and Xu, B. (2012) The Effect of Oxygen-Fuel Ratio and Temperature of Mixed Gas on Soot and NO_x Emission and Combustion Path of Oxygen Enriched Diesel Engine. *Combustion Science and Technology*, **18**, 491-498.

[8] Perez, P.L. and Boehman, A.L. (2009) Experimental Study of Oxygen-Enriched Diesel Combustion Using Simulated Exhaust Gas Recirculation. *Journal of Engineering for Gas Turbines and Power*, **131**, Article ID: 042802.

[9] Udayakumar, R. and Meher, A.K. (2005) Use of Oxygen Enriched air in a Direct Injection Diesel Engine. Institution of Engineers (India), Kolkata, 156-159.

[10] Marr, W.W., Sekar, R.R., Cole, R.L., Marciniak, T.J. and Longman, D.E. (1993) Oxygen-Enriched Diesel Engine Experiments with a Low-Grade Fuel. SAE Paper 932805.

[11] Curran, H.J., Gaffuri, P. and Pitz, W.J. (1998) A Comprehensive Modeling Study of *n*-Heptane Oxidation. *Combustion and Flame*, **114**, 149-177. http://dx.doi.org/10.1016/s0010-2180(97)00282-4

[12] Curran, H.J., Gaffuri, P. and Pitz, W.J. (2002) A Comprehensive Modeling Study of Iso-Octane Oxidation. *Combustion and Flame*, **129**, 253-280. http://dx.doi.org/10.1016/S0010-2180(01)00373-X

[13] Wang, H., Reitz, R.D., Yao, M.F., Yang, B.B., Jiao, Q. and Qiu, L. (2013) Development of an *n*-Heptane-*n*-Butanol-PAH Mechanism and Its Application for Combustion and Soot Prediction. *Combustion and Flam*, **160**, 504-519. http://dx.doi.org/10.1016/j.combustflame.2012.11.017

[14] Huang, H.Z., *et al.* (2014) Effect of Inlet State on PAHs Generation of Low Temperature Combustion of Butanol-Diesel. *Journal of Internal Combustion Engine*, **32**, 401-406.

[15] Zhou, J.J. (1990) Numerical Calculation for Working Process of Diesel Engine. Dalian University of Technology Press, Dalian.

[16] Zhang, W., Shu, G.Q., Shen, Y.G., Zhao, W. and Xu, B. (2012) Use Contain PAH n-Heptane Simplified Mechanism of the Numerical Simulation of the Oxygen-Enriched Combustion on Diesel Engine. *Journal of Internal Combustion Engine*, **30**, 296-304.

26

Fuzzy Estimation Analysis of Photovoltaic Model Parameters

Helal Al-Hamadi

Department of Information Science, College of Computing Sciences and Engineering, Kuwait University, Kuwait, Kuwait
Email: helal3113@hotmail.com, helal.alhamadi@ku.edu.kw

Abstract

One of the challenges in accurately estimating Photovoltaic (PV) cell electric performance is the uncertainty of the model equivalent circuit parameters. The parameters considered in the study are the series resistance, shunt resistance, photo current, saturation current, and diode ideality factor. Parameter estimation for the PV cell equivalent circuit model is challenging due to the implicit transcendental relationship of the I-V characteristics of the cell. This paper presents a fuzzy logic based study for estimating the uncertainty of the cell parameters. The model parameters change with temperature and irradiance, are the source of uncertainties. Mathematical programming is used to estimate the fuzzy parameters. The approach is performed on practical data and the results of the analysis provide the estimation of the PV cell parameters. Results of this research yielding better estimated parameters compared with other methods using the Absolute Mean Error (AME).

Keywords

Photovoltaic Cell, Parameter Sensitivity, Fuzzy Logic, Diode Model, Modeling

1. Introduction

Renewable energy sources are gaining more interest in recent years and will be an increasingly important part of power generation in the coming years [1]. The photovoltaic (PV) system provides an alternative renewable energy resource due to its environment friendly natural source of energy. It generates electricity on-site in remote and isolated areas without transmission losses or greenhouse gas emissions. The electrical properties of the PV modules determine the cost and performance of PV plants. The use of accurate PV model parameters representing effective PV models is necessary for accurate performance evaluation and control. Due to degradation and ageing of the PV modules, the actual the actual values of module parameters dynamically differ from those provided by the manufacturer. Moreover, the PV cell is a semiconductor material which exhibits a non-linear transcendental characteristic together with the unpredictable of the operating conditions complicates the modeling process. An exact equivalent circuit and its associated mathematical model is essential for providing accurate I-V characteristics of the PV module. The estimation of the PV module parameters helps to design more efficient PV modules and predicting their performance and efficiency. Several models have been proposed to describe the solar cell characteristics [2]-[8]. A Double-diode solar cell model is considered in this project with the same identity factor. A lumped parameter equivalent circuit with to model the solar cell. The main parameters to simulate the I-V terminal characteristics of the solar cell are, photo current, saturation current, shunt resistance, series resistance, and diode identity factor. The electrical current produced by a solar cell depends on the intensity of the incident light and on its intrinsic properties. An accurate estimation of the cell parameters required for accurate performance evaluation [3].

Several methods are proposed in the literature to estimate model PV parameters. In [4] PV parameters estimation of the equivalent model current voltage characteristics using least squares fitting method is effectively applied with minimal errors. A Simulated annealing based approach is proposed in [5] for optimal estimation of solar cell model parameters. In [6], a numerical curve fitting procedures are used to estimate the cell model parameters: a non-linear two-point interval division was applied to extract local parameters from the current-voltage data at search measurement point. In [7] a methodology based on radial basis function neural networks to estimate the electrical characteristic curves of PV modules from the environmental operating conditions is presented. Particle swarm optimization (PSO) was effectively applied to extract the solar cell parameters in [9] [10]. A genetic algorithm utilizing isolated points obtained experimentally is used to determine the PV external current-voltage characteristic curve of a solar cell is presented in [11]. Using data commonly provided by panel manufacturers, measured environmental parameters, and semi-empirical equations, a theoretical approach is presented in [12] to evaluate the uncertainty of the PV cell model parameter. A new parameters extraction method based on the differential evolution technique is introduced [13]. The method fits the I-V curve better than genetic algorithm having a lower fitness value and faster execution time.

In [14], the adaptive neuro-fuzzy inference systems as an AI technique which combines the advantages of fuzzy logicand adaptive networks, is applied for PV parameters estimation. The adaptive neuro-fuzzy inference systems is also used in [15] for parameter estimation of one diode equivalent circuit model of PV modules.

The present paper introduces a fuzzy-based methodology for the parameters estimation of PV cell. Using the of the PV equivalent circuit, **Figure 1**, the PV cell five equivalent circuit parameters: series resistance, shunt resistance, diode photo current, diode reverse saturation current and diodeideality factor, are assumed to be a fuzzy parameters. Measurement I-V data of the PV cell are used to form a set of equations with fuzzy parameters. Least square methods are used to obtain the optimum parameters. The proposed technique could be applied to estimate the parameters of PV modules in order to account for ageing, performance degradation, and changes of operating conditions. The results obtained are compared with other techniques in the literature.

2. PV Modeling

Accurate estimation of the equivalent circuit parameters is important for control and performance evaluation of the PV cell. Manufacturer datasheet parameters are nominal values that are measured in factory under specific temperature, illumination, and weather conditions. The parameters values changes with time due to aging and the nonlinear transcendental nature of the PV cell. Therefore, for PC cells in different weather and degradation

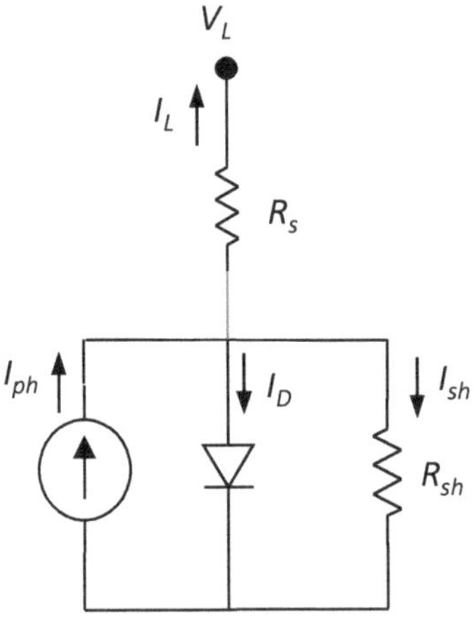

Figure 1. Equivalent circuit for a single diode lumped circuit.

conditions the parameters must be estimated to obtain more accurate values than that given in the nominal manufacturing datasheet.

Solar photovoltaic power generation employs solar panels also called modules that composed of a number of solar cells containing semi-conductor photovoltaic diode(s) which converts solar radiation into electric current. Mathematically, the solar photovoltaic cell is modeled by current voltage relationship (I-V) which exhibits a non-linear relationship due to the semi-conductor behavior of the cell. This (I-V) characteristic of the solar cell can be presented by a single diode model [2], [3] and [5] shown in **Figure 1**, and mathematically presented by Equation (1) and Equation (2).

$$I_L = I_{ph} - I_D - I_{sh} \tag{1}$$

$$I_L = I_{ph} - I_{SD}\left[e^{\left(\frac{q(V_L + I_L R_s)}{nkT}\right)} - 1\right] - \left[G_{sh}\left(V_L + I_L R_s\right)\right] \tag{2}$$

where $I_{ph}, I_{SD}, n, R_s, G_{sh} = (1/R_{sh})$ being the photocurrent, the diode saturation current, the diode identity factor, the series resistance and the shunt conductance, respectively. IL and VL are the terminal Current and voltage respectively; (q/kT) is the inverse thermal voltage, where k is Boltzmann's constant ($1.3806503 \times 10^{-23}$ J/K); q is the electronic charge ($1.602176565 \times 10^{-19}$ C) and T is the cell absolute temperature in Kelvin.

3. Fuzzy Logic Algorithm

The research idea is based on the fact that system parameters could be deduced if their values are bounded by using a fuzzy logic algorithm. Using measured data together with fuzzy logic a set of equations are formed with fuzzy input fuzzy parameters. Then Least squares are used to obtain the optimal estimated parameters based on the formulation in [18] and [19]. In this algorithm, PV input current and voltage has been taken as input, and error between the estimated values of the input current and the calculated one has been calculated as output. The flow chart of **Figure 2** represents the basic concept of the algorithm.

The equation of error and signal is given below as follows:

$$\Delta I_L = \left| I_L - \hat{I}_L \right| \tag{3}$$

where, I_L is the measured PV terminal current, and $\hat{I}_L$ is the estimated current.

A triangular shaped membership function is used for the input parameters. The range of the signal has been selected appropriate to each variable. **Figure 3** represents the graphical view of the membership function for R_s.

4. Results

Estimation of the five parameters of the PV module is illustrated in this section. Real photovoltaic solar module terminal data (V_{Li}, I_{Li}) for $i = 1, \ldots, N$ are considered in this testing [5]. The resulting parameters with comparison with previous work using other techniques are presented in **Table 1**. The comparison is based on the Absolute Mean Error (AME). For each method the estimated resulting parameters along with current the measured terminal are used to calculate the terminal voltage. The AME of the calculated and measured voltages are computed and presented in **Table 1**.

5. Conclusion

The paper presents a fuzzy-logic based technique to estimate the equivalent circuit parameters of PV cell. The PV cell five equivalent circuit parameters are modeled to be fuzzy parameters. Measurement I-V data of the PV

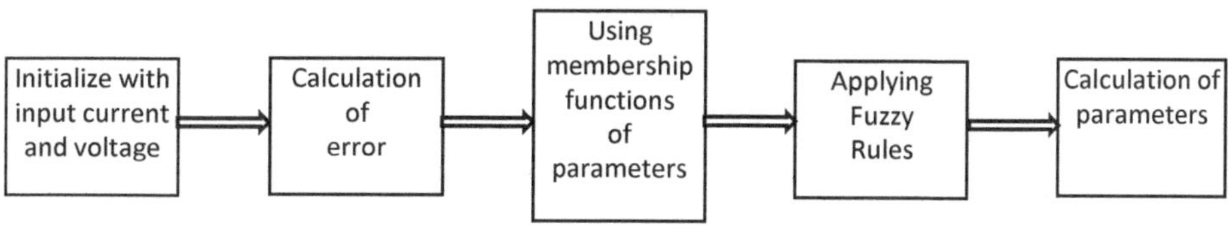

Figure 2. Fuzzy logic algorithm.

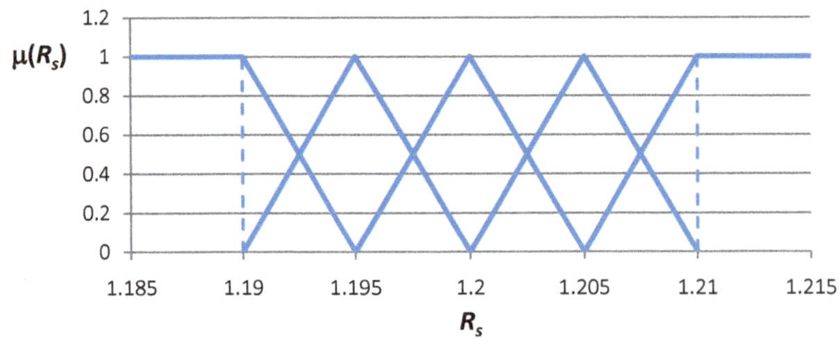

Figure 3. Graphical views of the membership function for R_s.

Table 1. Comparisons of estimated parameters by this technique and previous work.

	Ref. [16]	Ref. [17]	Ref. [5]	This work
I_{ph}(A)	1.0313	1.0339	1.0331	1.0341
I_{sd} (μA)	3.1756	3.076	3.6642	3.6391
R_s (Ω)	1.2053	1.203	1.1989	1.2019
G_{sh} (Ω^{-1})	0.0014	0.0018	0.0012	0 .0015
n	48.2889	48.1862	48.8211	48.1849
Error(MAE)	0.175367	0.183851	0.026887	0.016638

cell are used to form a set of equations with fuzzy parameters. Least square methods are used to obtain the optimum parameters. The proposed technique could be applied to estimate the parameters of PV modules in order to account for ageing, performance degradation, and changes of operating conditions. The results obtained are compared with other techniques in the literature yielding better estimated parameters compared with other methods using the Absolute Mean Error (AME).

Acknowledgements

The research is commissioned and financially supported by Kuwait University Research Sector under grant WI01/13.

References

[1] Renewables 2010 Global Status Report. 2010. http://www.rona.unep.org/documents/news/Renewables-GlobalStatusReport2010_pre-release_en_full.pdf

[2] Adamo, F., Attivissimo, F., Di Nisio, A. and Spadavecchia, M. (2011) Characterization and Testing of a Tool for Photovoltaic Panel Modeling. *IEEE Transactions on Instrumentation and Measurement*, **60**, 1613-1622. http://dx.doi.org/10.1109/TIM.2011.2105051

[3] Villalva, M.G., Gazoli, J.R. and Filho, E.R. (2009) Comprehensive Approach to Modeling and Simulation of Photovoltaic Arrays. *IEEE Transactions on Power Electronics*, **24**, 1198-1208. http://dx.doi.org/10.1109/TPEL.2009.2013862

[4] Ikegami, T., Maezono, T., Nakanishi, F., Yamagata, Y. And Ebihara, K. (2001) Estimation of Equivalent Circuit Parameters of PV Module and Its Application to Optimal Operation of PV System. *Solar Energy Materials and Solar Cells*, **67**, 389-395. http://dx.doi.org/10.1016/S0927-0248(00)00307-X

[5] El-Naggar, K.M., Al Rashidi, M.R., Al Hajri, M.F. and Al-Othman, A.K. (2012) Simulated Annealing Algorithm for Photovoltaic Parameters Identification. *Solar Energy Journal*, **86**, 266-274.

[6] Munji, M.K., Okullo, W., Van Dyk, E.E. and Vorster, F.J. (2010) Local Device Parameter Extraction of a Concentrator Photovoltaic Cell under Solar Spot Illumination. *Solar Energy Materials and Solar Cells Journal*, **94**, 2129-2136. http://dx.doi.org/10.1016/j.solmat.2010.06.041

[7] Bonanno, F., Capizzi, G., Graditi, G., Napoli, C. and Tina, G.M. (2012) A Radial Basis Function Neural Network

Based Approach for the Electrical Characteristics Estimation of Aphotovoltaic Module. *Applied Energy*, **97**, 956-961. http://dx.doi.org/10.1016/j.apenergy.2011.12.085

[8] Xiao, W., Lind, M.G.J., Dunford, W.G. and Capel, A. (2006) Real-Time Identification of Optimal Operating Points in Photovoltaic Power Systems. *IEEE Transactions on Industrial Electronics*, **53**, 1017-1026. http://dx.doi.org/10.1109/TIE.2006.878355

[9] Ye, M., Wang, X. and Xu, Y. (2009) Parameter Extraction of Solar Cells Using Particle Swarm Optimization. *Journal of Applied Physics*, **105**, 094502-094508. http://dx.doi.org/10.1063/1.3122082

[10] Qin, H.S. and Kimball, J.W. (2011) Parameter Determination of Photovoltaic Cells from Field Testing Data Using Particle Swarm Optimization. *IEEE Power and Energy Conference at Illinois* (*PECI*), 1-4.

[11] Petcuţ, F.M. and Dragomir, T.L. (2010) Solar Cell Parameter Identification Using Genetic Lgorithms. *Journal of Control Engineering and Applied Informatics*, **12**, 30-37.

[12] Attivissimo, F., Di Nisio, A., Savino, M. and Sapada Vecchia, M. (2012) Uncertainty Analysis in Photovoltaic Cell Parameter Estimation. *IEEE Transactions on Instrumentation and Measurement*, **61**, 1334-1342.

[13] Ishaque, K., Salam, Z., Taheri, H. and Shamsudin, A. (2011) Parameter Extraction of Photovoltaic Cell Using Differential Evolution Method. *IEEE Applied Power Electronics Colloquium* (*IAPEC*), 10-15.

[14] Kalika, S., Rajaji, L. and Gupta, S. (2012) Intelligent Technique Based Modeling for PVPS. *International Journal of Engineering and Innovative Technology* (*IJEIT*), **2**, 211-215.

[15] Kulaksız, A.A. (2011) ANFIS-Based Parameter Estimation of One Diode Equivalent Circuit Model of PV Modules. *Proc. IEEE International Symposium on Computational Intelligence and Informatics* (*CINTI*), 415-420.

[16] Easwarakhanthan, T., Bottin, J., Bouhouch, I. and Boutrit, C. (1986) Nonlinear Minimization Algorithm for Determining the Solar Cell Parameters with Microcomputers. *International Journal of Solar Energy*, **4**, 1-12. http://dx.doi.org/10.1080/01425918608909835

[17] Bouzidi, K., Chegaar, M. and Nehaoua, N. (2007) New Methods to Extract the Parameters of Solar Cells from Their Illuminated I-V Curve. 4*th International Conference on Computer Integrated Manufacturing Setif*, Algeria.

[18] Sherali, H.D. and Alameddine, A. (1992) A New Reformulation-Linearization Technique for Bilinear Programming Problems. *Journal of Global Optimization*, **2**, 379-410. http://dx.doi.org/10.1007/BF00122429

[19] Sherali, H.D. (2001) On Mixed-Integer Zero-One Representations for Separable Lower-Semicontinuous Piecewise-Linear Function. *Operation Research Journal*, **28**, 155-160. http://dx.doi.org/10.1016/s0167-6377(01)00063-3

27

Modeling of Criteria Air Pollutant Emissions from Selected Nigeria Petroleum Refineries

Temitayo Oladimeji[1], Jacob Sonibare[2], Moradeyo Odunfa[3], Augustine Ayeni[1]

[1]Chemical Engineering Department, Covenant University, Ota, Nigeria
[2]Chemical Engineering Department, Obafemi Awolowo University, Ile-Ife, Nigeria
[3]Mechanical Engineering Department, University of Ibadan, Ibadan, Nigeria
Email: temitayo.fatoki@covenantuniversity.edu.ng

Abstract

The dispersion models were used to estimate or predict the concentration of air pollutants or toxins emitted from sources such as industrial plants, vehicular traffic or accidental chemical releases. In this study, the Industrial Source Complex Short Term (ISCST3) emission dispersion model was used to measure the ground level concentration of criteria air pollutants within 50 km radius of location. This model considered emissions from major point sources of pollutants in four existing and twenty-three proposed Nigeria petroleum refineries. The obtained ground level concentration for 24-hr averaging periods of the criteria air pollutants at sensitive receptor around each of the refineries was compared with the National Ambient Air Quality Standards (NAAQS) of Nigeria, World Bank and World Health Organization (WHO) to determine their level of compliance. The highest ground level concentration predicted to be 450 - 1875 μg/m^3 for 24-h averaging period was obtained at Tonwei Oil Refinery, Ekeremor Local Government, Bayelsa State, while the lowest ground level concentration predicted to be 0.0099 - 0.1 μg/m^3 for 24-h averaging period was obtained at Amakpe International Refinery, Eket Local Government, Akwa Ibom State. Percentage set limits of criteria air pollutants ranging from 0.02% to 94.5% are within the set standard limits and no health risk is associated with areas around the plant's locations while percentage set limits of criteria pollutants ranging from 1.1 to 55.6 folds of the standards exceed the maximum permitted limits, hence affecting areas around the plants. The air quality standards guiding petroleum refinery emissions must be strictly considered, in order to ensure that the ground level concentrations do not exceed the required standard limits and prevent the adverse effects of air pollution in the Nigeria airshed.

Keywords

Air Quality Standards, Ground Level Concentration, Nigeria Petroleum Refinery

1. Introduction

Each year industrial facilities discharge pollutants into the environment, releasing large amounts of pollutants leading to respiratory, neurological and reproductive disorders, and cancers [1]. Nigeria is presently a major emitter of criteria air emissions: carbon monoxide (CO), oxides of nitrogen (NO_x), particulate matter less than 10.0 microns in diameter (PM_{10}), sulphur dioxide (SO_2) and volatile organic compounds (VOCs) which comes from number of sources within a petroleum refinery including equipment leaks (from valves or other devices), high-temperature combustion processes, etc [2]. The Industrial Source Complex Short Term (ISCST3) air dispersion modeling tool is a steady state Gaussian plume model which can be used to assess pollutant concentrations, and/ or deposition fluxes from a wide variety of sources are associated with an industrial source complex [3]. It was used to predict the ground level concentrations of criteria air pollutants from the four major existing and twenty-three proposed petroleum refineries in Nigeria, within 50 km radius of location from point sources of emission in the refineries, for the prevailing meteorological conditions assuming a "worst case scenario" *i.e.* 100% operation rate. The model is used to determine whether existing or proposed new industrial facilities are or will be in compliance with the National Ambient Air Quality Standards (NAAQS). Air pollution models constitute a set of formulae that take into account sources of pollution in a given area, the amounts of pollutants emitted by each source, chemical reaction transformations, different meteorological conditions and topographical features, and other factors that affect dispersion of pollutants. [4] defined the process of computer modeling of air pollution as the ability to use computers to solve basic equations that described the dynamics of meteorological and air pollution phenomena, and ultimately, the adverse effects of air pollution. The basis for most of those models was the complete equation for Gaussian Dispersion Modeling of continuous and buoyant air pollution plumes [5]. Previous studies on investigation on emissions using ISCST3 include [5]-[7]. This study is carried out to investigate the atmospheric conditions of the petroleum refineries in Nigeria and identify the environmental impacts of emissions of criteria pollutants in the area of influence. Determinations of their ground level concentrations and comparison with the National Ambient Air Quality Standards (NAAQS) of Nigeria, World Bank and World Health Organization (WHO) will indicate the level of air quality management required around Nigerian petroleum refineries.

2. Study Areas

As shown in **Figure 1**, Nigeria has 4 existing refineries; Kaduna Refining and petrochemical company (KPRC) located at Chikun Local Government of Kaduna State operating at 110,000 bbl/day; Warri Refining and petrochemical company (WRPC) located at Warri southwest local government of Nigeria's Delta State with an installed capacity of 125,000 bbl/day; Port Harcourt Refining Company (PHRC) has two refineries at Alesa-Eleme, near Port Harcourt in Nigeria's Southern Province, Rivers State. Port Harcourt I refinery has installed capacity of 60,000 bbl/day while Port Harcourt II refinery is the complex new refinery with installed capacity of 150,000 bbl/day [8] [9]. Due to scarcity of petroleum products, 23 proposed refineries have been licensed in Nigeria to ensure availability of the products (**Figure 1**), which includes the following with their installed capacities; Resources refinery and petrochemical Limited located in Ikot Abasi local government area of Akwa Ibom state (100,000 bbl/day); Amakpe International Refinery, Eket local government of Akwa Ibom state with production capacity of (12,000 bbl/day); Rehoboth Natural Resources Limited (12,000 bbl/day), Immingiri, Yenekoa local government of Bayelsa state; Amexum Corporation (100,000 bbl/day), Akpabuyo local government of Cross river state; Antonio Oil (27,000 bbl/day), Ipokia local government of Ogun state; Gasoline Associates International Limited Refinery (100,000 bbl/day), Ipokia local government of Ogun state; Ologbo Refinery Company Nigeria Limited (12,000 bbl/day), Ikpoba Okha local government of Edo state; Ode Aye Refinery Limted (100,000 bbl/day), Okiti pupa local government area of Ondo state; Orient Petroleum Resources Limited (55,000 bbl/day), Anambra West local government area of Anambra state; Qua Petroleum Refinery Limited (100,000 bbl/day), Ibeno local government of Akwa Ibom state; ChaseWood Consortium Nigeria Limited (70,000 bbl/day), Eket local government area of Akwa Ibom state; Clean Water Refinery (60,000 bbl/day), Eleme local government area of Rivers state; Niger Delta Refinery & Petrochemical Limited (100,000 bbl/day), Warri-South local government area of Delta state; NSP Refineries and Oil Services Limited (120,000 bbl/day), Andoni local government area of Rivers state; RIVGAS Petroleum and Energy Limited (30,000 bbl/day), Port Harcourt local government area of Rivers state; Sapele Refinery Limited located in Okpe-Sobo, Sapele local government area of Delta state (100,000 bbl/day); Starrex Petroleum Refinery (100,000 bbl/day), Eleme local government area of

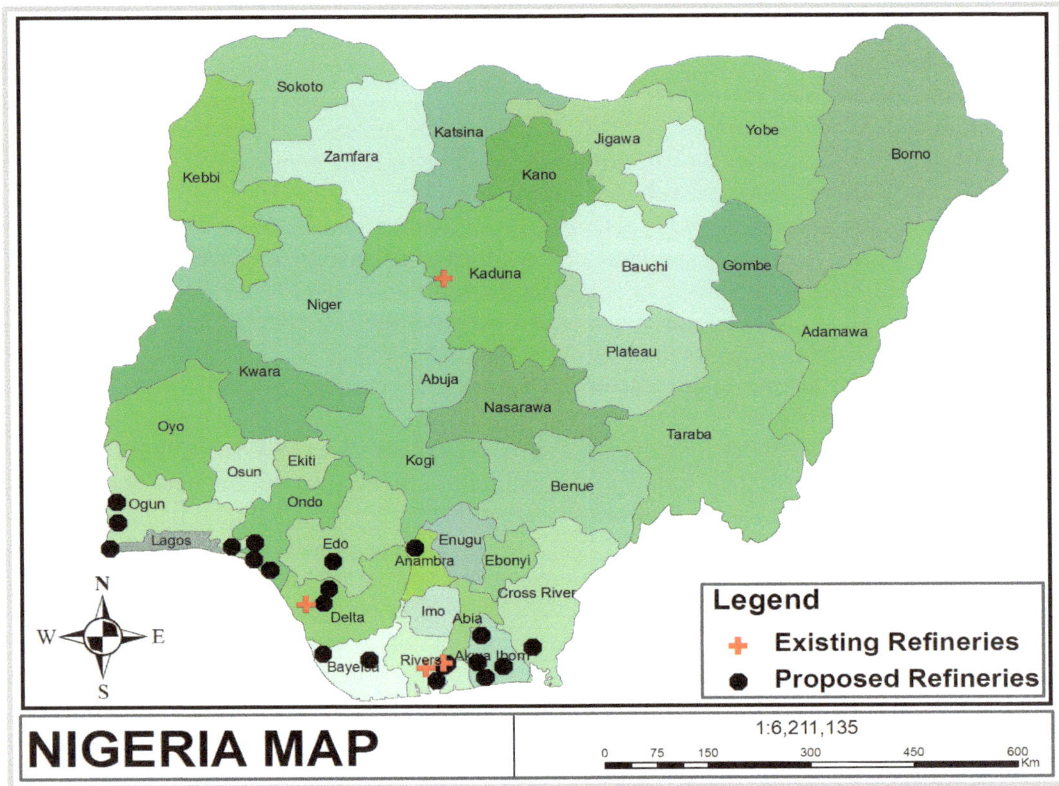

Figure 1. Map of Nigeria showing the existing and proposed Refineries.

Rivers state; Total Support Limited (12,000 bbl/day), Calabar Mnincipal local government area of Cross River state; Union Atlantic Petroleum Limited (100,000 bbl/day), Badagry local government area of Lagos state; Tonwei Oil Refinery (200,000 bbl/day), Ekeremor local government area of Bayelsa state; Ilaje Refinery and Petrochemicals (100,000 bbl/day), Ilaje local government area of Ondo state; Southwest Refinery and Petrochemical Company (100,000 bbl/day) is located in Ogun Waterside local government area of Ogun state; Owena Oil and gas Limited (100,000 bbl/day), Ilaje local government area of Ondo state [8]-[10].

3. Methodology

ISCST3 model uses steady state Gaussian plume equation to model emissions from point sources, including flares, process heaters, steam boilers and fluid catalytic cracking unit. This model considered emissions from the stacks of existing refineries and proposed petroleum refineries in Nigeria. Criteria air pollutants modeled for the ground level concentration include CO, NO_X, PM_{10}, SO_2 and VOC. The standards of ambient air quality as reported by NAAQS,WB and WHO is shown in **Table 1**. Major sources of emissions and parameters used as input to the air dispersion modeling for each of the refineries are shown in **Table 2** and **Table 3**. Some proposed refineries have similar operation rate, hence will have the same emission sources and dispersion model parameter. The ISC-AERMOD View air dispersion modeling tool developed specially for Microsoft Windows and runs under Windows 95/98/Me/NT/2000 and XP was employed in this study. This is a complete and powerful air dispersion modeling package which incorporates the popular U.S. EPA models into one interface: AERMOD, ISCST3, and ISC-PRIME. ISCST3 dispersion model interfaces [1]. The ISCST3 interface was used in this study which uses pathways that compose the run stream file as the basics for its functional organization. These pathways include:

i) Control Pathway (CO): Where the modeling scenario (worst case scenario *i.e.* 100% operation rate), the pollutants (CO, NO_X, SO_2, VOC and PM) and the averaging periods (24-hr) are specified.

ii) Source Pathway (SO): Where the sources of pollutant emissions (point sources of pollutants in the refinery-process heaters, fluid catalytic cracking units, steam boilers and flares) are defined.

Table 1. Standards of ambient air quality.

S/N	Pollutants	Averaging Period	Maximum Concentration ($\mu g/m^3$)		
			NAAQS	World Bank	WHO
1	CO	24-Hr	11,400	-	-
2	NO_x	24-Hr	75	150	-
3	SO_2	24-Hr	26	-	20
4	PM	24-Hr	250	80	50
5	VOC	24-Hr	6000	-	-

Source: FEPA (1991) [11]; World Bank (1998) [12]; WHO (2005) [13].

Table 2. Emission sources and dispersion model parameters for existing refineries.

Plant Name	Stack	Discharge Temperature (k)	Stack Diameter (m)	Exit Velocity (m/s)	Release Height (m)	[c]Emission (g/s)				
						PM_{10}	SO_2	NO_X	VOC	CO
Kaduna Refining & Petrochemical Company (KRPC) 110,000 BPSD	[a]Flare 1	1540	6.6	4.52	66	-	-	0.001	0.003	0.005
	Flare 2	1540	6.6	4.52	66	-	-	0.001	0.003	0.005
	Flare 3	1540	6.6	4.52	66	-	-	0.001	0.003	0.005
	[b]FCCU	330	2.75	23.47	67	-	54.5	7.9	24.3	1510.4
	[b]Boiler 1	550	2.18	9.05	30	-	304.1	1420.1	-	106
	Boiler 2	550	2.18	9.05	30	-	304.1	1420.1	-	106
	Boiler 3	550	2.18	9.05	30	-	304.1	1420.1	-	106
	[b]Heater 1	660	2.01	8.03	43	2.1	45.1	173.2	1.0	15.8
	Heater 2	660	2.01	8.03	43	2.1	45.1	173.2	1.0	15.8
Port Harcourt Refining Company I (PHRC) 60,000 BPSD	[a]Flare 1	840	1.1	2.47	36	-	-	0.0003	0.0006	0.002
	Flare 2	840	1.1	2.47	36	-	-	0.0003	0.0006	0.002
	[b]FCCU	180	1.5	12.8	37	-	12.5	2.0	5.6	347.7
	[b]Boiler 1	300	1.19	4.94	16	-	59.6	278.7	-	20.8
	Boiler 2	300	1.19	4.94	16	-	59.6	278.7	-	20.8
	[b]Heater 1	360	1.1	4.38	24	0.6	13.4	51.6	0.3	4.7
	Heater 2	360	1.1	4.38	24	0.6	13.4	51.6	0.3	4.7
Port Harcourt Refining Company II (PHRC) 150,000 BPSD	[a]Flare 1	2100	2.75	6.17	90	-	-	0.002	0.004	0.01
	Flare 2	2100	2.75	6.17	90	-	-	0.002	0.004	0.01
	Flare 3	2100	2.75	6.17	90	-	-	0.002	0.004	0.01
	Flare 4	2100	2.75	6.17	90			0.002	0.004	0.01
	[b]FCCU	450	3.75	32.01	92	-	103.5	14.9	46.3	2877
	[b]Boiler 1	750	2.97	12.35	41	-	267.8	1250.7	-	93.4
	Boiler 2	750	2.97	12.35	41	-	267.8	1250.7	-	93.4
	[b]Boiler 1	750	2.97	12.35	41	-	267.8	1250.7	-	93.4
	Boiler 2	750	2.97	12.35	41	-	267.8	1250.7	-	93.4
	[b]Heater 1	900	2.75	10.95	59	3.7	80.2	308.1	1.7	28
	Heater 2	900	2.75	10.95	59	3.7	80.2	308.1	1.7	28

Continued

Warri Refining & Petrochemical Company (WRPC) 125,000 BPSD	[a]Flare 1	1750	2.29	5.13	75	-	-	0.001	0.003	0.007
	Flare 2	1750	2.29	5.13	75	-	-	0.001	0.003	0.007
	Flare 3	1750	2.29	5.13	75	-	-	0.001	0.003	0.007
	Flare 4	1750	2.29	5.13	75			0.001	0.003	0.007
	[b]FCCU	375	3.13	26.68	76	-	67.3	9.7	30	1870.1
	[b]Boiler 1	625	2.48	10.29	34	-	228.8	1068.3	-	79.8
	Boiler 2	625	2.48	10.29	34	-	228.8	1068.3	-	79.8
	Boiler 3	625	2.48	10.29	34		228.8	1068.3	-	79.8
	[b]Heater 1	750	2.29	9.13	49	2.9	61.3	253.3	1.3	21.4
	Heater 2	750	2.29	9.13	49	2.9	61.3	253.3	1.3	21.4

Source: [a]Abdul-Wahab *et al.* (2001) [14]; [b]Stell (2000) [15]; [c]USEPA 1989 [16].

Table 3. Emission sources and dispersion model parameters for proposed refineries.

Plant Name	Stack	Discharge Temperature (k)	Stack Diameter (m)	Exit Velocity (m/s)	Release Height (m)	[c]Emission (g/s)				
						PM_{10}	SO_2	NO_X	VOC	CO
Resource Refinery & Petrochemical, Onne 100,000BPSD	[a]Flare 1	1400	1.83	4.11	60	-	-	0.001	0.002	0.004
	Flare 2	1400	1.83	4.11	60	-	-	0.001	0.002	0.004
	Flare 3	1400	1.83	4.11	60	-	-	0.001	0.002	0.004
	[b]FCCU	300	2.5	21.34	61	-	34.7	5.0	15.6	965.9
	[b]Boiler 1	500	1.98	8.23	27	-	165.7	773.9	-	57.8
	Boiler 2	500	1.98	8.23	27	-	165.7	773.9	-	57.8
	[b]Heater 1	600	1.83	7.3	39	1.7	37.3	143.2	0.8	13
	Heater 2	600	1.83	7.3	39	1.7	37.3	143.2	0.8	13
NSP Refineries & Oil services Limited, Andoni 120,000BPSD	[a]Flare 1	1680	2.2	4.93	72	-	-	0.001	0.003	0.006
	Flare 2	1680	2.2	4.93	72	-	-	0.001	0.003	0.006
	Flare 3	1680	2.2	4.93	72	-	-	0.001	0.003	0.006
	Flare 4	1680	2.2	4.93	72			0.001	0.003	0.006
	[b]FCCU	360	3.0	25.61	73	-	50.1	7.1	22.3	1390.9
	[b]Boiler 1	600	2.38	10.0	33	-	238.6	1114.4	-	83.2
	Boiler 2	600	2.38	10.0	33	-	238.6	1114.4	-	83.2
	[b]Heater 1	720	2.2	8.76	47	2.5	53.7	206.1	1.1	18.7
	Heater 2	720	2.2	8.76	47	2.5	53.7	206.1	1.1	18.7
Clean Water Refinery, Eleme (60,000BPSD	[a]Flare 1	840	1.1	2.47	36	-	-	0.0003	0.0006	0.002
	Flare 2	840	1.1	2.47	36	-	-	0.0003	0.0006	0.002
	[b]FCCU	180	1.5	12.8	37	-	12.5	2.0	5.6	347.7
	[b]Boiler 1	300	1.19	4.94	16	-	59.6	278.7	-	20.8
	Boiler 2	300	1.19	4.94	16	-	59.6	278.7	-	20.8
	[b]Heater 1	360	1.1	4.38	24	0.6	13.4	51.6	0.3	4.7
	Heater 2	360	1.1	4.38	24	0.6	13.4	51.6	0.3	4.7

Continued

RIVGAS Petroleum & Energy Limited, Port Harcourt (30,000BPSD)	[a]Flare 1	420	0.55	1.23	18	-	-	0.00007	0.00002	0.0004
	Flare 2	420	0.55	1.23	18	-	-	0.00007	0.00002	0.0004
	[b]FCCU	90	0.75	6.4	18	-	3.1	0.4	1.4	87
	[b]Boiler 1	150	0.59	2.47	8	-	14.9	69.6	-	5.2
	Boiler 2	150	0.59	2.47	8	-	14.9	69.6	-	5.2
	[b]Heater 1	180	0.55	2.19	12	0.2	3.4	12.9	0.07	1.2
	Heater 2	180	0.55	2.19	12	0.2	3.4	12.9	0.07	1.2
Towei Oil Refinery 200,000 BPSD	[a]Flare 1	2800	3.66	8.22	120	-	-	0.11	0.20	0.70
	Flare 2	2800	3.66	8.22	120	-	-	0.11	0.20	0.70
	Flare 3	2800	3.66	8.22	120	-	-	0.11	0.20	0.70
	Flare 4	2800	3.66	8.22	120	-		0.11	0.20	0.70
	[b]FCCU	600	5.0	42.68	122	-	2158.7	312.85	962.9	60,006
	[b]Boiler 1	1000	3.96	16.46	54.3	-	23,040	107605.4	-	8,033.4
	Boiler 2	1000	3.96	16.46	54.3	-	23,040	107605.4	-	8,033.4
	Boiler 3	1000	3.96	16.46	54.3	-	23,040	107605.4	-	8,033.4
	Boiler 4	1000	3.96	16.46	54.3	-	23,040	107605.4	-	8,033.4
	[b]Heater 1	1200	3.66	14.6	78.6	229.86	5183.0	19904.66	107.7	1,811.1
	Heater 2	1200	3.66	14.6	78.6	229.86	5183.0	19904.66	107.7	1,811.1
Orient Petroleum Resource Limited 55,000 BPSD	[a]Flare 1	770	1.01	2.26	33	-	-	0.001	0.03	0.04
	Flare 2	770	1.01	2.26	33	-	-	0.001	0.03	0.04
	[b]FCCU	165	1.38	11.74	34	-	368.48	55.62	163.3	10,167
	[b]Boiler 1	275	1.09	4.53	15	-	1741.5	8137.76	-	608.33
	Boiler 2	275	1.09	4.53	15	-	1741.5	8137.76	-	608.33
	[b]Heater 1	330	1.01	4.02	22	17.38	392.81	1505.19	10.43	139.05
	Heater 2	330	1.01	4.02	22	17.38	392.81	1505.19	10.43	139.05
Amakpe Internatna Refinery 12,000 BPSD	[a]Flare 1	168	0.22	0.49	7		-	0.0004	0.001	0.003
	Flare 2	168	0.22	0.49	7		-	0.0004	0.001	0.003
	[b]FCCU	36	0.3	2.56	7		13.70	2.43	10.43	483.19
	[b]Boiler 1	60	0.24	0.98	3		560.05	2602.24	-	209.24
	Boiler 2	60	0.24	0.98	3		560.05	2602.24	-	209.24
	[b]Heater 1	72	0.22	0.88	5	1.05	1.05	73.00	0.35	0.69
	Heater 2	72	0.22	0.88	5	1.05	1.05	73.00	0.35	0.69
Chase wood Consortiu Nigeria Limited 70,000 BPSD	[a]Flare 1	980	1.28	2.88	42	-	-	0.01	0.03	0.07
	Flare 2	980	1.28	2.88	42	-	-	0.01	0.03	0.07
	[b]FCCU	210	1.75	14.94	43	-	590.95	83.42	264.1	13.90
	[b]Boiler 1	19	350	1.39	19	-	2930.4	13678.8	-	983.8
	Boiler 2	19	350	1.39	19	-	2930.4	13678.8	-	983.8
	[b]Heater 1	420	1.28	5.11	28	31.28	636.14	2436.81	13.90	222.48
	Heater 2	420	1.28	5.11	28	31.28	636.14	2436.81	13.90	222.48

Source: [a]Abdul-Wahab *et al.* (2001); [b]Stell (2000); [c]USEPA (1989).

iii) Receptor Pathway (RE): where the receptors which determine the air quality impact at specific locations. Specifically, 50 km radius within each refinery was given adequate attention.

iv) Meteorological Pathway (ME): Where the atmospheric condition of the areas being modeled is defined, so it can be taken into account when determining the distribution of air pollution impacts for the area. Due to lack of upper air observations in Nigeria airport, meteorological database for the project location was combined using data from Cotonou meteorological observations. Cotonou has winds having prevalence for a southwesterly direction which is consistent with the winds observed in the refineries.

v) Terrain Grid Pathway (TG): Where the option of gridded terrain data to be used in calculating dry depletion in elevated or complex terrain is specified. Elevated terrain is relevant to this study.

(vi) Output Pathway (OU): Where you define which output results are necessary to meet the needs of the air quality modeling analysis. The resultant output is the ground level concentration of pollutants as shown in **Figure 2** and **Figure 3**.

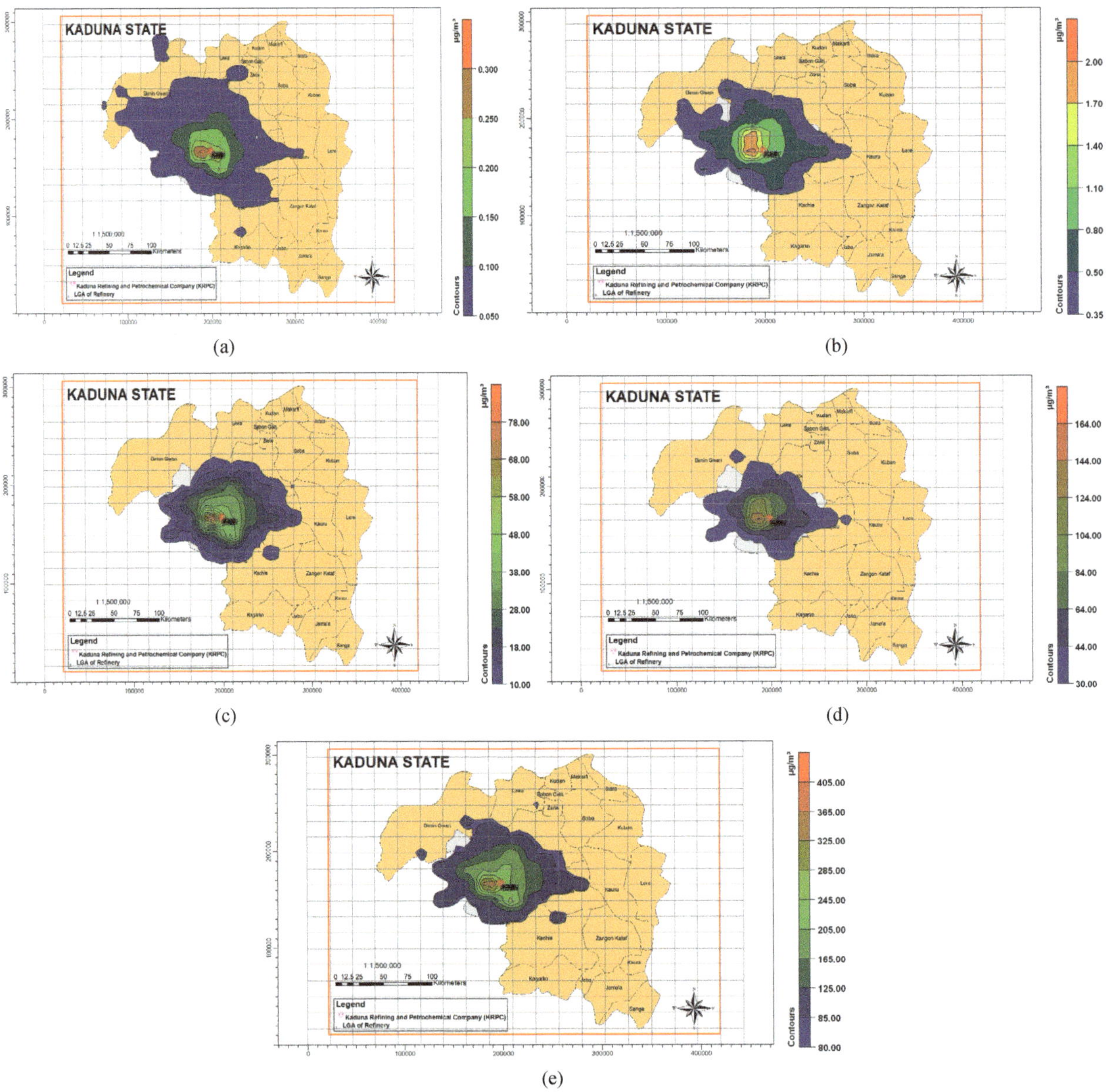

Figure 2. Predicted 24-hr Ground Level Concentrations of Criteria Pollutants from Kaduna Refining and Petrochemical Company, Kaduna State. (a) Predicted 24-hr PM_{10} Concentrations; (b) Predicted 24-hr VOC Concentrations; (c) Predicted 24-hr SO_2 Concentrations; (d) Predicted 24-hr NO_X Concentrations; (e) Predicted 24-hr CO Concentrations.

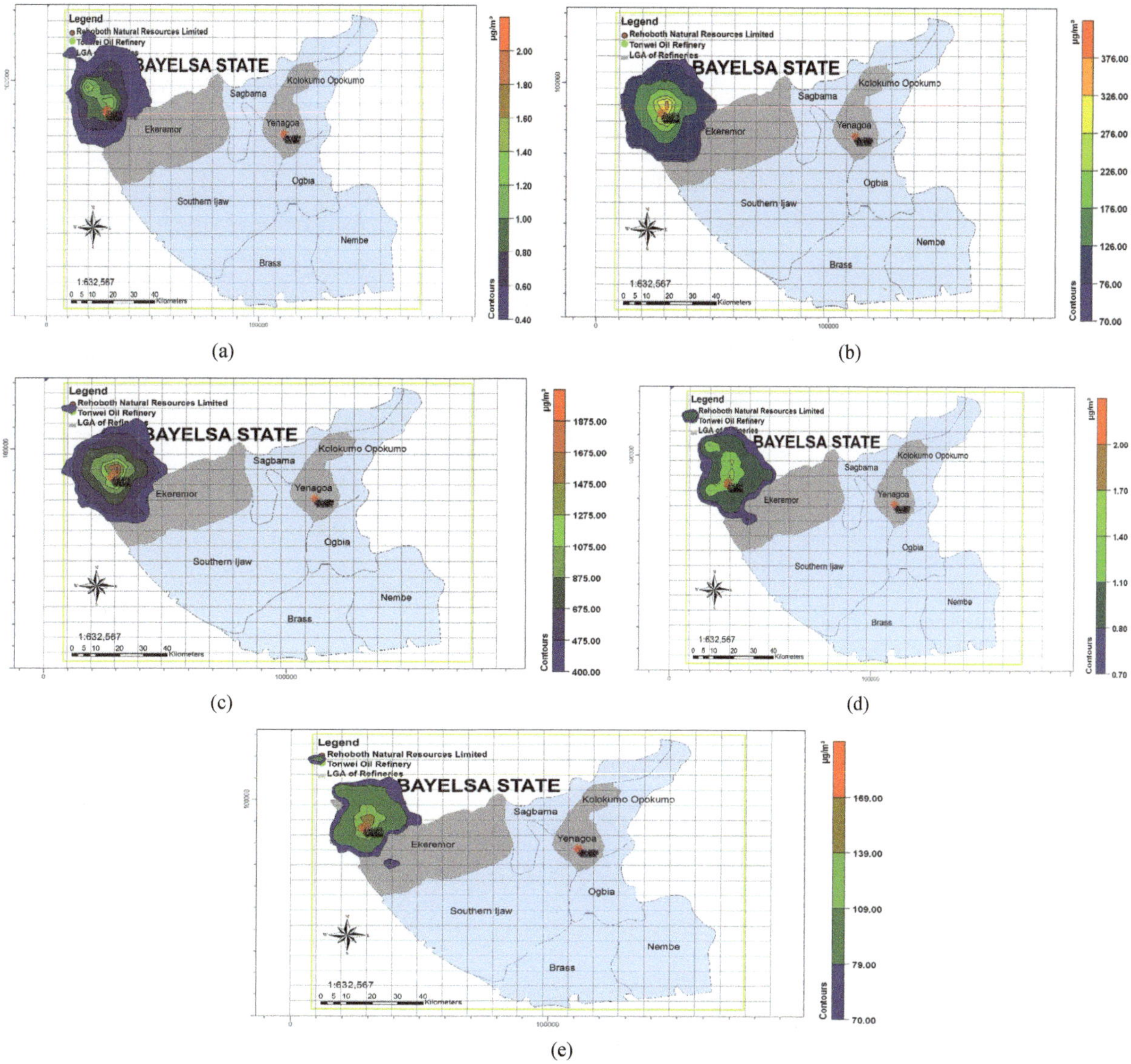

Figure 3. Cumulative 24-hr Ground Level Concentrations of Criteria Pollutants from Bayelsa state refineries. (a): Cumulative 24-hr PM_{10} Concentrations; (b) Cumulative 24-hr SO_2 Concentrations; (c) Cumulative 24-hr NO_X Concentrations; (d) Cumulative 24-hr VOC Concentrations; (e) Cumulative 24-hr SO_2 Concentrations.

From the existing and proposed refineries, the ground level concentration (glc) of the criteria air pollutant were computed and compared with the standards of ambient air quality derived from National Ambient Air Quality Standards (NAAQS) of Federal Ministry of Environment (FMENV), that of the World Bank environmental guidelines and World Health Organization (WHO) standards to determine their level of compliance (**Table 4** and **Table 5**).

4. Results and Discussion

An air quality impact on refineries is important to assess the impacts of emissions from petroleum refinery operations on their environment. **Table 2** and **Table 3** shows the emission sources and dispersion model parameters for the existing and proposed refineries these include the emission estimate of criteria pollutants from point sources (FCCU, Process heaters, Flares, Steam boilers) of emissions from the refineries. The dispersion modeling outputs from the ISC-AERMOD View (for the existing and proposed refineries in Nigeria) for 1-hour, 8-hour, 24-hour and annual averaging periods for the criteria pollutants are shown in **Figure 2** and **Figure 3**

Table 4. 24-hr averaging period maximum ground level concentration of criteria pollutants and % set limits in existing refineries.

S/N	Refinery	Air Pollutant	Maximum glc ($\mu g/m^3$)	Maximum Impact Local Government Area	% Set Limit of FMENV, WB, WHO
1	Kaduna Refining & Petrochemical Company (KRPC) 110,000 BPSD	PM_{10}	0.4	Chikun	0.2
		SO_2	82.7	Chikun	3.2[a]
		NO_x	404.9	Chikun	5.4[a]
		CO	164.1	Chikun	1.4
		VOC	2..3	Chikun	0.04
2	Port Harcourt Refining Company I (PHRC) 60,000 BPSD	PM_{10}	0.2	Eleme	0.08
		SO_2	26.3	Eleme	1.02[a]
		NO_x	121.4	Eleme	1.6[a]
		CO	73.0	Eleme	0.6
		VOC	1.0	Eleme	0.02
3	Port Harcourt Refining Company II (PHRC) 150,000 BPS	PM_{10}	0.7	Eleme	0.3
		SO_2	111.9	Eleme	4.3[a]
		NO_x	3,589.1	Eleme	4.7[a]
		CO	214.9	Eleme	1.9
		VOC	17.6	Eleme	0.3
4	Warri Refining & Petrochemical Company (WRPC) 125,000 BPSD	PM_{10}	1.7	Warri S/W	0.7
		SO_2	444.4	Warri S/W	1.7[a]
		NO_x	2,145.8	Warri S/W	2.9[a]
		CO	479.2	Warri S/W	4.2
		VOC	0.9	Warri S/W	0.1

[a]folds of % set limits of the standard *i.e.* above the set standard limits.

(surface plots available on request) while the Tables showing the results of maximum ground Level Concentration of air pollutant within 50 km from each of the refineries are shown in **Table 4** and the maximum ground level concentration impact from 50 km receptor locations obtained when the ground level concentration of the criteria pollutants from the refineries was compared with the set limits of Nigeria's National Ambient Air Quality Standard (NAAQS), World Bank and World Health Organization (WHO) are shown in **Table 5** and **Table 6**. The results predicted the total emissions of criteria pollutant emitted per annum in the existing and proposed refineries where NO_X has the highest rate of emissions. Also, from the existing refineries, Port Harcourt 2 refinery has the highest criteria pollutant emission rate due to its high installed capacity (150,000 bpsd) while from the proposed refineries Tonwei refinery has the highest rate of emissions of the criteria pollutant due to its high installed capacity (200,000 bpsd). It was observed that some refineries exceeds their area of influence *i.e.* the emission dispersion exceeds the plant location to the neighbouring local government area e.g. CO emissions for Rehoboth for 1 h averaging periods dispersed from Yenegoa to Ogbia. From **Table 5** and **Table 6**, criteria pollutants of ranges 0.02% - 94.5% of the standards falls within the set standards limit which means there is no health risk in the receptor locations while criteria pollutants of ranges 1.1 - 55.6 folds of the standards have exceeded the set standard limits and it means there is health risk in the receptor locations.

5. Conclusions

The ISCST3 dispersion model has been used in this study to evaluate the ground level concentrations from the

Table 5. 24-hr averaging period maximum ground level concentration of criteria pollutants and % set limits in proposed refineries.

S/N	Refinery	Air Pollutant	Maximum glc ($\mu g/m^3$)	Maximum Impact Local Government Area	% Set Limit of FMENV, WB, WHO
1	Amakpe International Refinery (12,000 bpsd)	PM_{10}	0.04	Ikot Ekpene	0.02
		SO_2	4.4	Ikot Ekpene	16.9
		NO_x	22.6	Ikot Ekpene	30.1
		CO	10.7	Ikot Ekpene	0.04
		VOC	0.2	Ikot Ekpene	0.002
2	Amexum Corporation (100,000 bpsd)	PM_{10}	0.5	Akpabuyo	0.2
		SO_2	6.7	Akpabuyo	0.4
		NO_x	314	Akpabuyo	0.4[a]
		CO	215	Akpabuyo	1.9
		VOC	4.1	Akpabuyo	0.1
3	Antonio Oil (27,000 bpsd)	PM_{10}	0.5	Ipokia	0.2
		SO_2	99.4	Ipokia	3.8[a]
		NO_x	413	Ipokia	1.5
		CO	37.5	Ipokia	0.3
		VOC	1.8	Ipokia	0.03
4	Chase Wood Consortium Nigeria Limited (70,000 bpsd)	PM_{10}	0.5	Eket	0.2
		SO_2	210	Eket	0.8[a]
		NO_x	1500	Eket	2.0[a]
		CO	170	Eket	0.01
		VOC	26	Eket	0.4
5	Clean Water Refinery (60,000 bpsd)	PM_{10}	0.5	Eleme	0.2
		SO_2	39.5	Eleme	1.3[a]
		NO_x	185.5	Eleme	2.0[a]
		CO	102.4	Eleme	0.9
		VOC	1.6	Eleme	0.02
6	Gasoline Associate & International Limited Refinery (100,000 bpsd	PM_{10}	0.5	Ipokia	0.2
		SO_2	58.6	Ipokia	2.3[a]
		NO_x	281.5	Ipokia	3.8[a]
		CO	139.8	Ipokia	1.2
		VOC	2.1	Ipokia	0.04
7	Ilaje Refinery & Petrochemicals (100,000 bpsd)	PM_{10}	0.8	Ilaje	0.3
		SO_2	61	Ilaje	0.2[a]
		NO_x	298	Ilaje	3.9[a]
		CO	192	Ilaje	1,7
		VOC	3.0	Ilaje	0.05

Continued

8	Niger Delta Refinery & Petrochemical Limited (100,000 bpsd)	PM_{10}	0.5	Warri South	0.2
		SO_2	72.7	Warri South	2.8[a]
		NO_x	347.3	Warri South	4.6[a]
		CO	417.1	Warri South	3.7
		VOC	6.7	Warri South	0.1
9	NSP Refineries and Oil Services Limited (120,000 bpsd)	PM_{10}	0.9	Andoni	0.4
		SO_2	395.0	Andoni	1.5[a]
		NO_x	1903.0	Andoni	1.3[a]
		CO	375.0	Andoni	3.3
		VOC	6.0	Andoni	0.1

10	Ode Aye Refinery Limited (100,000 bpsd)	PM_{10}	0.4	Okiti pupa	0.2
		SO_2	83.5	Okiti pupa	32.1[a]
		NO_x	440	Okiti pupa	5.9[a]
		CO	113.8	Okiti pupa	1.0
		VOC	2.0	Okiti pupa	0.03
11	Ologbo Refinery Company Nigeria Limited (12,000 bpsd)	PM_{10}	0.05	Ikpoba-oha	0.02
		SO_2	0.9	Ikpoba-oha	3.5[a]
		NO_x	8.9	Ikpoba-oha	11.8
		CO	5.0	Ikpoba-oha	0.04
		VOC	0.1	Ikpoba-oha	0.002
12	Orient Petroleum Resources Limited (55,000 bpsd)	PM_{10}	0.7	Anambra West	0.3
		SO_2	0.9	Anambra West	3.5
		NO_x	8.5	Anambra West	11.3
		CO	219.6	Anambra West	1.9
		VOC	3.0	Anambra West	0.01
13	Owena Oil & Gas Limited (60,000 bpsd)	PM_{10}	0.2	Ilaje	0.08
		SO_2	34.5	Ilaje	13.3[a]
		NO_x	159.4	Ilaje	2.2[a]
		CO	78	Ilaje	0.7
		VOC	1.2	Ilaje	0.02
14	Qua Petroleum Refinery Limited (100,000 bpsd)	PM_{10}	1.0	Ibeno	0.4
		SO_2	142.5	Ibeno	24.0[a]
		NO_x	623	Ibeno	8.3[a]
		CO	423	Ibeno	3.7
		VOC	6.5	Ibeno	0.1

Continued

15	Rehoboth Natural Resources Limited (12,000 bpsd	PM_{10}	0.05	Yenegoa	0.02
		SO_2	2.4	Yenegoa	92.3
		NO_x	12.1	Yenegoa	16.1
		CO	9.7	Yenegoa	0.1
		VOC	0.2	Yenegoa	0.003
16	Resources Refinery & Petrochemical Limited (100,000 bpsd)	PM_{10}	1.3	Ikot Abasi	0.7
		SO_2	191.2	Ikot Abasi	8.6[a]
		NO_x	897.3	Ikot Abasi	14.8[a]
		CO	399.4	Ikot Abasi	5.8
		VOC	6.0	Ikot Abasi	0.2
17	RIVGAS Petroleum & Energy Limited (30,000 bpsd)	PM_{10}	0.3	Port Harcourt	0.1
		SO_2	0.4	Port Harcourt	1.5
		NO_x	112.3	Port Harcourt	1.5[a]
		CO	3.9	Port Harcourt	0.03
		VOC	1.8	Port Harcourt	0.02
18	Sapele Refinery Limited (100,000 bpsd)	PM_{10}	0.3	Sapele	0.1
		SO_2	48.9	Sapele	18.8[a]
		NO_x	255.2	Sapele	3.4[a]
		CO	48.9	Sapele	0.4
		VOC	1.6	Sapele	0.03
19	South West Refinery & Petrochemical Company (100,000 bpsd)	PM_{10}	0.7	Ogun water side	0.3
		SO_2	59.4	Ogun water side	22.8[a]
		NO_x	270	Ogun water side	3.6[a]
		CO	149.8	Ogun water side	1.3
		VOC	2.4	Ogun water side	0.04
20	Starrex Petroleum Refinery (100,000 bpsd)	PM_{10}	1.0	Eleme	0.4
		SO_2	0.4	Eleme	1.5
		NO_x	112.3	Eleme	1.5[a]
		CO	3.9	Eleme	0.03
		VOC	1.8	Eleme	0.03
21	Towei Oil Refinery (200,000 bpsd)	PM_{10}	1.5	Ekeremor	0.6
		SO_2	376	Ekeremor	1.0[a]
		NO_x	1875.4	Ekeremor	2.5[a]
		CO	168.9	Ekeremor	1.5
		VOC	1.6	Ekeremor	0.03

Continued

22	Total Support Limited (12,000 bpsd)	PM_{10}	0.7	Calabar Municipal	0.3
		SO_2	2.9	Calabar Municipal	6.2[a]
		NO_x	16	Calabar Municipal	21.3
		CO	22	Calabar Municipal	0.2
		VOC	0.5	Calabar Municipal	0.01
23	Union Atlantic Petroleum Refinery (100,000 bpsd)	PM_{10}	1.8	Badagry	0.7
		SO_2	224	Badagry	86.2[a]
		NO_x	116.8	Badagry	1.6[a]
		CO	658	Badagry	5.8
		VOC	10.3	Badagry	0.2

[a]folds of % set limits of the standard (above the set standard limit).

Table 6. Cumulative ground level concentration impact on ambient quality of multiple refineries within each state.

S/N	Refinery	Air Pollutant	Cumulative Maximum glc (μg/m^3)	% Set Limit of FMENV, WB, WHO
1	Rehoboth Natural Resources Limited & Towei Oil Refinery Bayelsa State	PM_{10}	1.5	0.6
		SO_2	376	14.5[a]
		NO_x	1875.4	25.0[a]
		CO	168.9	1.5
		VOC	1.6	0.03
2	Total Support Limited & Amexum Corporation Cross River State	PM_{10}	0.5	0.2
		SO_2	7.0	2.7[a]
		NO_x	314.4	4.2[a]
		CO	215	1.9
		VOC	2.3	0.1
3	Antonio Oil, South West Refinery & Gasoline International Limited Refinery Ogun State	PM_{10}	4.6	1.8
		SO_2	107.4	4.1[a]
		NO_x	426.4	5.7[a]
		CO	149.8	1.7[a]
		VOC	2.6	0.04
4	Ode Aye Refinery Limited, Ilaje Refinery & Petrochemicals & Owena Oil & Gas Limited Ondo state	PM_{10}	0.8	0.3
		SO_2	84.5	3.3[a]
		NO_x	444.5	5.9[a]
		CO	199.5	1.8
		VOC	3.0	0.05
5	Warri Refining, Niger Delta Refinery & Sapele Refinery Limited Delta State	PM_{10}	1.7	0.7
		SO_2	443.3	17.1[a]
		NO_x	2145.8	28.6[a]
		CO	480.3	4.2
		VOC	7.0	0.1

Continued

6	Amakpe International Refinery, Chase Wood Consortium Nigeria Limited, Resources Refinery & Qua Petroleum Refinery Limited Akwa Ibom State	PM_{10}	8.0	3.2
		SO_2	501	19.3[a]
		NO_x	2214	29.5[a]
		CO	480.3	3.7
		VOC	7.0	0.4
7	Clean Water Refinery, NSP Refineries and Oil Services Limited, RIVGAS Petroleum & Energy Limited, Starrex Petroleum Refinery, Port Harcourt Refining Company I & Port Harcourt Refining Company II River state	PM_{10}	0.2	0.4
		SO_2	204	7.8[a]
		NO_x	957	12.8[a]
		CO	1.6	3.5
		VOC	0.1	0.1

[a] folds of % set limits of the standard *i.e.* above the set standard limit.

four existing refineries and twenty-three proposed refineries in Nigeria for the purpose of predicting the air quality impact. Emissions from minimum of seven and maximum of eleven elevated point sources of criteria pollutants with "worst case" scenario were considered. The 24-hr averaging period of ground level concentration was measured and compared with the Nigeria's ambient quality air set limits as set by National Ambient Air Quality Standard (NAAQS), World Bank and World Health Organization (WHO) within 50 km receptor locations.

From the modeling outputs, NO_x and SO_2 are the major pollutants that exceed the set limit, hereby causing harmful effects to the environment while PM, NOx and VOCs are within the set standard and have no effect on the environment. The highest ground level concentration predicted to be 450 - 1875 μg/m^3 for 24-h averaging period was obtained at Tonwei Oil Refinery, Ekeremor Local Government, Bayelsa State, proposed to operate at installed capacity 200,000 barrels/day capacity while the lowest ground level concentration predicted to be 0.0099 - 0.1 μg/m^3 for 24-h averaging period was obtained at Amakpe International Refinery, Eket Local Government, Akwa Ibom State, proposed to operate at installed capacity 12,000 barrels/day. According to Ambient Air Quality Standards (AAQS), percentage set limits of criteria air pollutants ranging from 0.02% - 94.5% are within the set standard limits and no health risk is associated with areas around the plant locations while percentage set limits of criteria pollutants ranging from 1.1 - 55.6 folds of the standards exceed the maximum permitted limits and health risk is associated with areas around the plant locations as stipulated by the National Ambient Air Quality Standard (NAAQS), World Bank and World Health Organization (WHO) emission set standards.

The study reveals that the contribution of emissions from criteria pollutant to its ambient air is significant. Possible impacts of anticipated emissions of criteria pollutants on humans [17], vegetations [18] soil and water [19] [20] are strong enough for necessary control measures to be implemented in both existing and proposed Nigerian petroleum refineries. Hence, the air quality standards guiding petroleum refinery emissions must be strictly considered, in order to ensure that the ground level concentrations do not exceed the required standard limits and prevent the adverse effect of air pollution in the Nigeria airshed.

References

[1] [USEPA] United State Environmental Protection Agency (1999) Proper Monitoring Essential to Reducing Fugitive Emissions under Leak Detection and Repair Programs. *Enforcement Alert*, **2**, 1-4.

[2] [WB] World Bank (1998) Petroleum Refining. Pollution Prevention and Abatement Handbook. *World Bank Group*, **3**, 377-380.

[3] [WHO] World Health Organization (2005) WHO Air Quality Guidelines for Particulate Matter, Ozone, Nitrogen Dioxide and Sulphur Dioxide Global Update. *Summary of Risk Assessment*, **23**, 1-20.

[4] Zannetti, P., Ed. (1994) Environmental Systems. *Proceedings of ENVIROSOFT 94 Conference*, Vol. II, San Francisco, November 1994, Computational Mechanics Publications, Southampton. http://www.witpress.com/

[5] Schnelle, K.B. and Dey, P.R. (1999) Atmospheric Dispersion Modeling Compliance Guide McGraw-Hill Professional. ISBN 0-07-058059-6.

[6] Elkamel, A., Abdul-Wahab, S., Bouhamra, W. and Alper, E. (2001) Measurement and Prediction of Ozone Levels around a Heavily Industrialized Area: A Neural Network Approach. *Environmental Research*, **5**, 47-49. http://dx.doi.org/10.1016/S1093-0191(00)00042-3

[7] Sonibare, J.A. and Ede, P.N. (2009) Potential Impacts of Integrated Oil and Gas Plant on Ambient Air Quality. *Energy & Environment*, **20**, 331-344. http://dx.doi.org/10.1260/095830509788066394

[8] [DPR] Department of Petroleum Resource (2001) Nigeria Oil Industry Statistical Bulletin. Department of Petroleum Resources, Ministry of Petroleum Resources, Lagos.

[9] [DPR] Department of Petroleum Resource (2004) Nigeria Oil Industry Statistical Bulletin. Department of Petroleum Resources, Ministry of Petroleum Resources, Lagos.

[10] [DPR] Department of Petroleum Resource (2010) Private Refineries and Petrochemical Plants Status. Nigeria Oil Industry Statistical Bulletin. Department of Petroleum Resources, Ministry of Petroleum Resources, Lagos.

[11] [FEPA] Federal Environmental Protection Agency (1991) National Interim Guideline and Standards for Industrial Effluents. *Gaseous Emissions and Harzadous Wastes*, **5**, 33-63.

[12] [WB] World Bank (1998) Petroleum Refining. In: World Bank, Ed., *Pollution Prevention and Abatement Handbook*, 1998: *Toward Cleaner Production*, World Bank Group, Washington DC, 377-380.

[13] [WHO] World Health Organization (2005) WHO Air Quality Guidelines for Particulate Matter, Ozone, Nitrogen Dioxide and Sulphur Dioxide. Global Update 2005, Summary of Risk Assessment, **23**, 1-14.

[14] Abdul-Wahab, S.A. and Al-Alawi, S. (2002) Assessment and Prediction of Tropospheric Ozone Concentration Levels Using Artificial Neural Networks. *Environmental Modelling Software*, **17**, 219-228. http://dx.doi.org/10.1016/S1364-8152(01)00077-9

[15] Chang, T. and Stell, J. (2000) Worldwide Refining Survey. *Oil Gas*, 66-120.

[16] [USEPA] United States Environmental Protection Agency (1989) Airs Facility Subsystem Source Codes and Emission Factor Listing for Criteria Air Pollutants. Research Triangle Park, N.C. U.S. Environmental Protection Agency.

[17] Gou, H., Lee, S.C., Chan, L.Y. and Li, W.M. (2004) Risk Assessment of Exposure to Volatile Organic Compounds in Different Indoor Environments. *Environmental Research*, **94**, 57-66. http://dx.doi.org/10.1016/S0013-9351(03)00035-5

[18] Rao, P.S., Gavane, A.G., Ankam, P., Ansari, M.F., Pandit, V.I. and Nema, P. (2004) Performance Evaluation of a Green Belt in a Petroleum Refinery: A Case Study. *Ecological Engineering*, **23**, 77-84. http://dx.doi.org/10.1016/j.ecoleng.2004.06.013

[19] Johnson, R.L., Thomas, T.B. and Zogorski, J.J. (2003) Effects of Daily Precipitation and Evapotranspiration Patterns on Flow and VOC Transport to Groundwater along a Watershed Flow Path. *Environmental Science Technology*, **37**, 4944-4954. http://dx.doi.org/10.1021/es026252c

[20] Kim, H., Annabel, M.D. and Rao, P.S. (2001) Gaseous Transport of Volatile Organic Chemicals in Unsaturated Porous Media: Effect of Water Partitioning and Air-Water Interfacial Adsorption. *Environmental Science Technology*, **35**, 4457-4462. http://dx.doi.org/10.1021/es001965l

28

Fuel Gas Production from Biomass Sources by Radio Frequency In-Liquid Plasma Method

Ismail Rahim[1,2*], Shinfuku Nomura[1*], Shinobu Mukasa[1], Hiromichi Toyota[1], Katsunori Kawanishi[1], Yoshihiko Makiura[3], Kazuhiko Kogoh[3], Kunihiro Ohshima[3], Susumu Katsuen[3]

[1]Department of Engineering for Production and Environment, Graduate School of Science and Engineering, Ehime University, Ehime, Japan
[2]Department of Automotive Engineering Education, State University of Makassar, Makassar, Indonesia
[3]KURABO Industries Ltd., Neyagawa, Japan
Email: [*]ismail_rahim@unm.ac.id, [*]nomura.shinfuku.mg@ehime-u.ac.jp

Abstract

Cellulose is a kind of saccharide that is the main component in cell walls of plants and therefore is the organic compound that exists in the largest amount in nature. The purpose of this experiment is to convert cellulose to a fuel. Radio frequency (RF) in-liquid plasma is generated in a cellulose distributed solution and a glucose solution, and the generation gas rate is measured. While hydrogen is the main gas generated by the plasma breakdown, carbon monoxide, carbon dioxide, and low-grade flammable gases are also produced. In the glucose water solution or the glucose distributed solution, the solution itself evaporates and decomposes inside the plasma but since the saccharides are non-volatile, they cannot penetrate into the plasma and are not decomposition. However, when the cellulose is at concentrations of 30 wt% or more, it becomes granular and can directly enter the plasma as a solid, where the plasma decomposes the cellulose itself, significantly increasing the amount of gas generated. In addition, the spectrometry of the plasma emission shows the solution after the creation of plasma has the ability to absorb ultraviolet light.

Keywords

In-Liquid Plasma, Radio Frequency Plasma, Saccharide, Hydrogen, Decomposition

[*]Corresponding author.

1. Introduction

Research has been conducted into feasibility of creating saccharides from biomass sources such as grass or wood and then breaking them down and converting them into chemical products [1]-[4] or biofuels [5]-[7] in order to move away from the use of fossil fuels, such as petroleum. Cellulose, which is found in these types of non-edible biomass materials, is the most common organic compound in nature. However, because it is a highly stable compound that includes the glycosidic bond, it is difficult to decompose [8]-[11]. A process using an enzyme called cellulase is most commonly used for its decomposition [12] [13]. However, it is really the energy from hydrogen that is expected to be the ultimate energy source to solve environmental and natural resource problems [14]-[16]. While there are many issues yet to be resolved, such as methods for storing and transporting hydrogen [17] [18], it exists in wide variety of raw as well as discarded materials, so the establishment of an effective recovery system is greatly desirable. If the natural hydrogen containing compound cellulose, which exists in abundance on our planet, could be used directly to create chemical-based products such as plastic or hydrogen, or if the byproducts could be used to simultaneously create value-added chemical products, such as nano-tech materials, or combustible gases, a transition could be made aimed at changing the reliance of making items from petroleum derivatives to one that is sustainable and more ecologically sound.

One effective method of treating waste materials is using a highly active energy field of plasma [19]-[21]. Recent papers have reported on the "in-liquid plasma method" in which gas bubbles are formed in liquids under high pressure, creating a chemical reaction that reaches 3500 K [22] [23]. This in-liquid plasma method is currently being widely used. This development enables liquids to be directly broken down by plasma, making it possible to decompose harmful substances without a catalyst [24] [25].

In this study, the cellulose conversion into fuel gas by plasma decomposition was performed using in-liquid plasma on water to which cellulose and glucose. A constituent monosaccharide of cellulose was added. The gases generated were investigated to identify the decomposition process. Artificial breakdown technologies for non-edible biomass will be presented in this experiment.

2. Decomposition of Saccharides Using a Batch-Type Breakdown Device

Figure 1 depicts the schematic of the apparatus used in this study, which comprises a 27.12 MHz high-frequency (HF) power source, matching box, reactor vessel and aspirator. As a visual window, a quartz glass container with an interior diameter of 55 mm, an exterior diameter of 60 mm and a height of 85 mm was used in the reactor

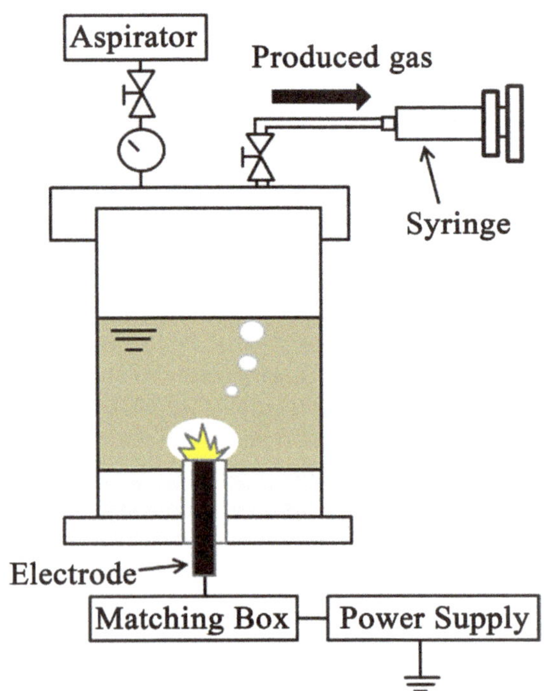

Figure 1. Batch-type plasma breakdown device.

vessel. A copper rod 3 mm in diameter was used as the electrode. The side surface of the electrode was covered with heat-resistant glass in order to reduce the contact surface between the electrode and the liquid. This prevents heat transfer from the electrode to the liquid, and by covering the dielectric body, which allows the HF field to be concentrated at the tip of the electrode [26] [27].

Before the start of the experiment, the gas discharge valve is closed and an aspirator is used to create negative pressure (0.01 - 0.02 MPa) inside the test reactor. In this state, the power output of the HF wave power source is increased while adjusting the impedance with the matching box resulting in plasma being generated in the liquid. The valve on the aspirator side is then closed and as the plasma is maintained in a stable condition, gases are generated which fill the reactor vessel and return it to atmospheric pressure. After atmospheric pressure has been reached, the exhaust valve is opened and the generated gas mixture is extracted.

Gas chromatography (GC) is used to analyze the components of the gas mixture. The gas production rate is calculated based on the time needed for obtaining the volume of a 50 ml syringe. The concentration of the glucose ($C_6H_{12}O_6$) and cellulose (Wako Pure Chemical Industries, Ltd., 38 μm powder) is changed and further analysis experiments are conducted. The amount of liquid was fixed at 120 g and the input energy was 150 W at atmospheric pressure.

3. Analysis of Generated Gas Mixture

The relationship between the concentration of the glucose and the gas mixture generated is shown in **Figure 2**. At the entire of concentrations, the glucose is completely dissolved in water. **Figure 2(a)** shows the volume of gases analyzed by GC, which is the amount of hydrogen gas in the generated gas mixture was 52% - 68%. Due to the presence of oxygen, CO and CO_2 gases are also generated, while hydrogen is the main gas generated by the plasma breakdown, and the flammable gas, CH_4, was also produced. Gas produced in plasma decomposition of clathrate hydrate also has similar gas contents [15] [28]. Besides, **Figure 2(b)** shows an increase in the concentration, which brings about an increase in H_2, CO and CO_2 at concentrations of 5 wt% or less. In contrast, the amount of O_2 generated decreases. At 10 wt%, the generated amount of H_2 gas stabilized.

On the contrary, the cellulose shown in **Figure 3** is insoluble, but two preparations are possible depending on the amount added. At concentrations of 27 wt% and below, the cellulose is distributed in a liquid form while at 30 wt% or more, the cellulose is in a powder form in the water as shown. In addition, as the concentration of the cellulose increases, the relative amount of water decreases, so the diameter of the powder grains becomes smaller.

As shown in **Figure 4**, up to a concentration of 5 wt%, the amount of H_2 generated decreases as the concentration in the liquid state increases, but at 5 wt% or more, this changes to an increase. From 10 wt% to 20 wt%, the generated H_2 settles into a stable value. At 30 wt%, the amount of each gas generated once again decreases. Then, at 40 wt% or more there is a change and a significant increase in gas production rate. The cellulose grains containing water at the reduced concentration of 30 wt% are pushed away from the region where plasma is generated by the high pressure it creates and because of the contact resistance, these larger diameter grains do not return to their original positions. Hence, air gaps form in the region of plasma generation and the plasma cannot reach the cellulose grains containing water.

Under the concentration of 40 wt% and above, even if the cellulose grains are pushed away by the pressure of the plasma, since the grains have smaller diameters there is less contact resistance, so even though some are pushed away, they circulate in the solution and some will sink back into the plasma region. The cellulose that returns is then directly decomposed by the plasma. By the reason of this, the gas production rate increases.

In addition, there was an increase of not only H_2, CO, CO_2 in the decomposition of the cellulose, but also of ethylene and acetylene, which were not detected in the glucose decomposition. At 50 wt%, the H_2 decreases and amounts of CO, CO_2 and CH_4 become greater than the volume of H_2. This could be due to the cellulose being directly broken down by the heat of the plasma.

The 0 wt% in **Figure 2** shows the result of plasma decomposition of pure water. In comparison, the glucose solution can generate up to approximately 1.5 times more hydrogen. In the case of cellulose in a liquid state, with the exception of the 27 wt%, all concentrations generated hydrogen that were below the value for thermal decomposition of water. With the intent of investigating the reaction mechanism in detail, the number of hydrogen, oxygen and carbon atoms making up the gas were calculated using the generation rate.

The change in the number of generated gas atoms in relation to the concentration for the glucose solution showed the same tendencies for each concentration (see **Figure 5(a)**). The generation of carbon atoms rises nearly proportionally to the concentration since carbon atoms will only be generated due to the glucose reaction, at low concentrations, but levels off at high concentrations. On the other hand, there were nearly no carbon atoms generated at 5 wt% or lower in the case of the cellulose as shown in **Figure 5(b)**, and the number of hydrogen and oxygen atoms generated fell as the concentration increased. This is because cellulose distributed through the liquid interferes with the reaction of the water and at low concentrations, only the water reacts. At concentrations of 10 wt% and above, not only the water, but also the cellulose reacts. In the concentrations of 40

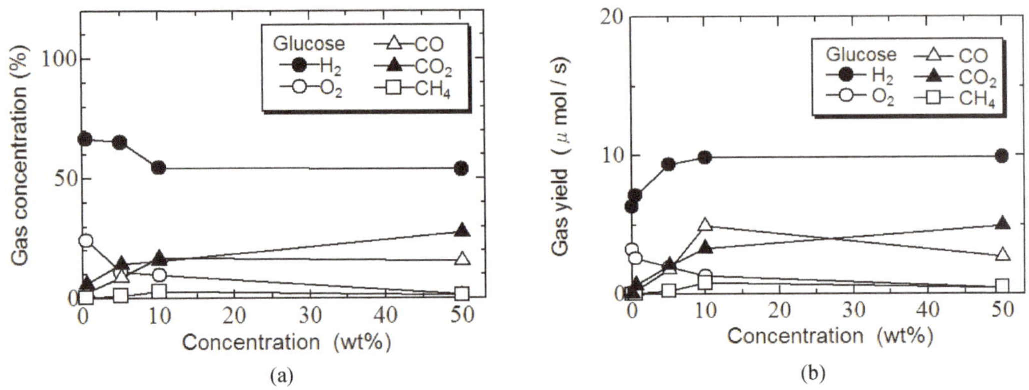

Figure 2. Plasma decomposition of glucose: (a) Volume of gas contents, (b) Gas production rate.

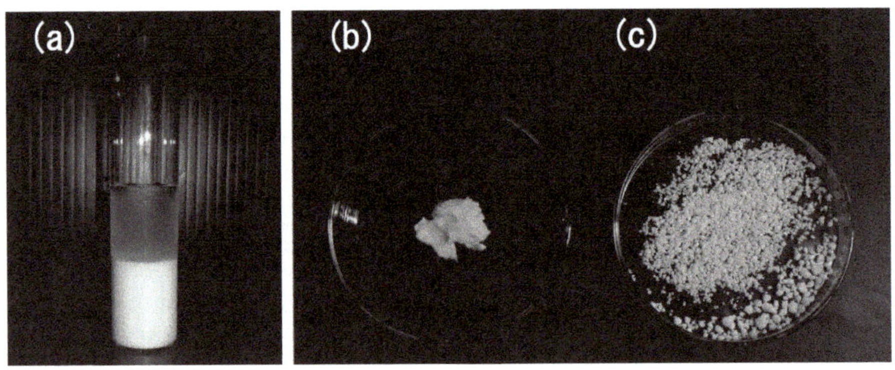

Figure 3. Images of cellulose soaked in water: (a) 10 wt%, (b) 30 wt%, (c) 50 wt%.

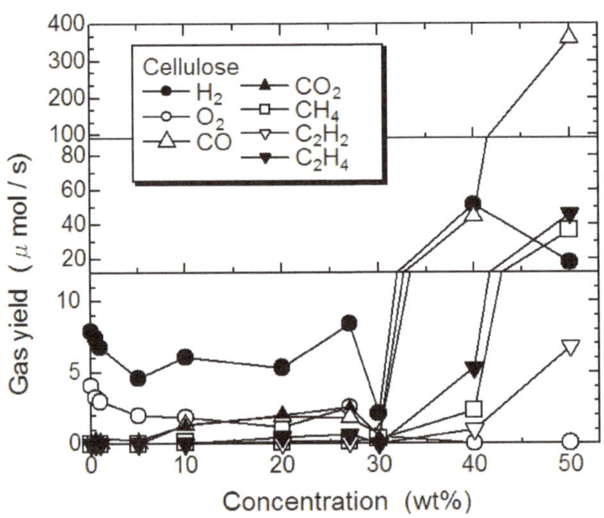

Figure 4. Gas production rate obtained by plasma decomposition of cellulose. The amount of gas generated is greatly changed by the concentration.

wt% or above, the ratio of C and O in relation to H in the gas mixture increases, intensifying the decomposition reaction by the plasma. Additionally, in concentrations of 40 wt% or less, the generation of H, O, C is nearly the same, but after the concentration reaches 50 wt%, the amount of O, and C generated become greater than H.

A related finding was that the reaction of glucose and cellulose in the liquid state is not due to a direct decomposition reaction by the plasma. Because of the In-liquid plasma is generated in the air bubbles created by the heat of the plasma evaporating the surrounding liquid. Thus, the glucose and cellulose cannot enter into these air bubbles due to its nonvolatile property.

Based on analyses of the values of plasma in pure water performed previously, the evaporated water in the air bubbles is decomposed by the plasma and the large volume of OH radicals that have been created are dispersed into the water [29] [30]. In addition, experiments have shown an increase in the concentration of hydrogen peroxide, which is thought to be formed from the OH radicals in the pure water [31]. OH radicals have a strong oxidizing property that indirectly breaks down the glucose and cellulose.

Spectrometry measurements were performed in order to investigate the types of radical formations in the plasma. **Figure 6(a)** shows the emission spectrum of pure water. H_α (656 nm), H_β (486 nm) OH, O (777 nm) and O (845 nm) are detected due to the presence of hydrogen and oxygen radicals resulting from the decomposition of water. **Figure 6(b)** shows the plasma emission spectrum for a glucose solution with a concentration of 50 wt%. Strong emission spectrums could be seen: 486 nm for H_β, 500 to 600 nm for Cu, 656 nm for H_α and 777 to 824 nm for O. Cu and Na (589 nm) are the materials used for the electrode and the heat-resisting glass. When the glucose and cellulose are directly broken down by the plasma, the expected spectrums of CO (250 to 400 nm) and C_2 ($\approx$516 nm) could be confirmed. However, the spectrum from the OH radicals seen at 309 nm during spectrometry of plasma in pure water was difficult to confirm. This is because the liquid after the plasma has been created has the ability to absorb ultraviolet light [32] [33].

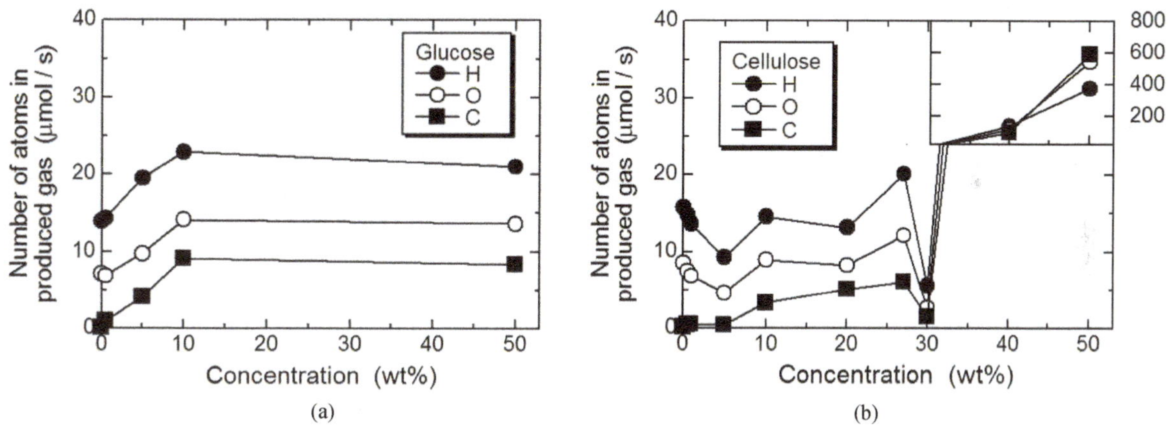

Figure 5. Number of hydrogen, oxygen and carbon atoms in the generated gas: (a) glucose, (b) cellulose.

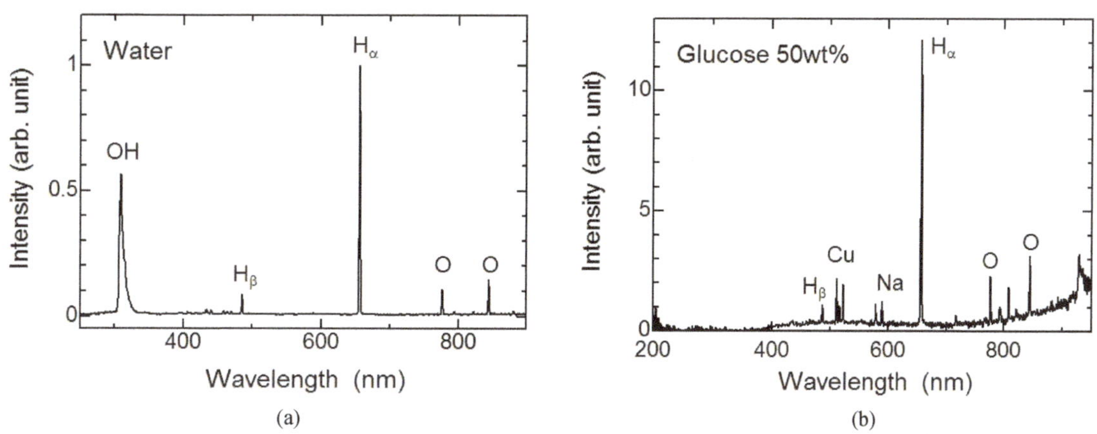

Figure 6. Plasma spectrometry measurements: (a) pure water, (b) glucose concentration at 50%.

4. Decomposition of Saccharides Using a Funnel Shaped Device for Breakdown

The problem of the cellulose not being able to enter into the plasma region was found using the experimental apparatus in the previous section. Therefore, as shown in **Figure 7**, the testing apparatus was improved by adding a taper to the bottom of the reactor vessel to form a funnel shape that makes it easier for the cellulose to accumulate at the bottom. The reactor vessel is funnel shaped and made of Pyrex glass. The liquid reservoir section has an inner diameter of 22 mm and an outer diameter of 25 mm. The section where the electrode is inserted has an inner diameter of 4 mm and an outer diameter of 6 mm. It is expected that the slope of the funnel will enable any cellulose that has settled to the bottom of the liquid reservoir to be efficiently broken down. A 3 mm diameter rod is used for the electrode. The top of the container is sealed with a silicone plug, making the container airtight. Before the start of the experiment, the gas discharge valve was closed and an aspirator was used to create negative pressure inside the vessel. After the pressure inside the device returned to atmospheric pressure, the gas was collected using the same procedure described in the previous section. The reaction liquid was a dissolved glucose solution in which powdered cellulose was dispersed.

Figure 8 shows the relationship between the concentrations of the gases generated for glucose and cellulose. As with the batch-type experiment, when the glucose concentration was 5 wt% or below, the amounts of H_2, CO, and CO_2 generated increased as the concentration increased while the amount of O_2 gas was reduced. In concentrations of 10 wt% and above, the amount of H_2 gas generated stabilized.

The amount of hydrogen generated was approximately twice the amount generated by the batch type device. In addition, the highest speed of hydrogen generation for cellulose in the funnel type device was approximately three times the highest value for the batch type in the liquid state. However, when compared to the concentration of 40 wt% shown in **Figure 2**, the amount of CO and CO_2 gas attained was smaller and direct decomposition of the cellulose by the plasma did not occur. How to introduce nonvolatile materials or fixed shaped particles into the plasma will be a topic for future research.

5. Conclusion

In this research, experiments for in-liquid plasma breakdown of glucose and cellulose were conducted and the components of the generated gas mixture were analyzed and the gas production rate was measured. Using plasma decomposition, a dissolved glucose-water solution is proposed, in which cellulose was dispersed to generate

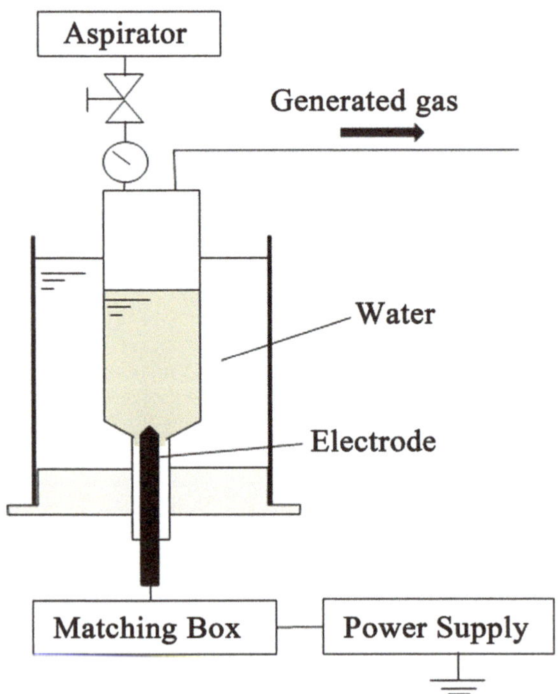

Figure 7. Plasma breakdown device using a funnel-shaped reactor.

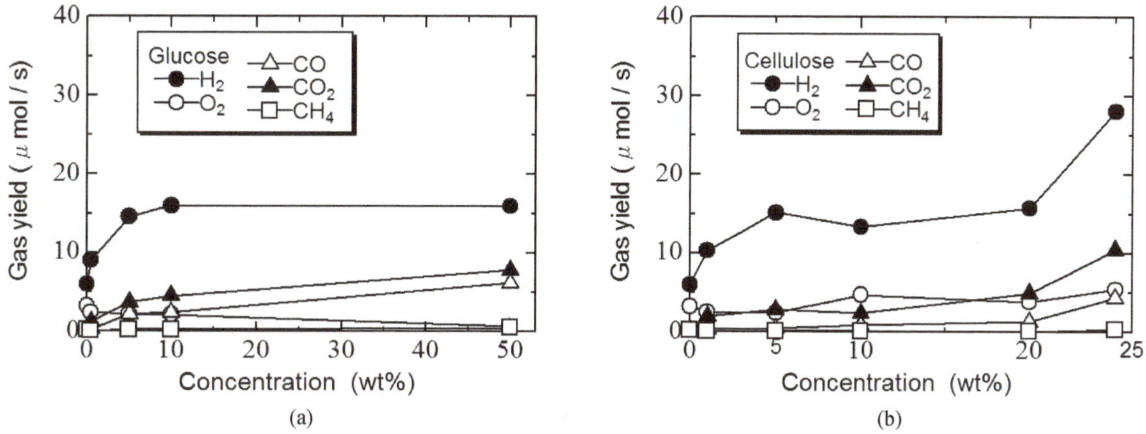

Figure 8. Plasma decomposition experiments using saccharides in a funnel-shaped reactor: (a) glucose, (b) cellulose.

a gas mixture that contains 50% - 66% of hydrogen gas. With the batch-type device, the cellulose became a powder at concentrations of 30 wt% and above. At concentrations 40 wt% and above, the generation of gas significantly increased because the powder entered directly into the plasma and the cellulose was directly broken down by thermal decomposition in the plasma. When the glucose-water solution or a solution with dispersed cellulose had concentrations of 27 wt% or less, the saccharides were not directly decomposed by the heat of the plasma, but rather were decomposed indirectly by active type radicals created by the plasma. With the funnel-shaped device, the generation speed of the hydrogen gas was greater than that of the batch-type device, but in this case as well, the direct heat from the plasma had little effect on the decomposition.

References

[1] De Souza, R.L., Yu, H., Rataboul, F. and Essayem, N. (2012) 5-Hydroxymethylfurfural (5-HMF) Production from Hexoses: Limits of Heterogeneous Catalysis in Hydrothermal Conditions and Potential of Concentrated Aqueous Organic Acids as Reactive Solvent System. *Challenges*, **3**, 212-232. http://dx.doi.org/10.3390/challe3020212

[2] Lai, L. and Zhang, Y. (2011) The Production of 5-Hydroxymethylfurfural from Fructose in Isopropyl Alcohol: A Green and Efficient System. *ChemSusChem*, **4**, 1745-1748. http://dx.doi.org/10.1002/cssc.201100489

[3] Tong, X., Ma, Y. and Li, Y. (2010) Biomass into Chemicals: Conversion of Sugars to Furan Derivatives by Catalytic Processes. *Applied Catalysis A*, **385**, 1-13. http://dx.doi.org/10.1016/j.apcata.2010.06.049

[4] Kazi, F.K., Patel, A.D., Serrano-Ruiz, J.C., Dumesic, J.A. and Anex, R.P. (2011) Techno-Economic Analysis of Dimethylfuran (DMF) and Hydroxymethylfurfural (HMF) Production from Pure Fructose in Catalytic Processes. *Chemical Engineering Journal*, **169**, 329-338. http://dx.doi.org/10.1016/j.cej.2011.03.018

[5] Kawashima, A., Matsubara, K. and Honda, K. (2008) Development of Heterogeneous Base Catalysts for Biodiesel Production. *Bioresource Technology*, **99**, 3439-3443. http://dx.doi.org/10.1016/j.biortech.2007.08.009

[6] Kawashima, A., Matsubara, K. and Honda, K. (2009) Acceleration of Catalytic Activity of Calcium Oxide for Biodiesel Production. *Bioresource Technology*, **100**, 696-700. http://dx.doi.org/10.1016/j.biortech.2008.06.049

[7] Brethauer, S. and Wyman, C.E. (2010) Continuous Hydrolysis and Fermentation for Cellulosic Ethanol Production. *Bioresource Technology*, **101**, 4862-4874. http://dx.doi.org/10.1016/j.biortech.2009.11.009

[8] Chundawat, S.P., Bellesia, G., Uppugundla, N., da Costa Sousa, L., Gao, D., Cheh, A.M., Agarwal, U.P., Bianchetti, C.M., Phillips Jr., G.N., Langan, P., Balan, V., Gnanakaran, S. and Dale, B.E. (2011) Restructuring the Crystalline Cellulose Hydrogen Bond Network Enhances Its Depolymerization Rate. *Journal of the American Chemical Society*, **133**, 11163-11174. http://dx.doi.org/10.1021/ja2011115

[9] Nishiyama, Y., Lngan, P. and Chanzy, H. (2002) Crystal Structure and Hydrogen-Bonding System in Cellulose Iβ from Synchrotron X-Ray and Neutron Fiber Diffraction. *Journal of the American Chemical Society*, **124**, 9074-9082. http://dx.doi.org/10.1021/ja0257319

[10] Xiang, Q., Lee, Y.Y., Pettersson, P.O. and Torget, R.W. (2003) Heterogeneous Aspects of Acid Hydrolysis of α-Cellulose. *Applied Biochemistry and Biotechnology*, **107**, 505-574. http://dx.doi.org/10.1385/ABAB:107:1-3:505

[11] Sakaki, T., Shibata, M., Miki, T., Hirosue, H. and Hayashi, N. (1996) Decomposition of Cellulose in Near-Critical Water and Fermentability of the Products. *Energy & Fuels*, **10**, 684-688. http://dx.doi.org/10.1021/ef950160+

[12] Yang, B., Dai, Z.Y., Ding, S.Y. and Wyman, C.E. (2011) Enzymatic Hydrolysis of Cellulosic Biomass. *Biofuels*, **2**, 421-450. http://dx.doi.org/10.4155/bfs.11.116

[13] Horn, S.J., Vaaje-Kolsad, G., Westereng, B. and Eijsink, V.G.H. (2012) Novel Enzymes for the Degradation of Cellulose. *Biotechnology for Biofuels*, **5**, 45. http://dx.doi.org/10.1186/1754-6834-5-45

[14] Lalaurette, E., Thammannagowda, S., Mohagheghi, A., Maness, P.C. and Logan, B.E. (2009) Hydrogen Production from Cellulose in a Two-Stage Process Combining Fermentation and Electrohydrogenesis. *International Journal of Hydrogen Energy*, **34**, 6201-6210. http://dx.doi.org/10.1016/j.ijhydene.2009.05.112

[15] Putra, A.E.E., Nomura, S., Mukasa, S. and Toyota, H. (2012) Hydrogen Production by Radio Frequency Plasma Stimulation in Methane Hydrate at Atmospheric Pressure. *International Journal of Hydrogen Energy*, **37**, 16000-16005. http://dx.doi.org/10.1016/j.ijhydene.2012.07.099

[16] Abbas, F.H. and Wan Daud, W.M.A. (2010) Hydrogen Production by Methane Decomposition: A Review. *International Journal of Hydrogen Energy*, **35**, 1160-1190. http://dx.doi.org/10.1016/j.ijhydene.2009.11.036

[17] Tozzini, V. and Pellegrini, V. (2013) Prospects for Hydrogen Storage in Graphene. *Physical Chemistry Chemical Physics*, **15**, 80-89. http://dx.doi.org/10.1039/C2CP42538F

[18] Ma, T., Isobe, S., Morita, E., Wang, Y., Hashimoto, N., Ohnuki, S., Kimura, T., Ichikawa, T. and Kojima, Y. (2011) Correlation between Kinetics and Chemical Bonding State of Catalyst Surface in Catalyzed Magnesium Hydride. *International Journal of Hydrogen Energy*, **36**, 12319-12323. http://dx.doi.org/10.1016/j.ijhydene.2011.07.011

[19] Heberlein, J. and Murphy, A.B. (2008) Thermal Plasma Waste Treatment. *Journal physics D*, **41**, Article ID: 053001.

[20] Choi, S., Park, D.W. and Watanabe, T. (2012) Thermal Plasma Decomposition of Fluorinated Greenhouse Gases. *Nuclear Engineering and Technology*, **44**, 21-32. http://dx.doi.org/10.5516/NET.77.2012.003

[21] Sakano, M., Tanaka, M. and Watanabe, T. (2001) Application of Radio-Frequency Thermal Plasmas to Treatment of Fly Ash. *Thin Solid Films*, **386**, 189-194. http://dx.doi.org/10.1016/S0040-6090(00)01641-2

[22] Nomura, S., Mukasa, S., Toyota, H., Miyake, H., Yamashita, H. and Maehara, T. (2011) Characteristics of In-Liquid Plasma in Water under Higher Pressure Than Atmospheric Pressure. *Plasma Sources Science and Technology*, **20**, Article ID: 034012. http://dx.doi.org/10.1088/0963-0252/20/3/034012

[23] Maehara, T., Miyamoto, I., Kurokawa, K., Hashimoto, Y., Iwamae, A., Kuramoto, M., Yamashita, H., Mukasa, S., Toyota, H., Nomura, S. and Kwashima, A. (2008) Degradation of Methylene Blue by RF plasma in Water. *Plasma Chemistry and Plasma Processing*, **28**, 467-482. http://dx.doi.org/10.1007/s11090-008-9142-2

[24] Nomura, S., Toyota, H., Mukasa, S., Yamashita, H., Maehara, T. and Kawashima, A. (2009) Production of Hydrogen in a Conventional Microwave Oven. *Journal of Applied Physics*, **106**, Article ID: 073306. http://dx.doi.org/10.1063/1.3236575

[25] Kurihara, H. and Yajima, T. (2007) Decomposition of Toluene by Liquid-phase Atmospheric-Pressure Microwave Plasma Generated Using Carbon Felt Pieces Impregnated with NaCl. *Chemistry Letters*, **36**, 870-871. http://dx.doi.org/10.1246/cl.2007.870

[26] Hattori, Y., Mukasa, S., Nomura, S. and Toyota, H. (2010) Optimization and Analysis of Shape of Coaxial Electrode for Microwave Plasma in Water. *Journal of Applied Physics*, **107**, Article ID: 063305. http://dx.doi.org/10.1063/1.3319616

[27] Hattori, Y., Mukasa, S., Toyota, H. and Nomura, S. (2013) Electrical Breakdown of Microwave Plasma in Water. *Current Applied Physics*, **13**, 1050-1054. http://dx.doi.org/10.1016/j.cap.2013.02.012

[28] Nomura, S., Putra, A.E.E., Mukasa, S., Yamashita, H. and Toyota, H. (2011) Plasma Decomposition of Clathrate Hydrates by 2.45GHz Microwave Irradiation at Atmospheric Pressure. *Applied Physics Express*, **4**, Article ID: 066201. http://dx.doi.org/10.1143/APEX.4.066201

[29] Mukasa, S., Nomura, S., Toyota, H., Maehara, T., Abe, F. and Kawashima, A. (2009) Temperature Distributions of Radio-Frequency Plasma in Water by Spectroscopic Analysis. *Journal of Applied Physics*, **106**, Article ID: 113302. http://dx.doi.org/10.1063/1.3264671

[30] Mukasa, S., Nomura, S., Toyota, H., Maehara, T. and Yamashita, H. (2011) Internal Conditions of a Bubble Containing Radio-Frequency Plasma in Water. *Plasma Sources Science and Technology*, **20**, Article ID: 034020. http://dx.doi.org/10.1088/0963-0252/20/3/034020

[31] Yang, Y., Cho, Y.I. and Fridman, A. (2012) Plasma Discharge in Liquid: Water Treatment and Applications. CRC Press, Boca Raton, 102.

[32] Sato, M. (2009) Degradation of Organic Contaminants in Water by Plasma. *International Journal of Plasma Environmental Science and Technology*, **3**, 8-14.

[33] Maehara, T., Nishiyama, K., Onishi, S., Mukasa, S., Toyota, H., Kuramoto, M., Nomura, S. and Kawashima, A. (2010) Degradation of Methylene Blue by Radio Frequency Plasmas in Water under Ultraviolet Irradiation. *Journal of Hazardous Materials*, **74**, 473-476. http://dx.doi.org/10.1016/j.jhazmat.2009.09.076

Permissions

All chapters in this book were first published in JPEE, by Scientific Research Publishing; hereby published with permission under the Creative Commons Attribution License or equivalent. Every chapter published in this book has been scrutinized by our experts. Their significance has been extensively debated. The topics covered herein carry significant findings which will fuel the growth of the discipline. They may even be implemented as practical applications or may be referred to as a beginning point for another development.

The contributors of this book come from diverse backgrounds, making this book a truly international effort. This book will bring forth new frontiers with its revolutionizing research information and detailed analysis of the nascent developments around the world.

We would like to thank all the contributing authors for lending their expertise to make the book truly unique. They have played a crucial role in the development of this book. Without their invaluable contributions this book wouldn't have been possible. They have made vital efforts to compile up to date information on the varied aspects of this subject to make this book a valuable addition to the collection of many professionals and students.

This book was conceptualized with the vision of imparting up-to-date information and advanced data in this field. To ensure the same, a matchless editorial board was set up. Every individual on the board went through rigorous rounds of assessment to prove their worth. After which they invested a large part of their time researching and compiling the most relevant data for our readers.

The editorial board has been involved in producing this book since its inception. They have spent rigorous hours researching and exploring the diverse topics which have resulted in the successful publishing of this book. They have passed on their knowledge of decades through this book. To expedite this challenging task, the publisher supported the team at every step. A small team of assistant editors was also appointed to further simplify the editing procedure and attain best results for the readers.

Apart from the editorial board, the designing team has also invested a significant amount of their time in understanding the subject and creating the most relevant covers. They scrutinized every image to scout for the most suitable representation of the subject and create an appropriate cover for the book.

The publishing team has been an ardent support to the editorial, designing and production team. Their endless efforts to recruit the best for this project, has resulted in the accomplishment of this book. They are a veteran in the field of academics and their pool of knowledge is as vast as their experience in printing. Their expertise and guidance has proved useful at every step. Their uncompromising quality standards have made this book an exceptional effort. Their encouragement from time to time has been an inspiration for everyone.

The publisher and the editorial board hope that this book will prove to be a valuable piece of knowledge for researchers, students, practitioners and scholars across the globe.

List of Contributors

Vaibhav Pal and R. Anand
Department of Mechanical Engineering, National Institute of Technology, Tiruchirappalli, India

P. Karthikeyan
Department of Automobile Engineering, PSG College of Technology, Coimbatore, India

Zhanhe Liu, Xufeng Xu and Elham Makram
Electrical and Computer Engineering Department, Clemson University, Clemson, USA

Hany A. Abdelsalam
Electrical and Computer Engineering Department, Clemson University, Clemson, USA
Electrical Engineering Department, Kafrelshikh University, Kafrelshikh, Egypt

Jamshed Khan
Goldwind Science and Technology Co. Ltd., Beijing, China

Usman Nasir
North China Electric Power University, Beijing, China

Al Kulasekara and Rarc Gopura
Department of Mechanical Engineering, University of Moratuwa, Katubedda, Sri Lanka

Ktmu Hemapala
Department of Electrical Engineering, University of Moratuwa, Katubedda, Sri Lanka

Néstor Ramos-García, Antoine Cayron and Jens Nørkær Sørensen
Department of Wind Energy, Fluid Mechanics Section, Technical University of Denmark, Lyngby, Denmark

Wen Zhong Shen and Wei Jun Zhu
Department of Wind Energy, Technical University of Denmark, Lyngby, Denmark

Hua Yang
School of Hydraulic, Energy and Power Engineering, Yangzhou University, Yangzhou, China

E. B. Elanchezhian, S. Subramanian and S. Ganesan
Department of Electrical Engineering, Annamalai University, Chidambaram, India

Leonidha Londo
Transmission System Operator, Tirana, Albania

Rajmonda Bualoti, Marialis Çelo and Nako Hobdari
Electric Engineering Faculty, Polytechnic University of Tirana, Tirana, Albania

Liping Zhang, Hehuan Peng, Qinling Yu, Ying Zhang, Zhijun Wang and Xianwei Tang
School of Engineering, Zhejiang A&F University, Linan, China

Wei Jun Zhu and Wen Zhong Shen
Department of Wind Energy, Technical University of Denmark, Lyngby, Denmark

Yan Gao and Cong Li
Shanghai University of Engineering Science, Shanghai, China

Momen Sughayyer
Mechanical Engineering Department, Palestine Polytechnic University, Hebron, Palestine

Gerhard Kramm
Engineering Meteorology Consulting, Fairbanks, USA

Gary Sellhorst
Geophysical Institute, University of Alaska Fairbanks, Fairbanks, USA

Hannah K. Ross
Department of Mechanical Engineering, University of Washington, Seattle, USA

John Cooney
Department of Atmospheric Sciences, Texas A & M University, College Station, USA

Ralph Dlugi
Arbeitsgruppe Atmosphärische Prozesse (AGAP), Munich, Germany

Nicole Mölders
College of Natural Science and Mathematics and Geophysical Institute, University of Alaska Fairbanks, Fairbanks, USA

Kennedy Izuchukwu Ogunwa
World Bank African Centre of Excellence in Oilfield Chemicals, Institute Of Petroleum Studies, University of Port Harcourt, Port Harcourt, Nigeria

Samuel Ofodile and Ozioma Achugasim
Department of Pure and Industrial Chemistry, University of Port Harcourt, Port Harcourt, Nigeria

Eduard A. Mikaelian
Gubkin Russian State University of Oil and Gas, Moscow, Russia

Saif A. Mouhammad
Physics Department, Faculty of Science, Taif University, Taif, Saudi Arabia

Ömer Gül and Tayfun Gündoğdu
Department of Electrical Engineering, Istanbul Technical University, Istanbul, Turkey

Luis Aromataris, Marcos Galetto and Fabian Rinaudo
Grupo de Análisis de Sistemas Eléctricos de Potencia, Facultad de Ingeniería, UNRC, Córdoba, Argentina

Sergio Preidikman
Grupo de Análisis de Sistemas Eléctricos de Potencia, Facultad de Ingeniería, UNRC, Córdoba, Argentina
Departamento de Estructuras, Facultad de Ciencias Exactas, Físicas y Naturales, Universidad Nacional de Córdoba, Córdoba, Argentina

Eduardo Toledo
Centro de Estudios de Energía para el Desarrollo (CEED), Facultad de Ingeniería, UNAM, Oberá, Argentina

Syed Mubashar Hassan
School of Engineering and Built Environment, Edinburgh Napier University, Edinburgh, UK

Syed Ali Hasnain
Centre of Excellence in Sustainable Building Design, School of Energy, Geosciences, Infrastructure & Society, Heriot-Watt University, Edinburgh, UK

Wichet Thipprasert and Prakasit Sritakaew
Department of Electrical, Faculty of Engineering, Rajamangala University of Technology Lanna Chiang Rai, Chiang Rai, Thailand

Barinaadaa Thaddeus Lebele-Alawa and Anthony Kpegele Le-ol
Department of Mechanical Engineering, Rivers State University of Science and Technology, Port Harcourt, Nigeria

J. Cheng and F. Choobineh
Department of Electrical and Computer Engineering, University of Nebraska-Lincoln, Lincoln, USA

Matias Sessarego, Néstor Ramos-García and Wen Zhong Shen
Department of Wind Energy, Technical University of Denmark, Lyngby, Denmark

Hani Albalawi and Elham Makram
Department of Electrical and Computer Engineering, Clemson University, Clemson, USA

Heng Wang and Weijun Liu
Automotive Engineering Institute, Shanghai University of Engineering Science, Shanghai, China

Helal Al-Hamadi
Department of Information Science, College of Computing Sciences and Engineering, Kuwait University, Kuwait, Kuwait

Temitayo Oladimeji and Augustine Ayeni
Chemical Engineering Department, Covenant University, Ota, Nigeria

Jacob Sonibare
Chemical Engineering Department, Obafemi Awolowo University, Ile-Ife, Nigeria

Moradeyo Odunfa
Mechanical Engineering Department, University of Ibadan, Ibadan, Nigeria

Ismail Rahim
Department of Engineering for Production and Environment, Graduate School of Science and Engineering, Ehime University, Ehime, Japan
Department of Automotive Engineering Education, State University of Makassar, Makassar, Indonesia

Shinfuku Nomura, Shinobu Mukasa, Hiromichi Toyota and Katsunori Kawanishi
Department of Engineering for Production and Environment, Graduate School of Science and Engineering, Ehime University, Ehime, Japan

Yoshihiko Makiura, Kazuhiko Kogoh, Kunihiro Ohshima and Susumu Katsuen
KURABO Industries Ltd., Neyagawa, Japan